*Physical Methods
in Modern Chemical Analysis*

Volume 2

PHYSICAL METHODS IN MODERN CHEMICAL ANALYSIS

Edited by

THEODORE KUWANA

Department of Chemistry
The Ohio State University
Columbus, Ohio

Volume 2

 1980

ACADEMIC PRESS

A Subsidiary of Harcourt Brace Jovanovich, Publishers

New York   London   Toronto   Sydney   San Francisco

ACADEMIC PRESS, INC.
111 Fifth Avenue, New York, New York 10003

United Kingdom Edition published by
ACADEMIC PRESS, INC. (LONDON) LTD.
24/28 Oval Road, London NW1 7DX

Library of Congress Cataloging in Publication Data
Main entry under title:

Physical methods in modern chemical analysis.

   Includes bibliographies and index.
   1. Chemistry, Analytic. I. Kuwana, Theodore.
QD75.2.P49   543   77–92242
ISBN 0–12–430802–3 (v. 2)

PRINTED IN THE UNITED STATES OF AMERICA

80 81 82 83   9 8 7 6 5 4 3 2 1

Contents

Analytical Aspects of Ion Cyclotron Resonance

Robert C. Dunbar

Refractive Index Measurement

Thomas M. Niemczyk

List of Contributors

Numbers in parentheses indicate the pages on which the authors' contributions begin.

William L. Davidson (171), 2381 Sarazen Drive, Dunedin, Florida 33528

Robert C. Dunbar (277), Department of Chemistry, Case Western Reserve University, Cleveland, Ohio 44106

S. W. Gaarenstroom (115), Analytical Chemistry Department, General Motors Research Laboratories, Warren, Michigan 48090

Thomas M. Niemczyk (337), Department of Chemistry, University of New Mexico, Albuquerque, New Mexico 87131

Donald J. Pietrzyk (1), Department of Chemistry, University of Iowa, Iowa City, Iowa 52242

N. Winograd (115), Department of Chemistry, The Pennsylvania State University, University Park, Pennsylvania 16802

Preface

The practitioners of chemistry today are faced with a multitude of increasingly complex problems concerned with *chemical analysis*. They are, for example, requested to find and identify trace amounts of materials in complex mixtures. Moreover, trace quantities, rather than being in the microgram range as thought of several years ago, are now being extended below the nanogram level to femtograms. The problem of identification is also nontrivial, extending from organic and inorganic compounds in various matrices to complex biological macromolecules. New tools often associated with sophisticated instrumentation are also constantly being introduced. Surface analysis is a good example of an area for which recent years have seen the advent of many new methods, and the abbreviations ESCA, SIMS, XPS, LEEDS, etc. are now common in the literature. These methods have made it possible to analyze and characterize less than monolayers on solid surfaces. Thus the demand upon a practicing chemist is to have a working knowledge of a wide variety of physical methods of chemical analysis, both old and new: the new ones as they are developed and applied, and the old ones as they are better understood and extended. It is the aim of "Physical Methods in Modern Chemical Analysis" to present a description of selected methodologies at a level appropriate to those who wish to expand their working knowledge of today's methods and for those who wish to update their background. It should also be useful to graduate students in obtaining a basic overview of a wide variety of techniques at a greater depth than that available from textbooks on instrumental methods.

"Physical Methods in Modern Chemical Analysis" will contain chapters written by outstanding specialists who have an intimate working knowledge of their subject. The chapters will contain descriptions of the fundamental principles, the instrumentation or necessary equipment, and applications that demonstrate the scope of the methodology.

It is hoped that these volumes continue the standard exemplified by the earlier volumes, "Physical Methods in Chemical Analysis," edited by Walter Berl in the 1950s and 1960s.

The patience and assistance of my wife Jane during the editing process are gratefully acknowledged.

Contents of Other Volumes

Volume 1

Selected Planned Chapters for Future Volumes

High Performance Liquid Chromatography

Donald J. Pietrzyk

Department of Chemistry
University of Iowa
Iowa City, Iowa

I. Introduction

A. Chromatography

1. Scope

Of all the different types of separation methods, chromatography has the unique position of being applicable to all types of problems and in all areas of science. It is perhaps the most accepted separation tool in the

1

modern analytical laboratory and is routinely used for purposes of purification, separation, identification, and quantitation.

Chromatography encompasses many different techniques. In this chapter one of these techniques, known as high performance liquid chromatography (hereafter referred to as HPLC), is discussed. No attempt has been made to survey all the literature in HPLC. Rather, the main purpose is to provide the reader with a discussion of the theory, instrumentation, and types of columns and eluting agents used in HPLC. It is hoped that this approach will provide a working background in HPLC, will facilitate the introduction of HPLC into laboratories where it is not now or only sparingly used, and will stimulate further development in laboratories where it is routinely used.

Modern liquid chromatography or HPLC has developed largely since about 1968. Its development and the increase in scientific applications parallels the surge observed in gas chromatography (GC) during the late 1950s and early 1960s.

Four major factors are probably responsible for the acceptance of HPLC as a companion to gas chromatography in related areas of organic, pharmaceutical, and biochemical analysis.

(1) HPLC provides a great reduction in analysis time in comparison with conventional (pre-1968) liquid chromatography. Furthermore, this reduction is accomplished over a wide range of unknown sample concentrations, including trace levels, with a high degree of precision and accuracy.

(2) HPLC is capable of generating high column efficiencies. That is, plate heights are extremely small in comparison with conventional liquid chromatography, and thus, a very large number of plates are possible per given column length. This means that short columns can be used, and/or that very complex mixtures can be separated. For example, complex mixtures such as urine, other biological fluids or tissue samples, environmental samples, and agricultural samples that were virtually impossible to separate several years ago can now be separated by HPLC.

(3) HPLC, unlike GC, does not require the sample to have an appreciable vapor pressure. Thus HPLC is applicable not only to the separation of lower molecular weight organic compounds, but also to macromolecular compounds. Furthermore, separations based on adsorption, partition, ion exchange, and size exclusion are possible. For these reasons, future applications of HPLC may be significantly greater than those of GC.

(4) Finally, the growth in the design and development of HPLC instrumentation has paralleled the demand for HPLC applications.

2. *Literature*

The theory and applications of chromatography have developed rapidly in the past 25 years, making chromatography one of the major techniques

TABLE I

Literature in Chromatography

Chromatography—General

Cazes, J., ed. (1977). "Liquid Chromatography of Polymers and Related Materials." Dekker, New York.

Deyl, Z., Macek, K., and Janak, J. (1975). "Liquid Column Chromatography: A Survey of Modern Techniques and Applications" (Journal of Chromatography Library—Vol. 3). Elsevier, Amsterdam.

Giddings, J. C. (1965). "Dynamics of Chromatography." Dekker, New York.

Heftmann, E., ed. (1975). "Chromatography," 3rd ed. Van Nostrand-Reinhold, New York.

Helfferich, F. (1962). "Ion Exchange." McGraw-Hill, New York.

Karger, B. L., Snyder, L. R., and Horvath, C. (1973). "An Introduction to Separation Science." Wiley (Interscience), New York.

Liteanu, C., and Gocan, S. (1974). "Gradient Liquid Chromatography." Halsted Press, New York.

Miller, J. M. (1975). "Separation Methods in Chemical Analysis." Wiley (Interscience), New York.

Perry, S. G., Amos, R., and Brewer, P. I. (1972). "Practical Liquid Chromatography." Plenum, New York.

Snyder, L. R. (1968). "Principles of Adsorption Chromatography" (Chromatographic Science—A Series of Monographs, Vol. 3). Dekker, New York.

Stahl, E. (1969). "Thin Layer Chromatography. A Laboratory Handbook," 2nd ed. Springer-Verlag, Berlin and New York.

Yau, W., Kirkland, J., and Bly, D. (1979). "Modern Size-Exclusion Liquid Chromatography." Wiley, New York.

Zweig, G., and Sherma, J., eds. (1972). "Handbook of Chromatography." CRC Press, Cleveland, Ohio.

Chromatography—HPLC

"Basics of Liquid Chromatography" (1977). 2nd ed. Spectra Physics, Santa Clara, California.

Baumann, F., and Hadden, N., eds. (1972). "Basic Chromatography." Varian, Walnut Creek, California.

Brown, P. R. (1973). "High Pressure Liquid Chromatography—Biochemical and Biomedical Applications." Academic Press, New York.

Dixon, P. F., Gray, C. H., Liu, C. K., and Stol, M. S. (1976). "HPLC in Clinical Chemistry." Academic Press, New York.

Done, J. N., Knox, J. H., and Loheac, J. (1974). "Applications of High Speed Liquid Chromatography." Wiley, New York.

Engelhardt, H. (1979). "High Performance Liquid Chromatography," trans. by G. Gutnikov. Springer-Verlag, New York.

TABLE 1 (*continued*)

Gruska, E., ed. (1974). "Bonded Stationary Phases in Chromatography." Ann Arbor Sci. Publ., Ann Arbor, Michigan.

Hawk, G. L., ed. (1979). "Biological/Biomedical Applications of Liquid Chromatography" (Chromatographic Science: A Series of Monographs—Vol. 10). Dekker, New York.

Huber, J. F. K. (1978). "Instrumentation for High Performance Liquid Chromatography" (Journal of Chromatography Library—Vol. 13). Elsevier, Amsterdam.

Johnson, E. L., and Stevenson, R. (1978). "Basic Liquid Chromatography." Varian, Walnut Creek, California.

Kirkland, J. J., ed. (1971). "Modern Practice of Liquid Chromatography." Wiley (Interscience), New York.

Knox, J. H., Done, J. N., Fell, A. F., Gilbert, M. T., Pryde, A., and Wall, R. A. (1979). "High-Performance Liquid Chromatography." Edinburgh University Press, England.

Lawrence, J. F., and Frei, R. W. (1976). "Chemical Derivatization in Liquid Chromatography" (Journal of Chromatography Library—Vol. 7). Elsevier, Amsterdam.

Parris, N. A. (1976). "Instrumental Liquid Chromatography" (Journal of Chromatography Library—Vol. 5). Elsevier, Amsterdam.

Pryde, A., and Gilbert, M. T. (1978). "Applications of High Performance Liquid Chromatography." Wiley (Interscience), New York.

Rajcsanyi, P. M., and Rajcsanyi, E. (1975). "High Speed Liquid Chromatography" (Chromatographic Science: A Series of Monographs—Vol. 6). Dekker, New York.

Rosset, R., Caude, M., and Jardy, A. (1976). "Practical High Performance Liquid Chromatography." Heyden, London.

Scott, R. P. W. (1976). "Contemporary Liquid Chromatography" (Techniques of Chemistry, A. Weissberger, ed., Vol. XI). Wiley, New York.

Scott, R. P. W. (1977). "Liquid Chromatography Detectors" (Journal of Chromatography Library—Vol. 11). Elsevier, Amsterdam.

Simpson, C. F., ed. (1976). "Practical High Performance Liquid Chromatography." Heyden, London.

Synder, L. R., and Kirkland, J. J. (1979). "Introduction to Modern Liquid Chromatography" 2nd ed. Wiley, New York.

Tsui, K., and Morozowich, W., eds. (1978). "GLC and HPLC Determination of Therapeutic Agents" (Chromatographic Science: A Series of Monographs—Vol. 9). Dekker, New York.

Unger, K. K. (1979). "Porous Silica—Its Properties and Use as a Support in Column Liquid Chromatography" (Journal of Chromatography Library—Vol. 16). Elsevier, Amsterdam.

Chromatography—Abstracts, Bibliography, Series

ASTM, "Bibliography on Liquid Exclusion/Gel Permeation Chromatography" (1977). AMD 40-S1 (AMD-40, 1974). Soc. Test. Mater., Philadelphia, Pennsylvania.

ASTM, "Liquid Chromatography Data Compilation" (1975). AMD 41. Am. Soc. Test. Mater., Philadelphia, Pennsylvania.

TABLE I (*continued*)

C.A. Selects, "High Speed Liquid Chromatography." Am. Chem. Soc., Chem. Abstr. Serv., Columbus, Ohio.[a]

C. A. Selects, "Ion Exchange." Am. Chem. Soc., Chem. Abstr. Serv., Columbus, Ohio.[a]

"Chromatographic Science: A Series of Monographs" (1965). Vol. 1. Dekker, New York.

Giddings, J. C., Grushka, E., Cazes, J., and Brown, P. R., eds. (1965). "Advances in Chromatography," Vol. 1. Dekker, New York.

Johnson, E. L. (1977). "Liquid Chromatography Bibliography." Varian, Walnut Creek, California.

"Journal of Chromatography Library" (1973). 1. Elsevier, Amsterdam.

Knapman, C. E. H., and Maggs, R. J. (1960). "Gas and Liquid Chromatography Abstracts." Butterworth, London.

Lederer, M., ed. (1959). "Chromatographic Reviews," Vol. 1. Elsevier, Amsterdam.

"Liquid Chromatography Literature—Abstracts and Index" (1972). Vol. 1. Preston Publ., Niles, Illinois.

Marinsky, J. A., and Marcus, Y., eds. (1966). "Ion Exchange and Solvent Extraction," Vol. 1. Dekker, New York.

Niederwieser, A., and Pataki, G. (1970). "Progress in Thin Layer Chromatography and Related Methods," Vol. 1. Ann Arbor Sci. Publ., Ann Arbor, Michigan.

Chromatography—Journals

Chromatographia

Journal of Chromatographic Science

Journal of Chromatography

Journal of Chromatography—Biomedical Applications

Journal of Liquid Chromatography

Separation Science and Technology

[a] This title is an ongoing series.

in the broad field known as "separation science." Many monographs, treatises, and review articles which document the advances in chromatography have appeared. Table I lists the major sources of information specific to HPLC and briefly surveys other key chromatographic literature allied to HPLC.

3. History

It is generally recognized that the discoverer of chromatography was the botanist Mikhail Tswett. Although there were previous reports in the literature of separations as a result of selective adsorption, it was Tswett who recognized and laid the foundations of the chromatographic process. In his classic experiments Tswett carried out adsorption chromatography. Many of the terms used to describe modern chromatography were introduced by Tswett (Magee, 1970).

As frequently happens when a new discovery is made, chromatography remained relatively neglected for many years. Starting in the 1930s, however, many major contributions were reported in the literature. These included the synthesis of high capacity ion exchangers to replace the previously used zeolites and, in a short period of time, the foundations of thin layer, partition, and gas chromatography were developed. In essence, these discoveries and their subsequent explosive growth have all played a major role in the development of modern liquid column chromatography or HPLC.

Probably the most decisive papers, aside from the introductory work of Tswett, were the reports by Martin and Synge (1941a,b) in which they introduced partition chromatography (Magee, 1970). Several major contributions can be traced to this work. First, they proposed a theoretical treatment of the chromatographic processes by considering the operating parameters of the column and expressed the efficiency of the separation in terms of the theoretical plate by analogy to distillation. Second, they described partition or liquid–liquid chromatography by recognizing that, by immobilizing one liquid on a suitable stationary phase, a separation could be achieved by passing a second liquid containing the sample mixture over the first.

Two other major contributions can be found in the work of Martin and Synge. These are best illustrated by considering the following quotations: "The mobile phase need not be a liquid but may be a vapour" (Martin and Synge, 1941a); "Thus, the smallest height equivalent to a theoretical plate should be obtainable by using very small particles and a high pressure difference across the length of the column" (Martin and Synge, 1941a). The first is the prediction of gas chromatography (introduced some 11 years later by James and Martin) and the second is the prediction of HPLC (introduced in the mid-1960s).

B. Liquid Column Chromatography

1. Phases

Chromatography in most of its forms is a method whereby the components of a mixture separate because of differences in their relative affinities for a stationary and a mobile phase. The notable exceptions are those chromatographic methods in which a difference in the molecular size of the components in the mixture is the key factor in determining the separation. The potential phase combinations are summarized in Table II. In this chapter

TABLE II

Types of Phases in Chromatography

Mobile phase	Stationary phase	Chromatographic mechanism
Liquid	Solid	Adsorption, ion Exchange, size Exclusion
Liquid	Liquid	Partition
Gas	Solid	Adsorption
Gas	Liquid	Partition

only the first two are considered. Furthermore, only liquid column chromatography employing modern columns and instrumentation will be discussed here.

2. Chromatographic Process

A physical picture of the chromatographic process in column chromatography is illustrated in Fig. 1. The stationary phase is packed into a cylindrical tube that is plugged at the bottom and is conditioned by passage of a mobile phase. The sample mixture is introduced at the top of the column into the mobile phase which carries it into the column. If the proper eluting agent is chosen as the mobile phase, the components of the sample mixture pass through the column and distribute themselves according to the differences in their affinities for the two phases.

In Fig. 1 it is assumed that the sample is a 2-component mixture and that the affinities of the phases for the two components are significantly different. The sequence of events that occurs in the chromatographic system as a function of time are shown in Fig. 1: (a) illustrates the column as it is equilibrated with passage of the mobile phase; (b), the sample is introduced above the column bed; (c), passage of the eluting mobile phase causes the sample to penetrate the column; (d), as the flow of the mobile phase is continued

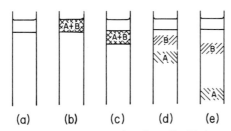

Fig. 1 Chromatographic processes in column liquid chromatography.

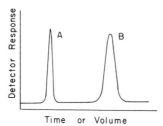

Fig. 2 A typical chromatogram in column liquid chromatography.

the two components begin to separate; and (e), with continued flow the distance between the two components increases. Eventually, the two components emerge from the bottom of the column. Monitoring the effluent of the column with a detector that responds to the components provides the chromatogram for the separation as shown in Fig. 2. For a fixed chromatographic system (a controlled flow rate with known mobile and stationary phases), the position of the chromatographic peak is characteristic of the component providing the peak, while the area under the peak is proportional to its concentration.

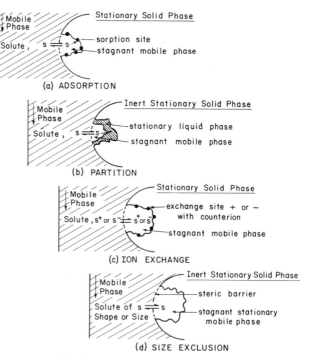

Fig. 3 Chromatographic mechanisms.

During the chromatographic process the solute is participating in mass transfer between the two phases at the interface (Fig. 3a) or in the bulk phase (Fig. 3b). The first is adsorption chromatography, and the latter is partition chromatography. Ion exchange and size exclusion mechanisms are illustrated in Figs. 3c and 3d, respectively. Since the system is dynamic, it is not at equilibrium throughout the column. However, the experimental conditions are usually chosen so that the system is as close to equilibrium as possible.

TABLE III

Intermolecular Forces That Influence
Adsorption and Partition Processes

Ionic forces
van der Waals forces
    (dipole–dipole, dipole–induced dipole,
    induced dipole–induced dipole)

Hydrogen bonding

Charge transfer

Many factors will influence the affinities resulting from adsorption and partitioning processes. To affect a separation, it is the intention of the operator to control, influence, and/or change these factors in some way so that the affinities of each component in the mixture will be different. The major factors are listed in Table III.

A quantitative discussion of the influence of the forces listed in Table III in chromatography is very complex particularly since the interactions occur between the solvent (mobile phase), solute, and stationary phase and are inter- and intramolecular. These interactions are summarized in Fig. 4.

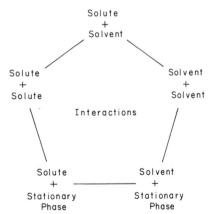

Fig. 4 Summary of the interactions contributing to the chromatographic process.

Many parameters that define a property of a given molecule or ion as it exists neat or in solution (usually aqueous solution) have been used in defining empirical relationships to account for the interactions. The more common ones are dipole moment, dielectric constant, polarizability, solubility, activity coefficients, equilibrium constants, boiling points, vapor pressures, electronegativity, and a variety of parameters associated with ionic properties.

Relationships accounting for these forces have been more successfully developed in adsorption column liquid chromatography (Snyder, 1962, 1968, 1974; Keller and Snyder, 1971) in comparison to partition column liquid chromatography (Martire and Locke, 1971). In general, however, most approaches in both areas tend to be limited because of the lack of availability of specific data and constants associated with the relationships.

Adsorption on a solid surface is as complex as partitioning when considering the system at the molecular level. Factors such as the distance between active sites relative to the size of the retained solute molecules, the degree and range of activity of the sites, the arrangement of the retained solute molecules, and the competition between the solute and solvent molecules for the site must be considered.

These types of interactions have been accounted for in a generalized equation developed by Snyder for adsorption chromatography on polar adsorbents such as silica gel and alumina

$$\log K = \log V_d + E_a(S^0 - A_s\varepsilon^0) \tag{1}$$

In this equation

(1) K is the adsorption constant for the solute;

(2) V_d and E_a, the adsorbent parameters, are the volume of the stationary phase and a surface activity function, respectively;

(3) S^0 and A_s are solute parameters which represent a dimensionless free energy of adsorption and the adsorbent area required by the solute, respectively; and

(4) ε^0 is a solvent strength parameter.

The details for the development of this expression are provided elsewhere (Snyder, 1962, 1968, 1974; Keller and Snyder, 1971).

Included within liquid–solid chromatography (Table II) are ion exchange and exclusion processes, see Figs. 3c and 3d. In the former, a reversible exchange of ions is possible between ions in the mobile phase and ions associated with oppositely charged ionic sites in the stationary phase. Thus, additional parameters to those listed in Table III are important. In exclusion processes molecular size of the solute in relation to pore sizes within the stationary solid phase is an additional parameter.

II. Theory

A. Fundamental Equations and Definitions in HPLC

1. The Equilibrium Constant

If the chromatographic system attains equilibrium, the ratio of the concentrations of the solute in the two phases is a constant and can be expressed as an equilibrium constant (assuming a linear isotherm). Thus, for a two component mixture of solutes 1 and 2

$$(K_D)_1 = C_s^1/C_m^1 \quad \text{and} \quad (K_D)_2 = C_s^2/C_m^2 \qquad (2)$$

where K_D is the distribution coefficient, and the solutes 1 and 2 are described by analytical concentrations C, in the stationary phase s, and in the mobile phase m. The sorption–desorption of 1 and 2 in the column at a given time is shown in Fig. 5, assuming the system is at equilibrium.

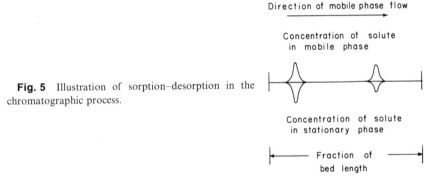

Direction of mobile phase flow

Concentration of solute
in mobile phase

Concentration of solute
in stationary phase

Fraction of
bed length

Fig. 5 Illustration of sorption–desorption in the chromatographic process.

2. Retention Times and Volumes

Consider the chromatogram in Fig. 6 where a constant, known flow rate is used in the elution. The appearance of the peaks in the effluent can be expressed in terms of time or volume. If the column dimensions (width and length) are known, the movement of the band can also be expressed by linear velocity. (Units in HPLC are usually cm/sec and the symbol U is used for linear velocity.)

In Fig. 6, t_{R_1} and t_{R_2} are the retention times for components 1 and 2, and t_m is the retention time for a solute that is not sorbed. Since this latter solute does not participate in the sorption–desorption process, it passes through the column at the same rate that the mobile phase is flowing, and is a measure of the total volume contained within the column. This is often referred to as the *dead* or *void volume* of the column. Corrected retention times, t_R' (also

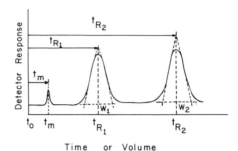

Time   or   Volume

Fig. 6   Illustration of chromatographic retention times and volumes.

referred to as the net retention time t_N) are given by

$$t_{R_1} - t_m = t'_{R_1} \tag{3}$$

$$t_{R_2} - t_m = t'_{R_2} \tag{4}$$

Since the chromatographic experiment is performed at a constant known flow rate, the position of each peak can be expressed as a retention volume or

$$V_{R_1} = t_{R_1} \times \text{flow rate} \tag{5}$$

$$V_{R_2} = t_{R_2} \times \text{flow rate} \tag{6}$$

$$V_m = t_m \times \text{flow rate} \tag{7}$$

where V_{R_1}, V_{R_2}, and V_m are the retention volumes for peak 1, peak 2, and a solute that is not sorbed, respectively. The corrected retention volumes, V_R' (also referred to as the net retention volume V_N) are given by

$$V_{R_1} - V_m = V'_{R_1} \tag{8}$$

$$V_{R_2} - V_m = V'_{R_2} \tag{9}$$

3. Separation Ratio

The ratio of the distribution coefficients for each of two components is called the separation ratio α, and is given by

$$\alpha = (K_D)_2/(K_D)_1 \tag{10}$$

By definition the component having the larger K_D value is designated as $(K_D)_2$. Therefore, the numerical value for α must be greater than one or equal to one if both components have the same K_D value.

If the chromatographic peaks are defined in terms of retention times as in Fig. 6 the separation ratio is given by

$$\alpha = \frac{t_{R_2} - t_m}{t_{R_1} - t_m} = \frac{t'_{R_2}}{t'_{R_1}} = \frac{(K_D)_2}{(K_D)_1} \tag{11}$$

For retention volumes α is

$$\alpha = \frac{V_{R_2} - V_m}{V_{R_1} - V_m} = \frac{V'_{R_2}}{V'_{R_1}} = \frac{(K_D)_2}{(K_D)_1} \tag{12}$$

The significance of the separation ratio is that the larger the value of α, the easier it should be to achieve a good separation. The actual values of the distribution coefficients for the components 1 and 2 are also important in determining the ease of separation. For example, both of the following provide the same α values

$$\alpha = 2/1 = 2, \qquad \alpha = 20/10 = 2$$

However, the latter separation would require a much longer analysis time.

4. *Isotherm*

Equation (2) states that the distribution coefficient is independent of concentration. Thus a plot of the concentration of the solute in the stationary phase versus the concentration in the mobile phase provides a linear distribution isotherm. From this kind of isotherm, a Gaussian shaped elution peak would be predicted and its retention time or volume should be constant as a function of sample size. This is illustrated in Fig. 7.

The ideal case is not always found. Figure 7 also illustrates two other types of isotherms. In both cases the distribution coefficient is dependent upon concentration, which causes the elution peak to be skewed and the retention or volume to change with sample concentration.

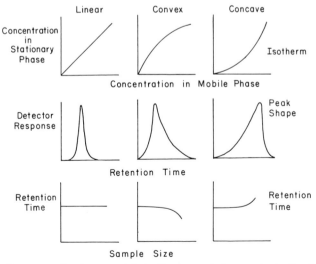

Fig. 7   Types of isotherms and their effect on peak shape and retention time.

The linear isotherm (actually all isotherms are found to be linear over a certain concentration range) and the convex isotherm are usually encountered in column chromatography. Much information about the nature of the interaction between the solute and the stationary phase can be obtained from isotherms particularly in liquid–solid chromatography (Snyder, 1962, 1968, 1974). For those that are nonlinear, the isotherms have been used to describe nonideal intermolecular attractions. For example, the convex isotherm is usually obtained where hydrogen bonding occurs between the solute and stationary phase. The first molecules sorbed cover up the most active sites and thus, additional sorption at increased concentration of solute is decreased. The concave isotherm is often characteristic of an end-on orientation of the solute molecule towards the stationary phase surface.

Another isotherm which can be encountered in chromatography is that which is influenced by chemisorption. In chemisorption attractive forces between solute and stationary phase are very strong and can be the result of a very stable hydrogen bond, charge-transfer interaction, or a chemical reaction. These forces, which are more chemical than physical, tend to be irreversible. Thus, the desorption kinetics are different than the sorption kinetics. Although usually undesirable in chromatography because of the very skewed elution peak, chemisorption can be used advantageously in situations where stripping is the main goal of the chromatographic application.

5. *Capacity Ratio*

Figure 8 illustrates the equilibrium distribution of the solute between the stationary and mobile phases. Consider a small section of the band that corresponds to the distance dx. The ratio of concentrations of the solute in the two phases is given by the column capacity ratio (also referred to as the

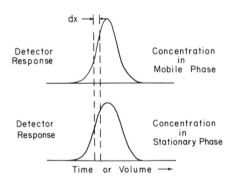

Fig. 8 Illustration of the equilibrium distribution of the solute between the mobile and stationary phase.

capacity factor) k'. Thus,

$$k' = \frac{q_s}{q_m} = \frac{C_s A_s \, dx}{C_m A_m \, dx} = \frac{C_s A_s}{C_m A_m} = K_D \frac{V_s}{V_m} \qquad (13)$$

where for the length dx, q is the number of moles of the solute in the two phases, A is the mean cross-sectional area of the two phases, V is the volume of the two phases, and s and m indicate the stationary and mobile phases, respectively. This is an important equation in chromatography since it relates the equilibrium distribution of the solute within the column to the thermodynamic properties of the column.

If one solute molecule were to be followed through the column it would spend a fraction of its time in the stationary phase and a fraction of its time in the mobile phase. Thus, the average fraction of time in the stationary phase is given by $q_s/(q_s + q_m)$, and the average fraction of time in the mobile phase is given by $q_m/(q_s + q_m)$. Combining with Eq. (13) gives

$$q_s/(q_m + q_s) = k'/(1 + k') \qquad (14)$$

and

$$q_m/(q_m + q_s) = 1/(1 + k') \qquad (15)$$

Assume that the mobile phase moves down the column at a constant velocity. U. When the solute molecule is in the mobile phase it moves at the velocity U, but when it is in the stationary phase it remains static. The average velocity of the band will, therefore, depend on the velocity U of the mobile phase, and the fraction of time the solute is in the mobile phase. The band velocity U_{band} is given by

$$U_{\text{band}} = U\left(\frac{1}{1 + k'}\right) \qquad (16)$$

The fractional rate R at which the band moves relative to the solvent front is given by

$$R = 1/(1 + k') \qquad (17)$$

and

$$k' = (1 - R)/R \qquad (18)$$

From Eq. (18) it can be seen that R is a function of the equilibrium distribution of the sample in the column. Thus, elution times (or volumes) are related to this quantity.

Using peak 1 in Fig. 6, the following can be written:

$$U_{\text{band}} t_1 = U_{t_m} \qquad (19)$$

which is

$$t_m/t_1 = U_{band}/U = 1/(1 + k_1') \tag{20}$$

and

$$t_m/t_1 = 1/(1 + k_1') \tag{21}$$

or

$$k_1' = (t_1 - t_m)/t_m \tag{22}$$

It can also be shown that the capacity ratio is given by

$$k_1' = (V_{R_1} - V_m)/V_m \tag{23}$$

It should be noted that k_1' is related to the distribution coefficient through Eq. (13). Also, the capacity factor is readily determined directly from the chromatogram.

6. The Chromatographic Equation

Combining Eq. (13) and (16) gives

$$U_{band} = UR \tag{24}$$

For conditions where the band does not move at all, $U_{band} = 0$, R must also be 0, while for the case where the band moves with the same velocity as the mobile phase, $U_{band} = U$, R must equal 1. Thus, $U_{band} = L/t_R$ and $U = L/t_m$, where L is the column length. Substitution into Eq. (24) gives

$$t_R = t_m/R \tag{25}$$

which, after conversion from retention times to retention volumes by Eqs. (2) and (22) gives

$$V_R = V_m/R \tag{26}$$

Combining Eqs. (13), (17), and (26) gives one of the fundamental equations of chromatography, or

$$V_R = V_m + K_D V_s \tag{27}$$

This equation is very useful in that it states that the total volume of mobile phase that flows while a given solute passes through the bed is the sum of two parts. The first part is V_m, which represents the dead or void volume through which every solute passes. The second part is $K_D V_s$ which represents the mobile phase that flows while the solute is immobile in the stationary phase. The latter term depends on the affinity of the solute for the stationary phase (K_D) and the amount of stationary phase (V_s). For partition chromatography the amount of stationary phase is expressed as V_s (volume of stationary phase); for adsorption chromatography it is expressed as A_s (the surface area of the adsorbent) or W_a (the weight of adsorbent); for ion exchange chromatography the capacity of the ion exchanger is expressed as V_s, and for exclusion chromatography the pore volume of the stationary phase is expressed as V_s.

7. Resolution

An indication of the separation of two chromatographic peaks is given by the resolution R_s. By definition it is the distance between the peak centers divided by the average band widths, where both are expressed in the same units. Thus, considering the two component system in Fig. 6,

$$R_s = \frac{t_{R_2} - t_{R_1}}{(W_2 + W_1)/2} = \frac{2\,\Delta t}{W_2 + W_1} \tag{28}$$

where W is expressed in the same time units as t or

$$R_s = \frac{V_{R_2} - V_{R_1}}{(W_2 + W_1)/2} = \frac{2\,\Delta V}{W_2 + W_1} \tag{29}$$

where W is expressed in the same volume units as V. For the case where the two peaks are close together, W_2 approaches W_1 and

$$R_s = \Delta t/W_2 = \Delta V/W_2 \tag{30}$$

It is important to note that resolution depends, not only on the distance between the peaks, but also on the narrowness of the peaks. The former is determined by column selectivity which is subject to the nature of the stationary and mobile phases. The latter is determined by the column efficiency which in turn is a function of column parameters such as flow rate, particle size, particle diameter, and packing of the column.

8. Efficiency

Column efficiency is expressed quantiatively by the number of theoretical plates N in the chromatographic bed. Since it is conveniently determined from the chromatogram its measurement will, in fact, account for the volume in connections between the column and the detector and for the volume within the detector itself. Thus, these volumes should be kept as small as possible.

When a solute band passes through the column several factors will contribute to broadening of the zone. The efficiency of the column describes numerically the extent to which the band is broadened and is dependent on a variety of column and kinetic parameters. These latter two factors will be considered later.

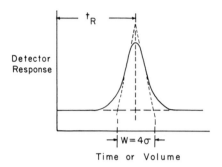

Detector
Response

Fig. 9 Illustration of the calculation of N from a typical chromatographic peak.

Consider the chromatographic peak in Fig. 9. The number of theoretical plates is given by

$$N = (t_R/\sigma)^2 = 16(t_R/W)^2 \tag{31}$$

where σ^2 is the variance of the peak in time units, t_R is the retention time for the peak, and W is the width of the peak at its base in time units. In units of volume Eq. (31) is

$$N = (V_R/\sigma)^2 = 16(V_R/W)^2 \tag{32}$$

where V_R is the retention volume, and W and σ are measured in volume units.

Column efficiency can also be expressed in terms of the plate height, H (HETP or height equivalent to a theoretical plate). Thus

$$H = L/N \tag{33}$$

where L is the length of the column bed. Since N is dimensionless, L and H must be expressed in the same units.

The number of plates and plate heights are a carryover from distillation theory and were introduced into chromatography through the historic work of Martin and Synge (1941a,b) in their development of the theoretical plate model

9. Factors Influencing Resolution

Resolution is related to efficiency, selectivity, and capacity factor, hence, control of resolution is possible by controlling these three fundamental parameters. The influence of the three are illustrated in Fig. 10a–d. Fig. 10a illustrates poor resolution. An increase in efficiency (narrowing of peaks) increases resolution as shown in Fig. 10b. An increase in resolution through the separation of the bands, Fig. 10c, is achieved by altering column selectivity so that peak 2 is sorbed to a greater extent. Resolution is obtained in Fig. 10c even though column efficiency has been reduced (peaks have been broadened). In Fig. 10d poor resolution is obtained because of low capacity factor in that neither peak is appreciably sorbed.

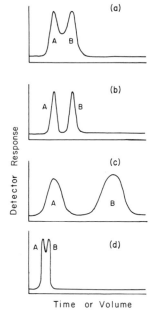

Fig. 10 Illustration of how efficiency, selectivity, and capacity factor affect resolution.

Combining Eqs. (11), (22), and (31) with (28), where $W_2 = W_1$, gives an equation relating resolution to N, the column efficiency, α, the column selectivity, and k', the capacity factor for the experimental condition, or

$$R_s = \left(\frac{1}{4}\right)\sqrt{N}\left(\frac{\alpha - 1}{\alpha}\right)\left(\frac{k'}{k' + 1}\right) \tag{34}$$

Since each of these variables can be varied independently, the influence of each on resolution can be evaluated.

The number of plates is related to the kinetics of the column and dependent on such readily controlled parameters as particle diameter, uniformity of the particle size, mobile phase velocity, and careful packing of the chromatographic bed. But changes in these factors which increase N will also increase R_s through a square root relationship. Since N is directly proportional to length of the column, it also follows that R_s is proportional to $\sqrt{L}$. Thus, if N or L are increased by a factor of 4, R_s increases only by a factor of 2.

While N deals with the chromatographic bed parameters, α and k' reflect those parameters which influence the interactions within the chromatographic system. Thus, variations in α and k' are achieved by changing the mobile phase, modifying the stationary phase, or by changing other parameters such as temperature and salt concentration.

If $\alpha = 1$ and $k_2' = k_1'$, then $R_s = 0$. However, for $\alpha > 1$ and for small changes in α, a large effect on resolution is experienced. A small change in k'

also contributes to a significant change in R_s. Although increasing k' can be desirable, too large of a k', or where $k'/(1 + k') \to 1$, will lead to long retention times and peaks which are broad; the larger the k' the larger the retention time and the broader the peaks. It is generally regarded that k' should not be larger than 5 to 10 if reasonable analysis times and narrow chromatographic peaks are desired.

The velocity in which an unretained solute migrates through the column is given by $U = L/t_m$. If this is combined with Eqs. (22) and (33) it can be shown that

$$N = Ut_R/H(1 + k') \tag{35}$$

Substitution for N in Eq. (34) gives the expression

$$t_R = 16R^2 \left(\frac{\alpha}{\alpha - 1} \right)^2 \frac{(1 + k')^3}{(k')^2} \frac{H}{U} \tag{36}$$

which states that the retention time, and subsequently the time for analysis, for the solute is a function of the resolution and the column operating conditions expressed in terms of selectivity, capacity, efficiency, and velocity of the mobile phase. If all other variables are held constant, a doubling of the resolution means that the retention time (time of analysis) will increase by a factor of 4. Doubling the velocity of the mobile phase should decrease the retention time by approximately a factor of 2 (H will also change with a change in mobile phase velocity and effect t_R). If the plate height is halved the t_R is halved. Finally, the largest effect is noted if α is changed; for example, if α increases from 1.1 to 1.2 the retention time decreases by a factor of 3 to 4.

As already noted large k' values produce excessive retention times. If t_R is plotted versus k' (see Fig. 11) a minimum is found. Furthermore, t_R varies little when k' is in the range of 1 to 5, provided k' has no effect on the other variables in Eq. (36). If Eq. (36) is differentiated with respect to k' and all variables are assumed to be constant, it is readily shown that the minimum is obtained when $k' = 2$.

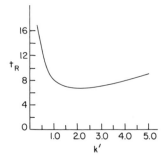

Fig. 11 Comparison of how retention time varies with capacity factor. (Reprinted by permission from C. F. Simpson, "Practical High Performance Liquid Chromatography," Heyden and Son Limited, London, 1976.)

B. *Chromatographic Models*

1. *Band Broadening*

As the components of a mixture travel through the column, each becomes diluted with the mobile phase and the concentration profile (or band) is broadened. If each of the components is traveling at a different rate, the band centers for adjacent bands will be separated and the completeness of the separation will be affected by the extent to which the adjacent bands are broadened. This section summarizes the highlights of the several approaches that have been taken to account for band broadening. Surveys as well as rigorous treatments have been provided elsewhere (see Grushka *et al.*, 1975; Giddings, 1965; see also reference books in Table I).

2. *Plate Model*

The first attempt to mathematically describe chromatography was the development of the plate model proposed by Martin, Synge, and co-workers (Martin and Synge, 1941a,b; James and Martin, 1950). Several assumptions were utilized in developing this theory. The major ones are the following.

(1)   The column is divided into a series of plates.
(2)   Distribution of the solute takes place between the stationary and mobile phase rapidly and reaches equilibrium within each plate.
(3)   The distribution coefficient for the solute is the same in all plates and is independent of concentration.
(4)   The diffusion in a longitudinal direction is considered to be negligible.

It is important to note that Martin and Synge fully realized the limitations of these assumptions. Specifically, they were aware that equilibrium is not reached at any point in the column. They were also aware of the fact that the plates are not discrete and discontinuous and that longitudinal diffusion does in fact take place.

Even with these simplifications the plate model has been a successful approximation in many aspects. It approximates band shapes in many cases and, perhaps its major contribution is that it has served as the fundamental basis for the development of more rigorous chromatographic band broadening theories and mathematical treatments (see, e.g., Grushka *et al.*, 1975; Giddings, 1965; Glueckauf, 1955).

The plate theory indicates the band shape to be a Poisson distribution which is Gaussian shaped at large plate numbers and is given by

$$C = C_0 \exp[-(V - V_R)^2/2\sigma^2] \tag{37}$$

and

$$C = C_0 \exp[-N(V - V_R)^2/2V_R{}^2] \tag{38}$$

where C is the concentration of the solute, C_0 is the concentration at the band maximum, V_R is the retention volume, V is the volume of mobile phase passed through the column, σ^2 is the band variance (in units of volume), and N is the number of plates. It can be shown further that

$$N = \frac{V_R^2}{\sigma_{(V)}^2} = \frac{t_R^2}{\sigma_{(t)}^2} = \frac{16 t_R^2}{W^2} = \frac{L^2}{\sigma^2} \tag{39}$$

since the standard deviation of the peak (square root of the variance) is directly proportional to the peak width and the peak width W is taken to be equal to 4σ. While N defines the number of plates, it also determines the distance a band spreads relative to the distance traveled. Thus, the larger the number of plates (greater efficiency) for a given length of column [see Eqs. (31) and (32)], the less the band spreads while passing through the column.

The plate height by definition is the rate of change in the band variance relative to the distance traveled or

$$H = d\sigma_{(L)}^2/dL \tag{40}$$

For a uniform column Eq. (40) becomes Eq. (32) ($H = L/N$) and

$$H = LW^2/16 t_R^2 = L/N \tag{41}$$

Since H is a measure of the band width, it also is an efficiency parameter; the smaller the H, the narrower the band and the more efficient the given column becomes.

It should be noted that if there are several processes contributing to spreading and these processes are independent of each other, statistically, the variance of the band will be the total sum of the variances for each of the processes. Thus, each chromatographic phenomenon must contribute to the efficiency parameters, N and H.

The plate model does not directly relate the various column processes to the band spreading even though Martin, Synge, and co-workers (Martin and Synge, 1941a,b; James and Martin, 1950) recognized these parameters. For example, mobile phase velocity, sorption phenomenon, temperature, diffusion processes, flow patterns, particle size of the stationary phase, solubilities of the sample, packing uniformity, particle size, and particle shape are not considered in the equations defining N and H. From a practical viewpoint this is the major limitation of the plate model theory since it is these same variables which can be readily manipulated by the investigator in carrying out HPLC separations.

3. Random Walk Model

The random walk model is a microscropic model in that it describes the behavior of individual solute molecules rather than that of the entire band.

This model was originated by Giddings and a rigorous discussion of it is provided elsewhere (Giddings, 1965).

In this model, the molecules are considered to be moving in a series of random steps as they advance down the column with the forward flow of the mobile phase. Each chromatographic process has a mean characteristic step length l and during residence in the column the solute will experience n such steps. Thus, after a certain time of movement down the column in a random travel, the solute molecules will become distributed about the central point. This positional distribution is given by

$$\sigma^2 = l^2 n \tag{42}$$

where the standard deviation (or width) of the distribution is the square root of Eq. (42). The random walk model according to Giddings (1965) defines these $n-l$ processes as broadening due to molecular diffusion and Eddy dispersion, and as broadening due to resistance to mass transfer in the stationary phase and in the mobile phase. The sum of these contributions provide the overall plate height expression.

Molecular diffusion of solute molecules introduced into the column as a small plug occurs in a longitudinal direction. The diffusion take places in both the mobile phase and stationary phase, with the former being much more significant. Since in HPLC a packed column is used, the broadening due to molecular diffusion, H_d, is given by

$$H_d = 2\gamma D_m/v \tag{43}$$

where γ is a tortuosity factor (<1) and is dependent on the nature of the packing of the column, D_m is the diffusion coefficient of the solute in the mobile phase, and v is the mobile phase linear velocity.

Equation (43) indicates that band spreading is decreased (efficiency increased) by using small uniform particles ($\gamma \to 1$), mobile phases of low diffusivity, and by operating at high flow rates. Since diffusion increases with the time the solute spends in the column, a high flow rate will decrease the residence time. Usually, with the flow rates and type of columns used in HPLC the effect of molecular diffusion on band broadening is the least significant.

Eddy dispersion occurs because flow velocity through a column varies for different paths through the column. Some molecules will travel more rapidly through open channels while others will lag behind because of slower flow through narrower channels. The average velocity determines the retention time, and the differing flow velocities making up the average contributes to broadening of the band about the average.

The band broadening due to eddy dispersion, H_E, is given by

$$H_E = 2\lambda d_p \tag{44}$$

where λ is a geometry factor related to the type of stationary phase particles used, and d_p is the particle diameter. Equation (44) indicates that the particles should be small, of uniform size, and homogeneously packed for a decrease in H.

Mass transfer describes the sorbing–desorbing process that occurs when the solute passes back and forth between the stationary and the mobile phase. Figure 12 illustrates why this process results in band broadening.

In Fig. 12a the upper band represents the distribution of the sample in the mobile phase and the lower one in the stationary phase at equilibrium. At a later instant movement of the mobile phase carries some of the sample down the bed and ahead of the mean for the band (see Fig. 12b). For equilibrium to be restored, some sample at the leading edge of the mobile zone must be transferred and sorbed to the stationary phase and at the tailing edge some of the sample must be transferred or desorbed from the stationary phase to the mobile phase. This mass transfer will occur as the sample migrates down the column and their sum contributes to the broadening of the zone.

The broadening due to resistance of mass transfer at the stationary phase, H_s, when using a stationary liquid phase is given by

$$H_s = qrd^2v/D_s \tag{45}$$

and for the mobile phase, H_m, by

$$H_m = \omega d_p{}^2 v/D_m \tag{46}$$

where q is a geometrical factor determined by the shape of the stationary phase particles, r is a parameter which is related to the fraction of molecules that spend their time in the two phases and is related to the distribution

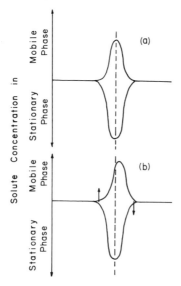

Fig. 12   Zone broadening due to mass transfer.

coefficient K_D and column capacity ratio, d is the thickness of the stationary phase penetrated by the solute, D_s is the diffusion coefficient for the solute in the stationary phase, D_m is the diffusion coefficient for the solute in the mobile phase, and ω is a column coefficient determined by the packing, the column diameter, and its shape.

Equation (45) indicates that band broadening is reduced by high diffusivity and by utilizing a stationary phase with a uniform configuration in which only a thin film of the phase is used in mass transfer. If the stationary phase is a solid, Eq. (45) is modified (Giddings, 1965). Equation (46) indicates that band broadening due to mass transfer effects in the mobile phase is reduced by uniform packing, avoiding column shapes that interfere in normal diffusion and mobile phase flow, by using mobile phase of low viscosity, and by using small uniform particles (affects plate height by the square power).

If each contribution, as suggested in the plate model, is independent of the others, the overall effect on H, according to Giddings (1965), is the sum of each contribution or

$$H = H_E + H_d + H_s + H_m \tag{47}$$

and for a stationary liquid phase H becomes

$$H = 2\lambda d_p + 2\gamma D_m/v + qrd^2v/D_s + \omega d_p^2 v/D_m. \tag{48}$$

Added to Eqs. (47) and (48) is a fifth term which accounts for mass transfer in the stagnant mobile phase within the packing particles.

Equation (48) has the form

$$H = A + (B/v) + C_m v + C_s v \tag{49}$$

which was previously derived for gas chromatography as the van Deemter equation (van Deemter et al., 1958). In Eq. (49), A and B are column constants and C_m and C_s are mass transfer terms for the mobile and stationary phases, respectively.

It is generally recognized that the various contributions to the overall expression of H are, in fact, not completely independent of each other. However, the conclusions regarding the factors that influence each contribution are valid, and Eq. (48) is a good qualitative description of those parameters that can be experimentally optimized in HPLC.

Several workers have improved the fundamental significance of the random walk model by a more detailed consideration of individual factors. For example, Giddings (1965) pointed out that the resistance to mass transfer in the mobile phase and in the eddy term are not independent of each other, but, in fact, are coupled. Equation (49) becomes the coupled plate height expression, or

$$H = (B/v) + C_s v + [1/(1/A + 1/C_m v)] \tag{50}$$

where the net result is that the coupling of the two terms is smaller than that due to their individual sums. Equation (50) is particularly more sound for HPLC since the B term (molecular diffusion term) is usually insignificant.

Other workers have considered the effect of interchannel and intraparticle contributions (Knox and Saleem, 1972) and the effect of the amount of stationary phase (Gruska et al., 1971) on resistance to mass transfer in the mobile phase. Stagnant mobile phase pockets can build up in the pores of the stationary phase particles. The influence of these pockets on mass transfer has been examined by Hawkes (1972).

4. The Nonequilibrium Theory

The nonequilibrium theory was initiated by Giddings and is rigorously treated elsewhere (Giddings, 1965). This approach, unlike the random walk approach, deals with gross processes of mass transfer and is used primarily for calculation of the C terms [see Eq. (49)] in the plate height expression. Briefly, it is a rigorous attempt to mathematically describe the chromatographic system indicated in Fig. 12 as it strives to reach equilibrium. The magnitude of the departure from equilibrium is, therefore, a measure of band broadening that the solute experiences.

The nonequilibrium theory, being more rigorous, becomes very powerful where H calculations are required for complex chromatographic systems (or for those where the random walk method fails) and for systems where particular attention is directed towards geometrical effects of the stationary phase particles. The major limitation in applying the theory is the difficulty in being able to identify and define all mass transfer processes mathematically.

5. Mass Balance Model

The mass balance model and its foundations have contributed much to chromatography (Grushka et al., 1975). For example, this approach was used to derive the van Deemter equation (van Deemter et al., 1958). More recently, this approach and its refinement have been used to identify information not only about the efficiency of the system but also about the shape of band profiles as they emerge from the column (Grushka, 1972). Perhaps this latter property is its main value since the band shape can provide much information about resolution and band capacity, as well as efficiency.

6. Plate Height Equations

Several variations of plate height equations have been derived and are reviewed elsewhere (Grushka et al., 1975; Waters et al., 1969; Snyder, 1969; Knox, 1976; Huber, 1969). In general, the principal value of these equations is to focus on specific chromatographic parameters and thus be able to optimize the column efficiency relative to that parameter and to be able to compare the properties of different columns. Often, these equations are modified versions or have their basis in the random walk model and the coupled plate height expression.

Although plate height and/or plate number are used to describe column efficiency, other defined parameters can also be used. Thus, approaches such as number of effective plates, number of plates generated in a given time, and the reduced parameter approach are becoming more widely used (Grushka et al., 1975; Knox, 1976, 1977). Probably, the main value of these are that they readily provide a basis for examining the many different chromatographic parameters and, subsequently, for optimizing the parameters. It appears that the reduced parameter approach (Knox, 1976, 1977) will prove to be particularly important in this respect.

C. Column Parameters

1. Scope

In designing a chromatographic separation three goals are usually sought. These are resolution, speed, and capacity. An optimum in all three cannot be achieved simultaneously, and a compromise is usually sought. Thus, any improvement in one is at the expense of either or both of the other two. In analytical HPLC, speed and resolution are usually desired and capacity is secondary. In preparative HPLC, the reverse is the goal; that is, capacity is the primary goal. Several parameters which influence these goals are readily controlled in the laboratory and these are briefly described in the following paragraphs.

2. Column Length

Column length is an important experimental parameter that controls the number of available plates. This in turn influences the resolution and the retention time.

The actual lengths of columns used in HPLC are usually determined by the resolution required, inlet pressures, and time of analysis. In some cases, longer columns are used because of the inability to detect the sample. This permits a larger sample to be used (the extra length compensates for approaching or actually overloading the column) so that it can be detected.

3. Column Diameter

Small diameter columns are used to minimize multiple path effects. Typically inner diameters of HPLC columns range from 2–5 mm. Since the sample capacity varies with the diameter (capacity increases with the square of the diameter), larger diameter columns are only used for preparative HPLC. If short columns are used they are usually used as linear columns. Longer columns can be made by connecting a series of shorter columns with dead volume connections. Coiling and other shapes have been considered, but these generally offer little or no advantage and can sometimes even provide a poorer performance (Barth *et al.*, 1972).

An optimum column diameter is predicted by considering the infinite diameter effect (Knox, 1966; Knox and Parcher, 1969; Laird *et al.*, 1974). If the sample enters the packed bed at the center of its cross section, lateral dispersion is given by

$$W_r^2 = 2.4d_p z + (32D_m/U)Z + W_i^2 \qquad (51)$$

where W_r is the radial peak width, W_i the initial band dispersion, d_p the particle size, Z the distance traveled in the column at velocity U, and D_m the diffusion coefficient of the solute in the mobile phase. In Eq. (51) the first term is due to stream splitting and the second to molecular diffusion. A typical peak spread illustrating an unfavorable and favorable condition is shown in Fig. 13. In the first case, Fig. 13a, the band reaches the wall edges long before it emerges from the column. Wall effects will prevail and contribute to a variation in flow rate across the column, particularly near the wall edge, and contribute to a further broadening and distortion of the peak. In the favorable case, Fig. 13b, wall effects are minimized. Thus, as a guideline

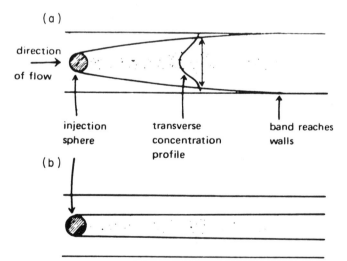

Fig. 13 Lateral band spreading of a sample injected centrally into a wide column. The particle-to-particle diameter ratio is 10 for (a) and 100 for (b). (Reprinted by permission from C. E. Simpson, "Practical High Performance Liquid Chromatography," Heyden and Son Limited, London, 1976.)

the column diameter, d_c, should be chosen so that

$$d_c > W_r \qquad (52)$$

Since central introduction of the sample is not always achieved, a safety factor of approximately 1 mm should be added to d_c.

From Eq. (51) two other parameters, namely column length and particle diameter are very important in developing and utilizing the infinite diameter column. In addition, it is important that columns be packed uniformly. Details about the relationship of particle size of the stationary phase to column diameter to produce an efficient infinite diameter column are provided elsewhere (Knox, 1966; Knox and Parcher, 1969; Laird *et al.*, 1974; Knox, 1976, p. 39).

4. *Particle Diameter*

Particle diameter and its uniformity in size is an important parameter that has a significant influence on the efficiency of the column. This is clearly pointed out in the band broadening theories outlined in the previous section.

Optimizing particle diameter in applications is complex because several practical factors are involved. Consideration must be given to pressure drop across the column, the availability of the packing material, the actual packing of the column, the nature of the instrumentation available (see Section III for a discussion of how several of the instrumental components can influence the observed efficiency) and the required resolution and analysis time. If the best efficiency is desired a uniformly packed column of spherical particles of uniform size (5 μm) at a column length of 100 mm is recommended assuming the available instrumentation is of favorable design (Knox, 1977). In practice, however, many quantitative separations are readily accomplished with particles of 37–50 μm size and at convenient analysis times. Furthermore, these particles are readily available, and columns of various lengths and diameters (consider infinite diameter) are packed in the laboratory by simple techniques.

For complex mixtures containing closely related components it is necessary to carefully design the column to achieve maximum efficiency so that adequate resolution is obtained. For simple mixtures that are readily resolved, there is usually a large margin in which to design a column for the separation.

5. *Types of Stationary and Mobile Phases*

Two of the main factors that determine resolution, efficiency, capacity, and time of analysis, and which are directly controllable, are the column packing (stationary phase) and the mobile phase. As previously indicated, adsorption, partition, ion exchange, and exclusion chromatography are readily carried out with HPLC techniques. The topics of stationary phases and mobile phases are treated in detail later in this chapter.

6. *Elution*

Elution in HPLC is done either isocratically or through the use of a gradient. For example mixtures or those which are easily resolved, isocratic elution is preferred. For more complex mixtures, particularly those that have a wide range in capacity factor values, optimum analysis time and resolution are achieved by employing a gradient.

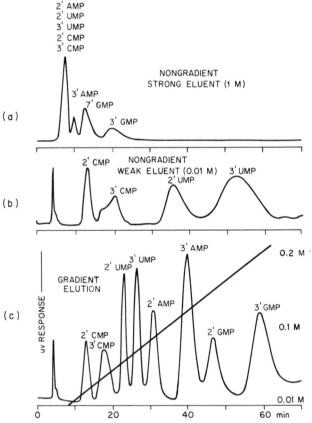

Fig. 14 Comparison of nongradient and gradient separation of a series of nucleotides. Separation (a) employs isocratic elution with a strong eluent of 0.1 M in electrolyte. Separation (b) employs isocratic elution with a weak eluent of 0.01 M in electrolyte. Separation (c) employs a linear gradient from 0.01–0.1 M in electrolyte. (Reprinted by permission from N. Hadden, F. Baumann, F. McDonald, M. Munk, R. Stevenson, D. Gere, F. Zamaroni, and R. Majors, "Basic Liquid Chromatography," Varian Aerograph, Walnut Creek, California, 1971.)

Although temperature and flow gradients can be employed in HPLC, almost all gradients used are a change in mobile phase eluting condition from one of weak eluting power to one of strong eluting power. Since using a gradient adds another degree of complexity to the separation it is essential that the gradient system provide reproducibility, versatility, rapid equilibration, and ease of operation. The first, reproducibility, is perhaps the most important, particularly, if the generated chromatogram is being used for qualitative or quantitative purposes. Slight changes in the gradient from separation to separation can have a significant effect on the retention time of the chromatographic peak (used for qualitative analysis) and on the area under the chromatographic peak (used for quantitative analysis).

Figure 14 illustrates the power of gradient elution. In Fig. 14 a strong isocratic eluting mixture of high salt concentration is used; only the more strongly retained components are resolved. In Fig. 14b a weak isocratic eluting mixture of low salt concentration is used; only components of low k' are easily resolved while those of large k' are obtained with broad peaks with large retention times. Using a gradient in which the salt concentration changes, Fig. 14c, the entire mixture is resolved in well-defined peaks and convenient elution time. Figure 14c illustrates the use of a linear gradient. Concave, convex, and stepwise gradients can also be employed.

Flow rates in analytical HPLC are used in the range of 0.5–5 ml/min, usually closer to the lower end of this range. Efficiency, and subsequently resolution, are favored by a low flow rate. In routine practice, however, a compromise is made and flow rates higher than those providing the best efficiency are used in order to reduce analysis time. Higher flow rates provide quicker elution, and consequently, more sample injections can be made per unit of time.

7. Inlet Pressure

Inlet pressures used in HPLC vary widely and depend on a variety of readily controlled experimental factors. In general, although operating inlet pressures may range from 200 to 5000 psi, a typical pressure is about 1000 psi. The higher pressures are not often needed, and frequently, if needed, they are an indication of a poorly packed column, fines in the column, plugging in the column, or swelling of the particles used as the stationary phase.

The pressure drop Δp across the column is given by

$$\Delta p = \phi \eta L u / d_p^2 \tag{53}$$

where ϕ is a dimensional parameter called the column resistance factor (Table IV lists typical ϕ values), η the viscosity of the mobile phase, L the column length, u the velocity of the mobile phase (given by L/t_m), and d_p the particle diameter.

TABLE IV

Typical Column Resistance Factors, ϕ

		Dry packed	Slurry packed
Solid-core particles	(spherical)	600–700	300–400
Porous particles	(irregular)	1000–2000	700–1500
	(spherical)	800–1200	500–700

Equation (53) states that the pressure drop increases as the particle size decreases by an inverse square function. Hence, it might be concluded that modern HPLC requires very high pressures because small particles are used. Although particle size is a contribution other reasons are as significant or even more significant parameters in determining the pressure drop (Knox, 1976).

8. Temperature

Temperature of the column in HPLC is a complicated parameter and can have a significant effect on the chromatographic result. Usually, ambient temperature is used for routine work, while for more fundamental studies, column temperature should be controlled.

Speed and resolution as well as operating conditions can be affected by increasing column temperature. The distribution coefficients of the solutes in the mixture between the two phases is sensitive to temperature effects. The pressure drop across the column is dependent on viscosity, see Eq. (53), which is affected by temperature. Because of the kinds of typical mobile phases used in HPLC, maximum temperatures used are seldom over 60°C.

D. Summary

Up to this point, the main considerations have focused on the column parameters in the chromatographic process and some of the practical aspects associated with these parameters. The actual types of mobile phases and stationary phases used in HPLC are considered in Section IV.A of this chapter.

It has been shown that resolution is optimized by alteration of α, N, or k'. The α' and k' are changed by altering the mobile phase, stationary phase, and temperature, while increasing N in HPLC is achieved by packing a more uniform column, reducing the flow rate, using a longer or multiple set of columns, reducing the particle diameter of the stationary phase used in the column, and using a less viscous mobile phase.

III. Instrumentation

A. The Chromatograph

1. Scope

Modern liquid column chromatography requires a careful control of all the chromatographic parameters and very sensitive detection if it is to be used effectively. Much of the instrumentation currently used has been developed or refined in the last 10 years and is still subject to improvement. The discussion in this section is one which should acquaint the reader with the instrumentation currently available and provide some insight into the advantages and disadvantages of the different components.*

The liquid chromatograph should be versatile and inert under the conditions of intended use. It should be capable of providing a wide range of flow rates and a reasonable precision for controlling flow, temperature, solvent composition, and detector response. Since high sensitivity detection is required, the signal to noise output should be large.

In addition to the above, the instrumentation should have low dead volume. In the previous sections the influence of broadening on resolution within the column was considered. Part of what is gained in the column through optimum design and control of column parameters can be lost if the instrumentation is poorly designed. That is, the broadening effects observed in the actual performing of a separation are not only due to those occurring within the column but also within the tubing, injector, and detector. Careful attention must be directed to proper sample injection, maintaining low volume tubing and low dead volume connectors, and utilizing proper detector geometry. Poorly moving or trapped mobile phase and excessive volume outside the column will reduce resolution. These factors are even more critical when using columns of very small internal void or deal volume.

2. A Liquid Chromatograph

Figure 15 shows a block diagram of a liquid chromatograph which indicates the main components. Complete LC instruments can be purchased or the user can buy or build individual components and fabricate a complete LC instrument. The following outlines the main features of the components. Additional information is readily available from the manufacturers and other sources (Parris, 1976; McNair and Chandler, 1974, 1976; McNair,

---

* The use of manufacturers' products here is intended to illustrate particular points and is not an endorsement of that particular component over all others. Often comparable products can be obtained from many manufacturers, and it is up to the user to make the final decision about which offers the best advantages for the types of problems faced by the user.

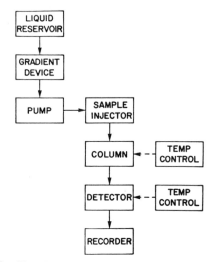

Fig. 15 Block diagram of a liquid chromatograph.

1978; Bombaugh, 1977; Wehrl:, 1974, Snyder and Kirkland, 1974; see also Table I).

3. *Mobile Phase Delivery System*

The mobile phase delivery system consists of a pumping system and a solvent supply. In general, the pump should be able to operate up to a maximum pressure of 3000 to 6000 psi and deliver the solvent at a known constant, reproducible flow through the column.

The requirements that a successful pump should satisfy are listed in Table V. The manufacturers of HPLC pumps have strived to satisfy these kinds of requirements, but no single pump meets them all, particularly when attempting to use one pump for both analytical and preparative scale HPLC. If a wide variety of separation problems are encountered in the chromatography laboratory, it is usually beneficial to have several different types of delivery systems available for use.

There are five types of pumps which comprise the two general classifications of pumps or (A) those that provide a constant pressure and (B) those that provide a constant flow. Table VI lists the type of pumps currently available.

Constant pressure pumps have the distinct advantage of being pulseless and thus, their operation will not affect the detector response. However, the flow rate can change if the permeability of the column or the viscosity of the mobile phase changes during the separation. This will affect the elution

TABLE V

Requirements for an HPLC Mobile Phase Delivery System

Sufficient pressure should be provided to maintain adequate flow rates with the selected column.

The flow rate should be reproducible.

A constant flow rate is necessary for qualitative and quantitative analysis.

The delivery should be pulse-free for minimum detector response at the required sensitivity.

A wide range of flow rates, 0.1 to 2 ml/min for analytical HPLC, 1 to 10 ml/min for small scale preparative HPLC, > 10 ml/min for large scale preparative HPLC, should be available.

Continuous operation is required.

The replacement or altering of the mobile phase should be possible with relative ease.

The delivery system should be adaptable to gradient elution.

The delivery system should not deteriorate in the presence of organic solvents nor should it corrode in the presence of electrolytes and buffers.

Degassing problems should be correctable.

TABLE VI

Types of Pumps Used in HPLC

(A) Constant pressure pumps

Liquid displacement by compressed gases

Liquid displacement of a piston driven by a compressed gas (pneumatic amplifier)

(B) Constant flow pumps

Piston or diaphragm driven by a moving fluid

A reciprocating piston

A syringe pump

volume and make it more difficult to use the data for quantitative and qualitative work.

Figure 16 illustrates a typical design for a liquid displacement pump driven by compressed gas (a) and by a piston driven by a moving fluid (b). Pumps of the first type are relatively inexpensive and utilize pressure from a gas tank. Safety features must be incorporated into the system so that the applied tank pressure can be controlled. Pressures are usually limited to 1500 psi, and it is necessary to avoid dissolving the driving gas in the mobile phase which enters the column (if it does dissolve serious degassing will

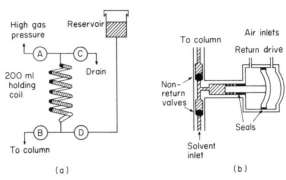

Fig. 16 Two examples of constant pressure pumps. (Reprinted by permission from J. N. Done, J. H. Knox, and L. Loheac, "Application of High Speed Liquid Chromatography," John Wiley and Sons, New York, 1974.)

occur in the column and/or detector). These pumps are restricted to isocratic elution.

The second type shown in Fig. 16b employs a compressed gas (air) at low pressure acting on a piston of large surface area which is connected directly to a hydraulic piston of low surface area. The ratio of these areas will determine the pressure amplification.

These kinds of pumps have a limited volume unless modified. One type of modification employs a rapid, automatic refill while the second employs two chambers, so that when one is emptying the other is being refilled. The pump can also be modified to deliver a constant flow and to deliver a gradient.

The liquid displacement pump employing pneumatic amplification is capable of producing the highest pressures available for HPLC ($>75\,000$ psi). Normally, inlet column pressures of this type should not be encountered in HPLC. If this kind of pressure is required to achieve a reasonable flow it is usually the result of a poorly packed column, particle fines in the column, or plugging at some point in the line. The availability of these high pressures, however, can be useful in HPLC for the packing of the stationary phase particles into the column, particularly when dealing with the micro size stationary phase particles.

Basic designs of the three types of constant flow pumps are illustrated in Fig. 17. A liquid driven diaphragm pump and a reciprocating pump are illustrated in Figs. 17a and 17b, respectively. The main difference between the two is that in the former a hydraulic fluid transmits the pumping action while in the latter the pumping action of the piston is directly on the solvent. These pumps have the undesirable characteristic of producing a pulsating flow, which can give rise to considerable baseline instability in flow sensitive

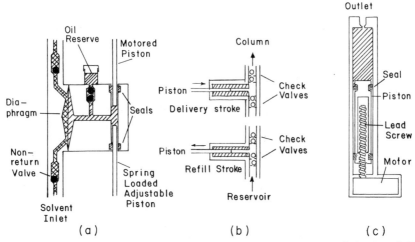

Fig. 17 Two examples of constant flow type pumps. (Reprinted by permission from J. N. Done, J. H. Knox, and L. Loheac, "Application of High Speed Liquid Chromatography," John Wiley and Sons, New York, 1974.)

detectors. Even with this limitation most pumps used in HPLC today are of the reciprocating piston type (Fig. 17b).

Several different kinds of piston pumps are commercially available and not all are designed for minimizing or elimination of the pulsation action. Those that do compensate for this are more expensive and are generally based on a dual pump system. There are variations in designs, but basically the dual piston pump works on the principle that when one pump is filling the other is emptying. Usually, a long restrictor coil of fine diameter tubing further dampens the pulse that remains in the delivered solvent. It is also possible to electronically program the emptying and filling of the two pump system by independently controlling the movement rate of the two pistons in order to maintain a uniform pressure throughout one complete cycle. A recent, and now commerically available, innovation is to use a three piston pump arrangement.

For the nondual type piston pumps the pulsating action can be minimized by adding a closed tube with an air space, a flexible bellows or tube, or a restrictor into the line after the pump. The first two are similar in that the air space or pressurized tube takes up some of the energy of the pulsation during the emptying stroke of the piston and then releases it during the refilling part of the stroke. This helps to maintain the pressure in the system. The third type can be a long coil of fine diameter tubing or a length of tube packed with fine diameter glass beads placed in the line after the pump.

In selecting a suitable piston pump it is important to consider the volume of the pump, its damping devices, and connecting tubes both from a point of view of convenience and practicality. Flushing the system can be time consuming, but more importantly, excessive volume makes it more difficult in utilizing gradient elution and recycle. Frequently, flushing valves are included in the pump design.

Piston type pumps are readily adapted to gradient elution. Usually, this is achieved either by premixing the solvents prior to pumping or by the addition of a second pump and an electronic module to the system. The module operates both pumps so that each delivers solvent from separate reservoirs to a mixing area in a known predetermined ratio.

The syringe pump shown in Fig. 17c is a constant flow pump that delivers the solvent pulse free. These kinds of pumps, which operate by a screw gear displacing a plunger through the reservoir, produce stable flow rates and are capable of developing high pressures. Syringe pumps are well suited to gradient operation since each solvent can be contained in separate syringe reservoirs and the speed of each syringe plunger can be electronically controlled to produce the gradient. The main disadvantages of the syringe pump are the inconveniences associated with changing the solvent system in the syringe and the fact that the syringe reservoir is limited in volume.

A suitable reservoir should be available for the nonsyringe type pumping systems. Ideally, it should have a capacity over 200 ml and be fitted with drying tubes if completely nonaqueous eluting mixtures are being used. An amber colored reservoir is preferred when using chlorinated solvents so that decomposition due to uv absorption does not take place.

When using the reciprocating and piston type pumps, it is necessary to make sure that the solvent coming from the reservoir does not contain dust particles. These kinds of pumps usually contain about $2-3$ μm filter disks to prevent dust from getting into the pump heads. Excessive dust will clog the filter and lead to poor performance. Furthermore, pump seals must be compatible with the solvents used to prevent excessive softening and subsequent breakdown of the seal.

Several of the pump systems indicated are readily modified so that they can be utilized in gradient elution. Many manufacturers have designed programmers that allow the operator to select the particular gradient desired for the separation. In general, they are electronic modules that will operate two or more pump systems that are connected to several reservoirs. Thus, the different solvents are introduced into a mixing area on the high pressure side according to how the operator wants the gradient to change. Typical designs may involve 2 or more identical high pressure pumps or one high pressure pump combined with a low pressure pump. It is also possible to generate gradients with simple mixing chambers and a single pump (Snyder, 1965).

Although inexpensive, this latter type of system does not generally offer the reproducibility provided by the commercial instrumentation. A recent useful design employs a single high pressure pump which switches from reservoir to reservoir by a switching valve according to a predetermined program selected by the operator. The solvents are mixed on the low pressure side in a mixing chamber prior to passage through the pump. This system can also be designed to produce ternary gradients.

Gradients involving temperature programming and flow programming are also possible (Scott, 1976). Although having potential in HPLC applications these kinds of gradients are not routinely used at present.

4. Introduction of the Sample

Introduction of the sample should be accomplished with minimum effort and the least possible interruption of the stabilized operating system. Since a suitable syringe offers accuracy and flexibility at sample volumes from a fraction of a microliter in analytical HPLC to several milliliters in preparative HPLC, development of HPLC injection systems utilizing syringes has taken place. Table VII cites the goals that an ideal injector should satisfy.

Samples can be introduced by one of three methods: (1) syringe–septum injection, (2) sample valve injection, and (3) stop-flow injection. The last of these three techniques for sample introduction is a carryover from classical liquid column chromatography and offers no advantage in HPLC. Even in preparative scale chromatography, where large samples are introduced, it offers no advantage since large scale injection can be accomplished by the other two techniques.

The main advantage of syringe injection is that the sample can be introduced directly into the column. If this is done diffusion of the sample prior to reaching the column is eliminated and little loss in efficiency (band broadening) is experienced in the injection step. Injection into the column

TABLE VII

Requirements for an HPLC Injector

1. On-stream injection
2. Usable at operating pressure
3. Solvent compatible and noncontaminating
4. Accurate and reproducible
5. Minimum contribution to band broadening
6. Application to analytical and preparative scale HPLC
7. No viscosity limit

packing, however, is hard on the syringe since the needles are fine bore and can be easily plugged by the stationary phase particles. Also, damage of the syringe tip to form burrs will shorten the life of the injection septum. One technique that is widely used to minimize these problems is to have a very thin layer of small glass beads at the head of the column, usually separated from the column packing by a thin stainless steel disc. Since particles will break off the septum during injection, this technique not only allows injection into the column and prevents needle plugging, but also facilitates cleaning up the contamination that forms at the head of the column without disturbing the column.

The choice of septum material is limited since the septum at some point will be in contact with the mobile phase. The better design for the injection system would be one in which the septum can be isolated from the mobile phase except during injection. This minimizes septum decomposition.

The life of a septum depends on its composition, the operator, and the design of the injector. It is a recommended procedure that one fine hole be introduced into the septum and that this hole be used repeatedly. Some workers have reported using a single septum for months while others have reported very short septum life times.

Several different valve injection systems have been designed. In general, all employ introduction of the sample into a sample loop by a syringe at atmospheric pressure. The valve mode is then changed and the sample is introduced into the line just above the column. Switching back to the original position allows the introduction of another sample. Although these valve injection systems have no septums, they still contain a syringe needle seal of some type and these seals are subject to wear.

Since the sample must pass through a loop and connecting tubing before reaching the column, broadening due to laminar flow may occur. With proper design and appropriate dead volume connections this broadening contribution is very small and comparable to that found in the better septum injection systems.

Aside from the differences in design approach taken by the manufacturers, the valve injection systems differ by whether they are fixed loop or variable loop type injectors. The general basic designs for valve type injectors are shown in Fig. 18.

The sample loop in the fixed loop type is at a fixed, accurately known, volume. In the load position, Fig. 18a, the syringe is introduced and the solution in the loop is replaced by continually injection of the sample until it appears in the drain vent. The sample size, therefore, corresponds to the loop size. Other sample sizes are obtained by changing the loop. Switching to the inject mode sends the sample onto the column with little or no sample loss.

In the variable loop injector, Fig. 18b, a similar approach is taken, except that the loop is large in volume (depending on its manufacture, it can be up

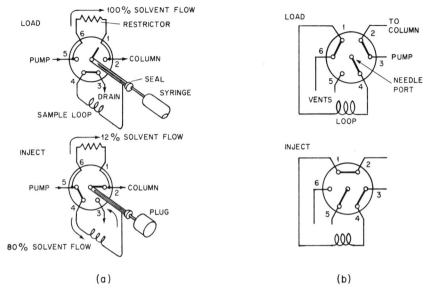

Fig. 18   Fixed loop (a) and variable loop (b) syringe injector. (Reprinted by permission from Rheodyne, Berkeley, California.)

to 2 ml) and the sample is introduced to the front of the loop nearest to the column. A corresponding volume of solution in the loop leaves the vent tube during injection. When the valve is switched to the inject mode a restrictor coil causes the major flow through the loop and sends the sample to the column. Depending on design details and its manufacture, the sample loss can be negligible to slightly more than found for the fixed loop injection system.

Whether using septum or valve type injection systems, specially designed high pressure syringes are generally not required. Normal syringes of appropriate accuracy are usually used. Also, it is important to use syringe needles that are free of burrs and have diameters that match the recommendation of the injector manufacture to avoid excessive wear (and leakage) of septums or seals. It is also important to keep in mind that the sample valve approach is readily automated for unattended HPLC operation.

B.   Detectors

1.   Scope

An ideal detector for modern HPLC should have a high sensitivity and a broad linear dynamic range. Other important performance characteristics that determine a detector's usefulness are its response and noise. Ideally, the detector should respond only to the solute and the response should be independent of the characteristics of the mobile phase. Thus, composition of the mobile phase can be altered, as in gradient elution, without influencing the background detector response.

Although individual HPLC detectors may be optimum in several of these requirements, no single detector is optimum in all operating parameters. Consequently, for the present several different detectors are required in the HPLC laboratory that is designed for a wide scope of applications.

The detectors used in HPLC can be classified in general terms as either bulk property detectors, solute property detectors, or specialty type detectors. The bulk type detector functions by measuring a bulk physical property of column eluent, while the solute type detector functions by measuring a physical and/or chemical property that is characteristic of the solute only. The specialty type detector is also a bulk property detector but differs in that the mobile phase plus solute is modified in some way prior to detection. This is not a precise classification since a solute property detector can provide a constant background signal due to the properties of the mobile phase. For example, a uv detector, which is one of the more widely used solute property detectors, will provide a signal due to the mobile phase if it is one that is chromophoric at the wavelength chosen for detection. On the average a bulk property detector will have a sensitivity of about 10^{-6} g/ml and the solute property detector about 10^{-9} g/ml.

Almost all detectors used in HPLC will provide a linear output or convert the output of the sensing device to a linear (voltage) output. This means that the output signal is directly proportional to the concentration of the solute in the mobile phase as it passes through the detector. The profile for the eluted peak, where solute concentration in the mobile phase is measured as a function of volume of mobile phase (or time at constant flow rate), will have or closely resemble a Gaussian shape. Such a shape, which depends on the linear response, minimizes the calibration and calculation procedures in quantitative analysis.

In order to evaluate or compare detectors for applications in HPLC it is imperative to consider their specifications. Table VIII lists a series of operational performance parameters by which detectors should be compared.

Several detector characteristics have a direct effect on band broadening. Since the column and elution parameters are designed to keep the individual solute bands narrow it is essential that the broadening due to the detector be kept at a minimum. If this is not the case, the resolution attained in the column can be lost in the detector.

There are several band broadening processes that are solely due to the detector and its electronics and each contributes to an overall detector band broadening effect. The broadening processes occur in the column detector

TABLE VIII

Specifications for Detectors

Specification	Definition
Dynamic range	Concentration range over which the dector will give a useful response
Response index	An indication of the detector's linearity
Linear dynamic range	Concentration range over which the detector's response is linear
Detector response	Can be either mass sensitive or concentration sensitive
Noise level	The amplitude of the detector output that is not related to an eluted solute
Detector sensitivity	The minimum concentration of solute in mass/unit volume passing through the detector that can be differentiated from the noise

connecting tubes and within the detector cell. Since the processes are random the Gaussian distribution of the solute is retained but broadened and this is added to the broadening occurring within the column.

Two electronic factors that can indirectly contribute to a broadened band are the result of the time constant of the detector amplifier and the recorder itself. These factors contribute to a recorder tracing showing the bandwidth to be larger and its peak height smaller than its actual shape within the detector. This broadening can be significant and the reader should consult the work of Sternberg (1966) for a detailed discussion of these factors.

Other factors which can influence band shape in the detector are turbulence within the connecting tubing or cell, adsorption effects in the detector, and the geometric shape of the detector. Turbulence that produces a convective mixing can actually reduce band broadening. However, the turbulence, if carried into the cell, can lead to excessive noise and subsequently reduce detector sensitivity. Designs to produce turbulence in the connecting tubing while minimizing it in the cell are possible, but often these create other problems and the advantages of increased turbulence is not fully realized.

Adsorption within the cell is a phenomena which is difficult to treat quantitatively and is one that is very dependent on the type of mobile phase employed in the elution. It is often encountered when the mobile phase is made up of two miscible solvents that differ widely in polarity. The more polar mobile phase component is adsorbed as a layer on the cell surface. When a solute passes through the cell it will compete with the adsorbed layer if it is of equal or greater polarity than the adsorbed layer. This results in a distorted peak shape, or in some cases a second peak appears. Modification of the eluting mixture will eliminate adsorption at the detector cell windows. It also follows that the composition of the cell will influence the adsorption.

It has been noted that the geometrical shape of the cell will influence the band broadening. Not all commercial detector cells have a cylindrical cross-section geometry. In general, those that do not, for example, a triangular or rectangular cross section, will contribute more to broadening than the cylindrical cross section assuming that the cells are of equivalent length and volume.

The problem of detection in HPLC is a significant one and although progress in detector design has taken place there still is no universal detector in HPLC that compares to the better universal detectors in gas chromatography. This is not to say that there are no effective detectors in HPLC, but rather, that there is much room for improvement. For example, cell design with respect to volume and sensitivity must be improved if the full potential of the efficiency generated in the column is to be realized. This is particularly important since a significant direction in stationary phase development is one whereby an improvement in efficiency is also accompanied by a reduction in loading capacities. At present it appears that the two most widely used HPLC detectors are the refractive index detector (a bulk property detector) and the uv detector (a solute property detector). In general, the latter is more sensitive and its versatility can be expanded if it is designed to provide a variable wavelength rather than a fixed wavelength.

Several reviews on detectors are available (Parris, 1976; McNair and Chandler, 1974, 1976; McNair, 1978; Bombaugh, 1977; Wehrli, 1974; Snyder and Kirkland, 1974; Scott, 1977a; Baumann, 1977; Bollet, 1977; see also Table I). The following discussion is devoted to only a selected number of detectors. Table IX lists other detectors used in HPLC that are not described in this chapter.

2. *Bulk Property Detectors*

A bulk property detector monitors a physical property of the eluent. In addition to measurement of refractive index, which was the basis for the development of the first useful type of HPLC detector, other common physical properties that are readily detected are changes in dielectric constant and electrical conductivity. Other types of bulk detectors are listed in Table IX.

Basically, all the bulk type detectors employ a transducer to provide a voltage–time output that is made proportional to the physical property being measured. The cell generally consists of a small volume, is cylindrical in shape, and contains suitable optics or electrodes as the sensing device. Any solute which passes through the cell will change the signal providing its

TABLE IX

Summary of Detectors for HPLC

Type of detector	Measurement
Bulk property detectors	
Refractive index	Changes in refractive index
Dielectric constant	Change in dielectric constant
Electrical conductivity	Change in conductivity
Density	Change in density
Thermal conductivity	Change in thermal conductivity
Interferometer	Change in an interference fringe pattern
Density balance	Differential flow across the base of two columns of a gas, one of which contains the solute
Vapor pressure	Change in vapor pressure
Solute property detectors	
uv	Absorption at a uv wavelength
Visible	Absorption at a visible wavelength
Fluorescence	Fluorescence
Electrochemical	Current under amperometric or Coulometric conditions
ir	Absorption at an ir wavelength
Heat of adsorption	Heat associated with absorption of a solute by a stationary phase
Radioactivity	Radio activity
Specialty type detectors	
Spray impact	Electric charge in an aerosol
Transport	Isolation of the sample on a moving wire, belt, or chain and subsequently vaporized and detected by a GC detector
Electron capture	Compounds that have the capability to absorb electrons

physical property, the one being detected, is sufficiently different from that of the mobile phase. Often a reference cell is part of the detector and contains only the mobile phase. Thus, the actual signal observed is a comparison of the measuring cell output to the reference cell output. The major advantage of dual cell operation is an increased sensitivity since the noise level is reduced and variations in the system are compensated for.

All bulk detectors, more or less, suffer from similar limitations. In general they have a limited lower level of sensitivity which is directly attributible to the physical property being detected. Although their dynamic range is relatively large, the linear range is small. Since the bulk detectors are susceptible to changes in column conditions, column flow rate, and mobile phase composition, they are not generally suitable for gradient, temperature, or flow programming. The use of the reference cell does not eliminate these factors. For these reasons the bulk property detectors are used in analytical HPLC under isocratic and isothermal conditions and are usually used if the solutes being separated are nonchromophoric in the range of the available uv detector.

The limited sensitivity of the bulk type detector, particularly for the refractive index detector, can be used advantageously in preparative scale HPLC where the main goal is to allow passage of the maximum amount of sample throughput per unit of time while still maintaining sufficient resolution to bring about purification. For this reason continual development of these detectors is important.

3. *Refractive Index Detector*

Figure 19 illustrates a typical example of a modern refractive index detector. This particular detector is a differential refractometer and measures the refraction of a light beam due to the difference in refractive index between the sample and reference channel. The sample channel contains mobile phase plus solute from the column while only the mobile phase passes through the reference channel.

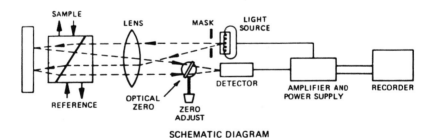

Fig. 19 Waters R401 refractive index detector fro HPLC. (Reprinted by permission from Waters Associates, Inc., Milford, Massachusetts.)

In the operation of the refractive index detector the lens collimates the light beam from an incandescent lamp and the parallel beam passes through the cells to the mirror. The mirror reflects the beam back through the cells and lens, which focuses it into a photocell. The angle of deflection arising from the difference in refractive index between the two cells causes the location of the beam to change as the angle changes. Thus, the photocell output is a measure of a beam change on it rather than one of a change in the beam intensity. The signal is then amplified and recordered.

This type of refractive index detector, which employs a deflection principle of refractometry, is able to cover the entire refractive index range from 1.00–1.75. It has a cell volume of 10 μl (analytical HPLC) or 70 μl (preparative scale HPLC). An optical block and heat exchanger are provided in order to bring the temperature of the column effluent to that of the cell.

The refractive index detector will detect all substances in the mobile phase that have a significant difference in refractive index in comparison to that for the mobile phase. Thus, the observed sensitivity will depend on this difference. In general, the refractive index detector can be considered to be one of the least sensitive detectors.

As indicated previously it is not useful for solvent gradient, flow, or temperature programming and is only used isocratically. Its greatest use is in size exclusion chromatography where polymers are being separated, in preparative scale chromatography, and as a companion backup detector to the uv detector to check column effluent for solutes emerging from the column that are nonchromophoric.

4. The Dielectric Constant Detector

In chromatography the presence of a solute in the mobile phase will change the dielectric constant of the mobile phase providing its polarity is significantly different than that of the mobile phase. If the column effluent passes through a sensing device that responds to changes in dielectric constant the sensing device can be used as an HPLC detector. Several types of detector designs have been described in the literature (Grant, 1959; Johansson and Karram, 1958; Vespalec, 1975; Poppe and Kunysten, 1972).

In general, the sensing device is in the form of a cyclindrical or parallel plate condenser and is designed to have a small volume. Since sensitivity of the detector is directly related to capacity of the condenser, the plates have to be as close together as possible.

A suitable circuit for making the measurement is an ac bridge in which the detector comprises one arm of the bridge. Useful bridge circuits are the Wein bridge (for detector cells that have large capacities of > 100 pF) and the Schering bridge (for detector cells that have small capacities of 1–10 pF). Another option is to make the capacitor containing the dielectric part of an oscillator circuit. These circuits are widely used. Discussions of these circuits are readily available in texts dealing with instrument design.

Although sensitivity is directly related to the differences in dielectric constant between the mobile phase and the mobile phase plus solute, the measurement of this or the circuit used is usually not the limiting factor in determining the overall detector sensitivity. The limitations ascribed to bulk detectors, in general, as outlined earlier are the limiting factors. It should also be pointed out that the general guideline, which suggests that the presence of a functional group or of several functional groups produces a molecule with greater polarity and, consequently, a larger dielectric constant, does have exceptions. For example, dioxane with two ether functional groups is polar yet has a very low dielectric constant.

5. *Electrical Conductivity Detector*

Any solute that is completely or partially ionized in the mobile phase can be detected with the electrical conductivity detector. In essence, the detector is one in which the detector cell consists of a small chamber containing two electrodes. An ac potential is applied across the electrodes and if ionic species are in the solution they can act as current carriers. When current flows through the solution between the electrodes, the solution is said to be conducting. The usual circuit used for this measurement is a Wheatstone bridge circuit or a variation of it. Since this is such a common circuit in instrumentation, it will not be discussed here.

The volume of the cell should be as small as possible and the electrodes, as in any conductance measurement, should be an inert conducting material such as platinum. The nature of the cell essentially forces the shape and distance between electrodes to be small. The former variable requires the solution to be higher conducting, while the latter variable permits the measurement of a lower conducting solution. Since the conductance of an ionic solution is dependent on temperature, thermostating the cell may be required.

Several commercial electrical conductivity detectors are available and have been shown to be very reliable. They have low volume cells and employ both a sample and reference cell with each being one arm of the Wheatstone bridge circuit. This allows a differential measurement since the conductance of the mobile phase plus solute is being measured in the sample arm and is compared to the measurement of the mobile phase only in the reference arm. Several workers have reported designs of cells for electrical conductivity detectors that have significantly smaller volumes (Scott, 1977a; Pescok and Saunders, 1968).

Since the conductivity detector responds to any solute that produces ions in solution, it has wide use as a detector for following separations of inorganic and organic acids, bases, or salts. Particularly significant is its applications for detecting inorganic compounds.

The detector can still be used where the mobile phase is ionic itself, such as the case when using a buffer in the eluting mixture. In this kind of application, the sensitivity usually experienced is approximately 10^{-6} g/ml and is typical of other bulk type detectors even if the differential mode is used. If a low conducting mobile phase is used, such as pure solvent or mixed solvent in the absence of electrolytes, a greater sensitivity is experienced and often approaches that of the solute property detector (approximately 10^{-8} g/ml).

6. Solute Property Detectors

A solute property detector responds to a specific property of a solute in the mobile phase that the mobile phase does not possess or has to a significantly lesser extent. These types of detectors are the most used in HPLC at present. They are the most sensitive (on the average about 10^{-9} g/ml), most versatile, and offer the widest linear dynamic range. Most are commercially available at several different levels of sophistication. Space limitations do not allow a discussion of all the solute property detectors. Only the uv, fluorescence, and electrochemical detectors are considered here; others are listed in Table IX.

7. The Ultraviolet Adsorption Detector

Of all the detectors available for HPLC the type most often used is the ultraviolet adsorption detector (Parris, 1976; McNair and Chandler, 1974, 1976; McNair, 1978; Bombaugh, 1977; Wehrli, 1974; Snyder and Kirkland, 1974; Scott, 1977a; Baumann, 1977; Bollet, 1977; Callmer and Nilsson, 1974; Baker et al., 1974). Three types of detectors are readily available. One type has a fixed wavelength at 254 nm (corresponds to a strong emission line from the Hg arc lamp) and perhaps an option through a filter accessory to provide 280 nm. The second is designed to use a series of filters and allows the selection of several wavelengths in addition to 254 and 280 nm. The third type, which offers the most versatility and usually the best sensitivity, is the variable wavelength detector. This type of detector employs a monochrometer which allows the operator to choose any wavelength, not only in the uv, but also in the visible region. An added feature of many of the commercially variable wavelength detectors is that the adsorption spectrum of the solute can be obtained when it is in the detector cell.

In order for a uv detector to be useful, it is necessary that the solute absorb within the active region of the detector. The detector cell usually has a cylindrical shape and a volume of about 8 μl and consists of a 1-cm path length. Energy from the source is focused onto the cell and the energy transmitted is detected by a photomultiplier tube. The electrical output of the photomultiplier tube is amplified, and with suitable electronics it is not unusual to achieve up to 0.005 absorbance units full scale with a noise level at about ± 0.0002 absorbance units. Other features that are often part of the detector are double beam operation (sample and reference cell) and thermostatic control.

This basic design is common to all commercial uv detectors. They differ, however, in detail as to how each function is achieved and it is not possible to outline all these features here.

All compounds containing π electrons will absorb in the uv region. However, the location of the absorption and the molar absorptivity at the absorption wavelength will vary widely. Since aromatic and simple conjugated compounds and their substituted derivatives absorb at or near 254 or 280 nm, these two wavelengths are common to the fixed wavelength detectors. Since there are many simple π-containing functional groups that absorb at other wavelengths, often lower than 254 nm the variable wavelength detector offers more versatility. Thus, compounds such as sugars, amino acids, peptides, aliphatic carboxylic acids, and monoolefinic, or acetylinic compounds can be detected with the variable wavelength detector using a wavelength below 210 nm. These compounds would not be detected with the fixed wavelength detector unless they are multifunctional or conjugated.

The sensitivity of the uv detector depends not only on the detector design but also on the molar absorptivity of the solute being detected. Since the variable wavelength detector allows selection at the wavelength of maximum absorption, highest sensitivity for a given solute is obtained with this detector, assuming all other factors are equal. If several different solutes are being eluted from the column, one wavelength is chosen for the measurement even though a variable wavelength detector is used since changes during the course of the chromatogram can cause a large shift in the baseline absorbance.

Although it would appear that the uv detector is ideal for gradient elution provided that nonabsorbing eluents are used, problems are often experienced. In general, this is due to a refractive index change in the mobile phase as the gradient is carried out. The shape of the drift in the baseline that is observed will, therefore, depend on the rate and magnitude of the change in refractive index. Several manufacturers have been able to minimize this effect through design of the cell and its entrance and exit slits. However, large refractive index changes are often still bothersome in these detectors.

8. *Fluorescence Detector*

A fluorescence detector responds to solutes that will fluoresce. In general, for an organic compound to fluoresce, the compound must have at least some conjugation as a minimum. However, not all conjugated compounds will fluoresce since fluorescence yields vary widely. Thus, the fluorescence detector is more selective than the uv detector (Parris, 1976; McNair and Chandler, 1974, 1976; McNair, 1978; Bombaugh, 1977; Wehrli, 1974; Snyder and Kirkland, 1974; Scott, 1977a; Baumann, 1977; Bollet, 1977; Callmer and Nilsson, 1974; Baker *et al.*, 1974; Johnson *et al.*, 1977).

   As molar absorptivity influences the sensitivity of the uv detector, fluorescence yield influences the sensitivity of the fluorescent detector. Often, the fluorescence detector is more sensitive than other detectors, with respect to concentration change. An added feature is that the detector is less sensitive to slight instrument variations since fluorescence is an emission and the measurement is one of a higher signal (fluorescence) superimposed on a low signal (background).

   Several fluorescence detectors are commercially available. A typical design includes a Hg lamp as an uv source for irradiation of the sample, a sample cell, and photocell appropriately placed to measure the fluorescent emission. Initial commercial detectors contained relatively large cell volumes ($>25\ \mu l$); however, more recent instruments have smaller volume cells.

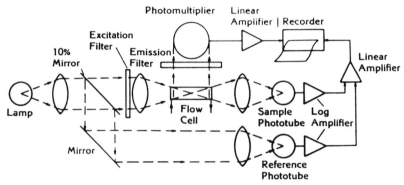

Fig. 20   Schematic of the Du Pont 836 Fluorescence–Absorbance HPLC Dector. (Reprinted by permission from Du Pont Instruments, Wilmington, Delaware.)

   One commercial detector is designed to measure absorption and fluorescence simultaneously. A diagram of this detector is shown in Fig. 20. In this design the mobile phase plus solute passes through the sample cell which is irradiated by the source. One detector is placed to detect transmission through the cell, while the other is placed at an angle to detect the fluorescent emission. Both signals are compared to a reference and plotted with a two-channel two-pen recorder. A typical chromatogram is shown in Fig. 21.

   Unlike the uv detector, the fluorescence detector is much more compatible with gradient elution. This is consistent with the nature of the instrumentation and principles involved in making the measurement.

   The limitations of the fluorescence detector are that it is a specific detector and limited only to those solutes that fluoresce. Furthermore, fluorescence yield is highly dependent on the solution environment. Thus, control of pH, concentration, and quenching, which can have a large effect on fluorescence, may require conditions that are not compatible with the chromatography.

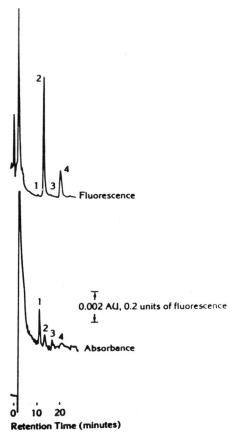

Fig. 21 Dual tracing using the Du Pont 836 fluorescence–absorbance HPLC detector for the separation of aflatoxins in 50 μl peanut butter extract. Operating conditions: Column, "Zarbax-" SIL 25 cm × 2.1 mm i.d., Mobile phase, 60% CH_2Cl_2 (50% H_2O saturated), 40% $CHCl_3$(50% H_2O saturated), 0.3% methanol; Pressure, 2000 psig; Flow, 0.7 ml/min; Temperature, ambient; Detector, uv photometer, 365 nm (0.02 AUFS); Fluorescence, excitation 365 nm; emission, Corning CS-3-72. Peak identity: 1, Alfatoxin B_1, 5 ppb; 2, Alfatoxin G_1, 1 ppb; 3, Alfatoxin B_2, 3 ppb; 4, Alfatoxin G_2, 1 ppb. (Reprinted by permission from Du Pont Instruments, Wilmington, Delaware.)

Since the detector is very sensitive, extra care is necessary to remove fluorescent impurities from the eluting agent.

9. *Electrochemical Detector*

The application of electrochemical detectors in HPLC is determined by the voltametric characteristics of the molecules being detected. Both oxidation and reduction processes can be followed. Usually, the electrochemical detector is operated on the limiting current plateau for the electroactive species. However, under certain circumstances it is advantageous to use a potential on the rising portion of the current–voltage curve; for example, detector selectivity can be increased.

Both amperometric (Kissinger *et al.*, 1973; Kissinger, 1977a; Maruyama *et al.*, 1977; Buckta and Papa, 1976) and coulometric detectors (Davenport and Johnson, 1974; Tjaden *et al.*, 1976) have been developed. It appears at present that the former offers more advantages in that this type of electro- chemical detector tends to be more sensitive and more simple in design. Only the amperometric type detector will be discussed here.

Figure 22 illustrates a typical design of an amperometric type detector which employs an electrode embedded in a rectangular cell (Kissinger *et al.*, 1973). In this design the electrode surface is part of the channel wall formed by sandwiching a fluorocarbon gasket (50–125 μm) between two machined blocks of a suitable plastic such as Kel T. Tubular cells have also been used. Cell volumes as small as 1 μl can be achieved and typical electrode materials are platinum, gold, glassy carbon, and carbon paste. Either classical two electrode measurements or three electrode measurements employing an auxiliary electrode as shown in Fig. 22 can be used. If the latter is used careful placement of the auxiliary electrode is necessary in order to maintain a wide linear response range.

Electrochemical detection has several limitations. It is necessary to con- sider the properties of the mobile phase and how they will influence the electrochemical measurement. Variations in the mobile phase which can

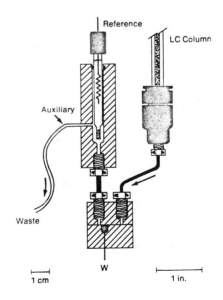

Fig. 22 Typical design for an amperomet- ric type electrochemical detector. (Reprinted with permission from P. T. Kissinger, *Anal. Chem.* **48**, 447A (1977). Copyright by the American Chemical Society).

improve electrochemical detection may not be compatible with those required for successful chromatography or vice versa. Thus, it is necessary to optimize column and detector performance simultaneously. More satisfactory electrochemical detection is favored by mobile phases composed of polar solvents containing electrolytes and buffers but is not suitable for nonpolar solvents. For this reason electrochemical detection is not recommended for normal phase chromatography and is most often used with reversed phase and ion exchange type stationary phases. Most applications have employed oxidation in the detection mainly because of the interference of trace reducibles such as dissolved oxygen, trace metal ions, and hydrogen ions. However, detection is not limited to oxidation. Although the dropping mercury electrode can be used (Scott, 1977a), it is often mechanically awkward relative to solid electrodes. (Recent improvement in mercury electrode detectors has taken place. For example, a mercury hanging drop detector is currently commercially available from Princeton Applied Research.) Several different solid electrodes can be used as indicated earlier and the choice of the electrode material is based on ruggedness and long-term stability. Particularly important is the latter property, since complex reactions are often part of solid electrode electrochemistry. These reactions can change the electrode surface during the chromatographic separation and consequently change its response. Even with these limitations, detection limits at and sometimes below the picomole level are readily achieved for detection of molecules that are readily oxidized or reduced.

10. *Specialty Type Detector*

A specialty type detector for purposes of classification is one in which a solute property is still being measured but only after modification of the mobile phase–solute mixture. Table IX lists the specialty type detectors that have been developed. Only the spray impact detector and the transport detector will be described here. Of these two, only the latter is commercially available.

11. *Spray Impact Detector*

When a liquid is dispersed to produce an aerosol, an electric charge is produced. This phenomenon, which is not yet completely understood, was used as a basis for an HPLC detector (Scott, 1977a; Mavery and Juvet, 1974). Figure 23 shows the design of the spray impact detector. The eluent from the column strikes a conducting target to form a spray and a potential and current are developed at the electrode. Small quantities of a solute in the mobile phase will have a large effect on the potential and this significant

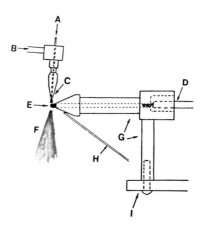

Fig. 23 Spray impact detector. Part identifications: A, LC column outlet; B, air inlet to aspirator; C, stainless steel aspirator jet; D, electrometer coaxial cable with expoxy seal; E, glassy carbon or gold target electrode; F, spent spray from target; G. PTFE body and mounting arm; H, glass capillary to aspirator vacuum line for charged droplet removal; I. laminated plastic mounting. (Reproduced from R. A. Mowery, Jr. and R. S. Juvet, Jr., *J. Chromatogr. Sci.* **12**, 687 (1974) permission of Preston publications, Inc.)

change is what is monitored. Sensitivity is very favorable, being of the order of 10^{-10} to 10^{-11} g/ml for inorganic and organic solutes where water is the mobile phase. Introduction of an organic solvent causes a decrease in the sensitivity; for example, for pure methyl ethyl ketone, sensitivities are typical of bulk physical property detector or about 10^{-6} g/ml.

The spray impact detector is still under investigation and is not currently commercially available: its true potential has not been realized at present (Mowery and Juvet, 1974).

12. Transport Detector

The principle of the transport detector (Scott, 1977a) is to remove the solute from the mobile phase so that the solute can be detected free of interferences from the mobile phase. This has been accomplished by using a carrier, such as a metal chain, wire, or disc, that passes through the column effluent continuously. A film of effluent is deposited on the carrier which subsequently passes through a chamber where only the mobile phase is removed, usually by evaporation. If a solute is present it remains on the carrier and passes into another chamber where it is detected. A typical detector would be the flame ionization detector (FID). The ultimate sensitivity of the transport detector would appear to be high and be that of the FID or whatever sensing device is used for detecting the isolated solute. However, the present design presents several limitations, hence, sensitivity, even when using the FID, is similar to that of other bulk property detectors. Also contributing to the decrease in sensitivity is the fact that the mobile phase must be completely removed and the solute is not lost during this removal step. This detector is not useful for detecting solutes with appreciable vapor pressures.

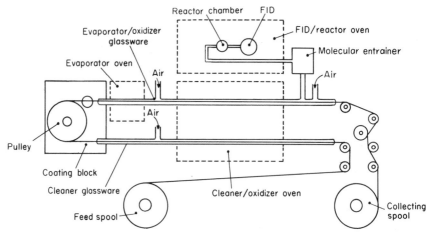

Fig. 24 Pye unicam wire transport detector. (Reprinted by permission of Pye Unicam, Ltd., Cambridge CB1 2PX, England.)

Figure 24 illustrates the design for the commercially available Pye Unicam transport detector. In this particular detector, a wire is used and the mobile phase is removed in the evaporator oven at 105°C. The solute is pyrolyzed in a N_2 stream and this stream sweeps the pyrolysis products directly into the flame ionization detector. The detector output is amplified and recorded. Depending on the sample, detector sensitivity of about 5×10^{-6} g/ml is achieved.

One application that perhaps is still to reach its potential is to use the transport detector as a means of introducing the sample into a mass spectrometer which then provides the mass spectrum of the solute. If the column effluent is split-streamed so that part of the effluent passes through a normal detector the chromatogram for the solute is also obtained. Several workers have contributed to the present development (Scott, 1977a; Arjine et al., 1974; Scott et al., 1974; Horning et al., 1973). Presently, one instrument based on a LC/MS design introduced by Finnigan Corporation is commercially available (McFadden et al., 1976).

C. Pre- and Postcolumn Derivatization

Conversion of the solute in the mobile phase to a species that is more easily detected is a technique that has been used extensively in classical liquid column chromatography. A typical example is the formation of the highly colored product that results from the reaction of ninhydrin and an amino acid. This reaction is forced to take place after the mobile phase

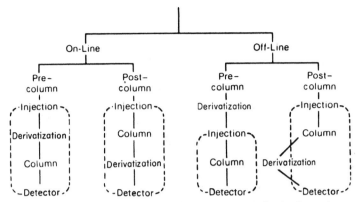

Fig. 25 Options in pre- and postcolumn usage of derivative formation.

containing the amino acid solute leaves the column and prior to entering the visible absorption detector by pumping ninhydrin into the line.

If postcolumn derivatization is employed, a suitable reaction time must be allowed; this is usually accomplished by increasing the time and consequently the volume between the column and the detector cell. This can contribute significantly to band broadening and therefore, some of the increased sensitivity through derivatization is lost. An alternate approach is to convert the solutes in the mixture into derivatives which are more easily detected and then proceed to separate the derivatives. If precolumn derivatization is used, the compound is chemically transformed into another structure and the original molecule may not be readily recovered after separation. Figure 25 summarizes the options available in using derivative formation in HPLC. Each approach offers advantages and limitations that are a function of the required reagents, type of derivative formation reaction, and particular suitability of the reaction to automation (Lawrence and Frei, 1976).

Pre- and postcolumn derivatization to aid detection usually involves an improvement in molar absorptivity, in fluorescence emission, or introduces radioactivity (Lawrence and Frei, 1976). Table X lists several of the more common reagents used in derivative formation.

As indicated previously, column effluent can be directed into the mass spectrometer by the transport detector so that the mass spectrum of the solute can be obtained. It is also possible to use direct sampling of the column effluent into the mass spectrometer (Arjino *et al.*, 1974). Other techniques that are useful in characterizing the solute include obtaining its uv-visible absorption spectrum directly in the detector cell, or collecting the peaks and proceeding with conventional approaches to chemical and instrumental methods for characterizing the sample responsible for the peak.

TABLE X

Reagents Useful for Enhancing uv
and Fluorescence Detection in HPLC

Functional group	Reagent
uv detection	
R—C(=O)—OH	o-p-Nitrobenzyl-N,N'-diisopropyl isourea
	p-Bromophenacyl bromide
R—OH	3,5-Dinitrobenzoyl chloride
R—NH$_2$	N-Succinimidyl-p-nitrophenyl acetate
R₂C=O	p-Nitrobenzyloxyamine hydrochloride
Amino acid	Ninhydrine
	2,4-Dinitro-fluorobenzene
	Phenylisothiocyanate
Fluorescence detection	
R—C(=O)—OH	4-Bromomethyl–7-methoxycoumarin
R—SH	7-Chloro–4-nitrobenzo–2-oxa-1,3-diazole
R—NH$_2$	1-Dimethylaminonaphthalene–5-sulfonyl chloride
Ar—OH	1-Dimethylaminonaphtalene–5-sulfonyl chloride
R(H)C=O	1-Dimethylaminonaphthalene–5-sulfonyl hydrazine

D. Chromatographic Automation

The direction of HPLC instrument design has not just focused on the development of individual components of the instrumentation. In recent years modification and development of modules that are readily microprocessor controlled has also taken place.

A typical microprocessor controlled HPLC instrument is the Spectra-Physics SP-8000 HPLC which is schematically illustrated in Fig. 26. The microprocessor has the capabilities of controlling all the critical functions of

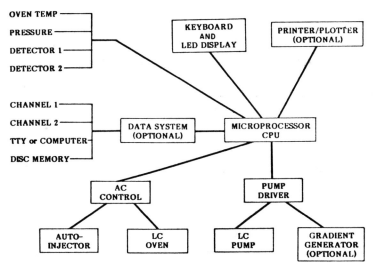

Fig. 26 Schematic diagram of the Spectra Physics Model SP-8000 illustrating micro-processor control of all functions and data handling.

the instrument and receives the data and produces the required analytical information. For example, the display will show on request the flow rate, solvent composition, oven temperature, column pressure, and time into run. A permanent record is provided by a printer plotter. It should be noted that this particular instrument is capable of providing a ternary gradient and has a dual channel capability that permits recording simultaneously with the chromatogram, either pressure, temperature, alternate detector output, or percentage of solvent A, B, or C.

Much of the time that is required to complete a chromatographic analysis is consumed by sample preparation. Many of these time consuming manual operations can be completed automatically. This includes various pretreatment techniques of the sample such as taking aliquots of liquid samples, the disintegration and dissolution of tablets or other solid samples, removal of inert material through filtration, dialysis, or even extraction, the drying and redissolving of the sample, and finally the automated introduction of the the sample into the chromatograph. Even pre- and postcolumn reactions to facilitate separation and/or detection are readily automated. This increased level of automation does not prevent the final data from still being handled by integrator or computer.

A typical automated system designed by the Technicon Instrument Corporation is schematically illustrated in Fig. 27. This instrumentation is used for assaying fat-soluble vitamins in pharmaceutical tablets. Vitamin D at about 10 μg and Vitamins A and E at about 1 μg are readily determined after two concentrating steps. In the first, a theoretical concentration gain of 4.14 is

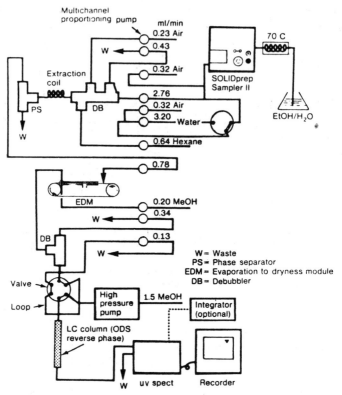

Fig. 27 Example of a totally automated HPLC. This instrument is available from Technicon Corporation and this particular design is used for fat soluble vitamin determinations in pharmaceutical tablets. The SOLID prep sampler II and EDM are Technicon modules. The first is used to sample solids, while the second is a solvent evaporator.

achieved in an extraction, while in the second, a theoretical gain of 2.75 is realized in a step whereby the sample is evaporated to dryness and then redissolved. A throughput of up to 10 samples per hour is possible and the peaks are readily quantified through modern data handling instrumentation.

IV. The Column

A. Mobile Phase

1. Scope

The heart of the chromatographic system is the column since it is in this area where the operator must make judgements as to the type of stationary and mobile phase to use to first bring about retention of the components of the sample and then to bring about their separation. In essence, the main objectives are: (1) to retain the sample (selectivity), (2) to have the sample components migrate through the column at different rates (resolution), and and (3) to have these processes occur in a reasonable time period (efficiency).

In order to meet these objectives successfully, knowledge about the chemical and physical interactions that take place between sample, mobile phase, and stationary phase is very useful. Figure 4 summarized these types of interactions.

In general, as outlined earlier in this chapter, four types of separation mechanisms account for chromatographic separation in HPLC. These are adsorption, partition, ion exchange, and size exclusion. Although much is known about these processes the choice of the correct chromatographic conditions cannot always be predicted with certainty. In most instances confirming experiments are necessary either by performing the experiments or by using analogy to literature results.

2. Role of the Mobile Phase

The mobile phase should be compatible with the sample, column, detector, and other instrument components while providing a particular selectivity to the system. Any combination of solvents can be used as long as the mobile phase retains suitable flow properties. Small amounts of other chemicals (salts for ionic strength control, salts of particular size and/or charge, and buffers for pH control) can be added to impart a special selectivity to the system.

In partition, adsorption, and ion exchange the mobile phase participates actively in chemical processes that influence the interactions summarized in Fig. 4 and subsequently leads to a separation. In contrast, in size exclusion the mobile phase has little effect upon the separation and serves mainly as a carrier of the sample. Hence, it is chosen because of its good solvent properties for the sample. Often the mobile phase used in the first three processes is described as interactive because of its influence on the chemical processes present in the system while the latter is said to be noninteractive because of its influence only on the physical forces.

Table XI summarizes the general requirements for selecting a mobile phase.

3. Selection of the Mobile Phase

Although the main role of the mobile phase is to move the sample through the bed of the stationary bed, its choice must be made on more than just the consideration of this property. The mobile phase will undergo interactions with the solute molecules and with the stationary phase itself. The strength and type of interactions involved will have a direct bearing on not only the

TABLE XI

General Requirements for Selecting a Mobile Phase in HPLC

The mobile phase must have suitable eluting power for the separation.

The mobile phase must have low viscosity so that it can be conveniently pumped.

If the mobile phase is used isocratically, its preparation must be reproducible if it is used as a mixture. If the mobile phase mixture is produced by gradient, the gradient must be reproducible.

The mobile phase must be a good solvent for the sample and permit easy recovery of the sample components after separation.

The mobile phase must have a high level of purity and stability.

The mobile phase must be compatible with the detector.

If the sample is to be recovered after separation, the mobile phase must be volatile or easily separated from the sample.

Attention should be directed towards the toxicity and flammability of the mobile phase.

---

resolution but also the efficiency of the separation (see the discussion of Eq. (34)].

Because of the type of interactions that are involved, certain properties are more important than others in selecting a mobile phase for adsorption, partition, ion exchange, and size exclusion chromatography. Furthermore, these are different for the four general types of HPLC.

In selecting the mobile phase, it is necessary to consider the role of the stationary phase. This will become more apparent in Section IV.A.4. Also chromatography can be performed in a normal phase or reversed phase mode. By convention the former employs a polar stationary phase and nonpolar, or one of lesser polarity, mobile phase, while the latter is just the opposite. In the following discussion on mobile phase selection for the four types of chromatography this consideration should also be kept in mind.

4.   *The Mobile Phase in Adsorption Chromatography*

In adsorption chromatography, the mobile phase molecules are in direct competition with the sample molecules for the adsorption sites. If the interactions between the sample molecules and stationary phase are strong then the interactions between the mobile phase molecules and the stationary phase must also be strong to achieve suitable elution. If they are weak the mobile phase–stationary phase interactions should be weak.

An approximate but yet very useful scale of eluting solvents can be established by consideration of the polarity of a series of solvents. This is done by empirically rating the solvents in an order which indicates the strength of adsorption on an adsorbent; it is referred to as an eluotropic series and it is

TABLE XII

Eluotropic Series for Alumina

Solvent	ε°	Solvent	ε°
n-Pentane	0.00	Tetrahydrofuran	0.45
Isooctane	0.01	Acetone	0.56
Cyclohexane	0.04	Methyl acetate	0.60
Carbon tetrachloride	0.18	Aniline	0.62
Xylene	0.26	Acetonitrile	0.65
Toluene	0.29	n-Propanol	0.82
Benzene	0.32	Ethanol	0.88
Ethyl ether	0.38	Methanol	0.95
Chloroform	0.40	Ethylene glycol	1.11
Methylene chloride	0.42	Acetic acid	Large

customary to list the solvents in an increasing order of solvent adsorption. A more quantitative approach to identifying solvent strength, which has been applied to polar adsorbents, is through the establishment of a "solvent strength parameter" (Snyder, 1968). A condensed list of solvents rated towards alumina is shown in Table XII. The order is also similar to one for silica as the adsorbent. For a nonpolar adsorbent, such as charcoal or a nonpolar organic copolymer adsorbent, the order of solvents will be the opposite, as an approximation, to that in Table XII. An exact eluotropic series is difficult to define since each type of adsorbent will influence the order, and furthermore, slight variations in eluotropic order are observed when comparing different samples, particularly when the structures of the samples are markedly different. This is particularly true for nonpolar adsorbents. Attempts have been made to establish more exact polarity scales by trying to account for the complex nature of the many types of interactions that can occur in adsorption (see Table III). Some success has been achieved and this is discussed in detail elsewhere (Snyder, 1962, 1974; Keller and Snyder, 1971; Snyder and Kirkland, 1974; Snyder, 1971; Saunders, 1977; Bakalyer et al., 1977; Scott, 1977b; Tjssen et al., 1976).

Selection of a solvent from Table XII provides a way of influencing the retention (k') of sample components on the adsorbent. That is, the larger the ε° value the smaller the k' value will be for the sample components. Intermediate ε° values can be readily achieved by mixing solvents and using isocratic elution or by a gradual mixing which is accomplished by gradient elution.

5. The Mobile Phase in Partition Chromatography

Traditional parition chromatography, where a liquid layer as a stationry phase is coated onto a solid inert support, is currently not used in HPLC as much as it once was. Reasons for this are that bleeding of the liquid layer during elution not only creates considerable difficulty in maintaining a uniform and reproducible column, but can also present serious limitations to detection of column effluent. In addition the mobile and stationary phase liquids must have a high degree of immiscibility. Perhaps the major one is that chemically bonded type stationary phases, which were a later development, are much more versatile and reproducible in HPLC applications. If traditional partition chromatography is to be employed, the eluotropic series, such as that shown in Table XII, can be used to predict solvent strengths to be used. For normal phase partition chromatography, choosing solvents with higher ε° value (Table XII) will decrease the k' value. In reversed phase, the opposite is used. The problem of dissolving the stationary liquid layer is usually more severe in traditional reversed phase partition chromatography because to decrease k' it is necessary to change to less polar solvents. It is these kinds of solvents which are most likely to have a high solubility for the stationary liquid phase.

Much of what was described here also applies to mobile phase selection when using chemically bonded stationary phases, which are available in normal and reversed phase modes. The chemically bonded phases are treated in more detail in Section IV.B.4.

6. The Mobile Phase in Ion Exchange and Size Exclusion Chromatography

Most mobile phases in ion exchange HPLC separations of organic samples are aqueous solutions containing salts to maintain ionic strength and/or buffers to maintain pH. In some cases small amounts of organic solvents are added to improve the eluting conditions and/or improve solubility of the sample. For organic acids, bases, and ampholytes pH control will determine whether the molecule is charged, neutral, single or multicharged, a cation, an anion, or even a zwitterion. Added salts will influence salting in and salting out phenomena and can be used to facilitate the HPLC separation of organic nonelectrolytes on ion exchange stationary phases. This approach has been used in separations by gravity flow ion exchange (Sherma and Rieman, 1958; Rieman, 1961) but has not been widely used in HPLC ion exchange separations, where almost all applications are for the separation of weak or strong organic electrolytes.

In size exclusion HPLC the major role of the mobile phase is to act as a carrier for the solute. Hence, the major criteria (others are listed in Table XI) are to choose a solvent that readily dissolves the sample while keeping the stationary phase in a solvent-wetted and swollen state.

Ideally, the interactions between the sample components and the mobile phase, the imbibed mobile phase, and the solid matrix should be the same. If not, a mixed interaction mechanism occurs and it is necessary to consider the influence of adsorption and partition phenomena when choosing the mobile phase.

B. Stationary Phase

1. Role of the Stationary Phase

The major role of the stationary phase is to exhibit some level of affinity for the sample components. For adsorption, partition, and ion exchange HPLC, chemical processes are involved (see Table III) and the system is considered to be interactive. For noninteractive chromatography, such as in size exclusion chromatography, affinity for the sample should be absent and only physical phenomena should be involved.

A summary of the general requirements for the stationary phase is more difficult to make than that for the mobile phase because of the contrasting properties in the four types of chromatography. For example, the sample capacity of the column is directly related to the quantity of available stationary phase and this is achieved differently in the four types of chromatography. In liquid–solid chromatography the capacity is proportional to the surface area of the adsorbent, in liquid–liquid chromatography to the volume of the stationary liquid phase, in ion exchange chromatography to the number of ion exchange sites, and in size exclusion chromatography to the volume of imbibed mobile phase in the pores of the packing. Many other stationary phase properties are different and are discussed in detail elsewhere, where the specific types of chromatographic processes are treated separately (see Table I).

Column parameters, such as column length and diameter and particle size of the stationary phase, were briefly discussed earlier in this chapter. One additional topic which is important to consider is the procedure for packing the stationary phase into a column.

Although almost all commercially available stationary phases can be purchased in prepacked columns, there are several situations where packing the column in the laboratory is preferred. Perhaps the major one is that self-packing of the column provides the user with the ability to control the performance of the column. However, the smaller and more uniform the particles of the stationary phase, the more critical the packing procedure is if efficient columns are to be made. Also, packing of gels (used in size exclusion) and ion exchangers (organic copolymer type matrix) must be done carefully because these types of stationary phases are not rigid and must be preswollen before packing.

There are two basic methods of packing columns with rigid stationary phase particles. One method involves dry packing and the second is a wet or slurry packing technique.

The major goal in packing a column is to produce a uniformly packed column. The more homogeneous the column is, the higher its efficiency.

Dry packing is generally restricted to larger particles (>25 μm). Smaller particles can, in some cases, be dry packed; however, efficiencies should be expected to be lower than that obtainable by slurry packing.

Perhaps the main advantage of dry packing is that it is a simple and rapid procedure in comparison to slurry packing. Also, in dry packing it is easy to establish the actual weight of the stationary phase used in the column.

The general technique of dry packing is to plug one end of the column with a suitable end fitting and slowly pour the powder into the open end while gently tapping the column. Problems arise from (1) voids being formed in the settling process, (2) larger particles settling more rapidly than smaller ones, (3) compression that occurs when the column is wetted if the particles readily swell, and (4) the particles agglomerating into larger masses.

If done properly slurry packing will minimize or eliminate many of the above problems and will provide the user with a compact, dense, highly efficient column. The basic technique is to suspend the stationary phase particles in a solvent or solvent mixture by vigorous stirring and transfer the slurry into the column with little or no settling of the particles.

Even though the slurry is agitated prior to entering the column, once in, the heterogeneity can begin to develop because of differences in settling. This is minimized by using a solvent system that has a density close to that of the packing, by using a viscous solvent, or by a combination of both. Table XIII lists densities and viscosities for several solvents.

Other factors to control are the flow rate and inlet pressure during introduction of the slurry. One approach uses a high constant flow; thus, the pressure builds up as the column is packed. Alternatively, a high constant inlet pressure is used (constant pressure pump) and the flow rate falls off as the bed height increases.

The solvent used in the slurry should be degassed. Although a wide variety of slurry concentrations have been used, a 10%–30% w/v slurry is usually appropriate. The experimental techniques for carrying out this procedure are listed in Table XIII.

It is useful to review the nature of a packed HPLC column prior to discussing the different types of stationary phases used in HPLC, since this plus the individual properties of the stationary phase will play a major role in determining band broadening or efficiency, resolution, selectivity, and

TABLE XIII

Densities and Viscosities for Selected Solvents
Used in HPLC Column Packing Procedures

Solvent	Density (g/ml)	Viscosity (cP), 20°C
Diiodomethane	3.3	2.9
1,1,2,2-Tetrabromethane	3.0	—
Tetrachloroethylene	1.6	0.9
Carbon tetrachloride	1.6	1.0
Chloroform	1.5	0.6
Dichloromethane	1.3	0.4
n-Propyl alcohol	0.8	2.3
Ethyl alcohol	0.8	1.2
Methyl alcohol	0.8	0.6
n-Heptane	0.7	0.4

Packing technique[a]	Slurry solvent
Balanced density	Tetrabromoethane, tetrachloroethylene, diiodomethane
Ammonia slurry	0.001 M Aqueous ammonia
Balanced viscosity	Cyclohexanol, polyethylene glyco 200
Nonbalanced density	Carbon tetrachloride, methanol, acetone, dioxane–methanol, THF–H$_2$O, isopropyl alcohol, chloroform–methanol

[a] See Majors (1977) for a survey of references describing each of these procedures.

capacity. In an ideal case small solid particles of uniform diameter are tightly packed together. Usually, the solid is the stationary phase or it is modified by coating or chemically bonding a liquid to the solid. If the solid particles are porous the solid contains interior pores that are accessible to the mobile phase. Some of the mobile phase is trapped in these pores and is immobile. This stagnant mobile phase is considered to be part of the stationary phase and its presence increases the complexity of the stationary phase.

The total volume of the packed bed, V_T, is given by

$$V_T = V_{SS} + V_S + V_M \tag{54}$$

where V_{SS}, V_S, and V_M are volumes occupied by the solid support, the stationary phase, and the mobile phase, respectively. For nonporous particles, V_M is the space between the stationary phase particles, while for porous particles, V_M accounts for this space as well as the internal volume of the particles. Thus, there are two definable porosities or

$$\text{total porosity} = \varepsilon_T = V_M/V_T \tag{55}$$

and

$$\text{interparticle porosity} = \varepsilon_\text{I} = \text{interparticle volume}/V_\text{T}. \qquad (56)$$

Thus, ε_T represents the fraction of dead space in the bed and ε_I is only that fraction of volume between the particles. For nonporous, solid-core type stationary phase particles only one type of porosity, ε_I, is possible and $\varepsilon_\text{T} = \varepsilon_\text{I}$,

If the stationary phase is not a solid but a liquid, it should be held immobile on an inert support as a uniform, thin film of liquid. In practice, small pools of liquid form within the pores and crevices of the inert stationary phase particles. Furthermore, pools can form between the individual particles. This concentration of the mobile phase in pools is undesirable and adds to the complexity of the true nature of the stationary phase. In general, pool formation in partition chromatography is difficult to eliminate and is promoted by using large amounts of stationary liquid phase, very porous inert supports, and irregularly shaped inert support particles.

Stationary phase particles used in HPLC are either totally porous or superficially porous. The totally porous particles are either spherical or irregularly shaped, have deep pores, and are available in a wide variety of particle sizes. They can be inorganic, such as silica, or organic copolymers, such as poly(styrene–divinylbenzene). Figure 28a illustrates this type of stationary phase particle.

The superficially porous type stationary phase particles are illustrated in Figs. 28b and 28c; these stationary phase particles are also referred to as being porous layer, controlled surface porous, solid core, or pellicular. The pore depth is controlled by the thickness of the active stationary phase. In general, the superficially porous particles are spherical or close to it.

The totally porous stationary phases have deep pores, are available in a wide range of particle sizes, have large surface areas, and high sample capacities. They are usually not useful in HPLC at particle sizes greater than 50 μm since diffusion into and out of the deep pores becomes a limiting factor and leads to a significant contribution to band broadening. Reducing the flow rate will only compensate for this to a small extent, while increasing the flow rate to achieve a faster analysis time will decrease the column efficiency and resolution. Shortening the diffusion paths or pore depth by reduction in particle size will increase column efficiency and resolution since

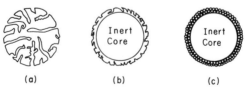

(a)   (b)   (c)

Fig. 28 Types of porous particles: (a) totally porous particle; (b) and (c) superficially porous particle.

mass transfer rates are increased. In the past the totally porous stationary phases were easier to pack in columns because they were most often used as larger particles (50-μm size) and hence, could be dry packed. Presently, they are commercially available at a nominal 5-μm and 10-μm size and must be slurry packed.

The superficially porous stationary phases provide an alternate way of decreasing the pore depth. As shown in Figs. 28b and 28c, the active chromatographic support is a thin porous outer shell. Mass transfer is rapid and good column efficiency and resolution is maintained at the high linear velocities used in HPLC. Typically, the outer layer is chromatographically active and can be inorganic, such as silica or alumina, or organic, such as one of a variety of polar, intermediate, or nonpolar type polymers which can be further modified. In most cases the inner core is glass and the layer thickness is typically 2%–3% of the diameter of the solid core. This leads to a reduced surface area and subsequently, a significantly lower loading capacity in comparison to a totally porous stationary phase of similar particle size. (On the average this difference is about a factor of 10.) The superficially porous particles are usually available only in a few specific size ranges. However, the particles are large enough so that they can be efficiently dry packed. The solid core not only eliminates inner porosity but also provides physical strength to the particle. The result is a densely packed bed of good column permeability that resists expanding or contraction as the mobile phase is changed.

Initially, the totally porous type stationary phase particles were readily available only at the large particle size (50-μm average) and this availability contributed to their wide use. Subsequently, the superficially porous beads, which were smaller and more uniform in size, were developed and offered many advantages over the totally porous particles. Thus, the superficially porous particles went through a period of extensive use, particularly in cases where sample loading was not a problem. Currently, this trend has reversed because of the wide availability of the totally porous particles at a uniform particle size of 5 and 10 μm; columns of these particles are classified as being microparticulate columns and in this chapter they are designated as microparticles. Table XIV compares the typical properties of the superficially porous beads to the currently used totally porous microparticles.

Much has been done to classify stationary phases in HPLC according to type, separation mode, and chromatographic properties and is discussed and reviewed in a variety of sources (Knox, 1976; Snyder and Kirkland, 1974; Snyder, 1965, 1971; Scott, 1976; Waltan, 1978; Grushka, 1974; Leitch and DeStefano, 1973; Majors, 1974a,b, 1977; see also Table I). Of particular interest, from a practical viewpoint are the series of reviews provided by R. E. Majors (Majors, 1974a,b, 1977). In these reviews Majors has carefully documented the commercially available stationary phases for HPLC at the

TABLE XIV

Comparison of Average Properties of Superficially Porus (SP)
and Microparticle (MP) Adsorbents

Property	SP	MP
Average particle size (μm)	30–40	5, 10
H values (mm)	0.2–0.4	0.01–0.03
Column lengths (mm)	50–100	10–30
Column diameters (mm)	2	2–5
Pressure drop (psi/cm)	2	20
Sample capacity (mg/g)	0.05–0.1	1–5
Surface area (m²/g)	10–15	400–600
Bonded phase coverage (% wt.)	0.5–1.5	5–20
Ease of packing	Easy, dry pack	Difficult, slurry pack

time of the review. The stationary phases are classified according to mode, functionality, quantity of phase, column dimensions, and chromatographically useful properties such as particle size, surface areas, porosity, and other distinguishing column parameters.

2. Liquid–Liquid HPLC

As previously indicated, in liquid–liquid chromatography or partition chromatography separation occurs because of differences in the distribution of the sample components between a stationary and mobile liquid phase.

In general, column efficiency and resolution is high but applications in HPLC over the years has decreased substantially for two main reasons. First, a useful column is prepared by coating particles of inert support with a liquid layer which serves as the stationary liquid phase. Improper choice of the liquids used for the mobile and stationary phase can result in a slow removal of the stationary liquid phase. If this occurs a continual change in the nature of the separation mechanism will take place which is undesirable. Presaturation of the mobile phase with the stationary phase liquid will minimize this effect. In general, the two phases are chosen so that they have little or no solubility in one another. Consequently, they tend to be quite different from one another in solvent properties and this may not be optimum in regards to affecting the separation.

The second limitation is that partition HPLC is best applied to the separation of compounds that have a low range of capacity factor values. Since the stationary liquid phase is chosen because it is a good solvent for the sample and a poor one for the mobile phase, increasing the eluting power of the mobile phase to elute compounds possessing higher capacity factor values will also tend to dissolve the stationary mobile phase. Presaturation at this higher mutual solubility is generally not satisfactory.

A better and more useful approach to solving the problem of separating compounds with large capacity factors by liquid–liquid HPLC is to use a reversed stationary phase mode. Both normal and reversed phase partition chromatography can be carried out. In the former the stationary liquid phase is polar and the mobile liquid phase is nonpolar. Thus, the nonpolar compounds elute first and are followed by the polar compounds. In reversed phase the stationary liquid phase is nonpolar and the mobile liquid phase is polar. Thus, the polar compounds are eluted first and are followed by the nonpolar compounds. An exact reversal in elution order is often observed in switching the phases. However, this should not be expected in all cases; slight variations to major changes in elution order have been observed. A typical normal phase liquid–liquid HPLC separation is shown in Fig. 29.

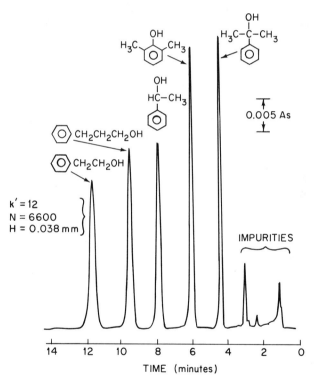

Fig. 29 Separation of hydroxylated aromatics by normal phase liquid–liquid HPLC using a liquid coated stationary phase. Column conditions: Column, 0.25 m × 3.2 mm i.d.; Packing, porous silica microspheres, diameter 5–6 μm, pore size 350 Å; Stationary phase, β,β'-oxydipropionitrile, approximate loading 30% by weight; Mobile phase, hexane, saturated with stationary phase; Flow rate, 1 ml/min; Inlet pressure, approximately 40 bars (600 psi); Temperature, 27 °C. (Reproduced from J. J. Kirkland, *J. Chromatogr. Sci.* **10**, 593 (1972) by permission of Preston Publications, Inc.)

These limitations in liquid–liquid chromatography coupled with the fact that most of the separations can be performed with solid stationary phases has led to the decline in use of liquid–liquid HPLC. However, partition HPLC can provide high resolving power for closely related compounds owing to its power to differentiate minute differences in solubility of the compounds. Thus, current applications usually deal with separations of these kind, particularly if they can not be easily performed on solid phases.

Table XV lists typical phase systems for liquid–liquid HPLC. The majority of normal phase separations have been accomplished with β,β'-oxydipro-pionitrile (BOP) and Carbowaxes as stationary liquid phases. For reversed phase, squalene, cyanoethylsilicone, and certain hydrocarbon polymers appear to be the most popular.

Typical support materials include inactive supports such as Zipax[R] (a controlled surface porous support), glass beads, organic copolymers of various types, and active supports such as silica and alumina. Most adsorbents used as stationary phases in liquid–solid chromatography are potentially useful as the support material. However, the more active the support material is, the greater the contribution of a mixed mechanism (partition

TABLE XV

Typical Phase Systems for Stationary
and Mobile Phase

Stationary phase	Mobile phase
Normal phase	
β,β'-Oxydipropionitrile	a–d
Carbowax 400–4000	a–d
Glycols (higher molecular weight)	a–d
Cyanoethylsilcone	a–d
Reversed phase	
Squalane	e–g
Zipax–HCP	e–g
Cyanoethylsilicone	e–g

^{a} Hexane, heptane, isooctane, and other saturated hydrocarbons.

^{b} Benzene, xylene, and other aromatic solvents.

^{c} Group **1** containing up to 10% methanol, ethanol, isopropanol, dioxane, THF.

^{d} Groups **1** or **2** containing up to 10% chlorinated hydrocarbons.

^{e} Water and water–alcohol mixtures.

^{f} Water–acetonitrile mixtures.

^{g} Acetonitrile.

and adsorption) to the chromatographic process. Coating procedures and modification of active supports to reduce their contribution are provided in many sources and will not be documented here (Knox, 1976; Snyder and Kirkland, 1974; Snyder, 1965, 1971; Scott, 1976; Leitch and DeStefano, 1973; Majors, 1974a,b, 1977; see also Table I). Table XVII in the next section lists those commercially available adsorbents which can be used in liquid–liquid HPLC, as well as typical loadings for the stationary liquid phase. It should be noted that totally porous adsorbents, including the microparticles, as well as superficially porous adsorbents can be used. The high level of efficiency achieved in Fig. 29 is in part the result of using microparticles as the solid support for holding the stationary liquid phase.

3. Liquid–Solid HPLC

Liquid–solid HPLC or adsorption HPLC is based on the interactions that take place between the fixed active sites on the solid stationary phase and the compounds being separated. The adsorbents are characterized by having large surface areas with a high degree of porosity and can be classified as being totally porous or superficially porous (see Fig. 28).

Typical polar adsorbents are silica and alumina while typical nonpolar adsorbents are charcoal and organic copolymers such as polystyrene–divinylbenzene. The former group is commercially available in a variety of surface areas, levels of porosity, as totally porous or superficially porous, in spherical or irregular shapes, and at a wide range of nominal sizes. The latter group, although very useful (Colin and Guiochon, 1976; Bebris et al., 1978; Chu and Pietrzyk, 1974; Pietrzyk and Chu, 1977; Kroeff and Pietrzyk, 1978a; Cantwell, 1976; Majors and MacDonald, 1973) is not yet at a comparable level of commercial availability.

The adsorbents are used in HPLC in three physical forms. The totally porous type is used in a large particle size range (50 μm average) and in the microparticle range (5 and 10 μm). The former group is a general purpose type and provides moderate efficiencies and resolution. The latter group is highly efficient and their usefulness extends to the separation of more complex mixtures. The third type or superficially porous type (20 μm) is comparable to the microparticles in general applications though somewhat less efficient.

Tables XVI and XVII list the commercially available microparticles and superficially porous particles, respectively, that are used for liquid–solid and liquid–liquid HPLC. The original reference should be consulted for a listing of the commercial suppliers and references that provide additional information about each of the packings.

In general, the microparticles will have efficiencies that are 10 times that for the superficially porous particles. On the average, best plate heights

TABLE XVI

Commercially Available Microparticle Type Packings and Prepacked Columns[a]

Type	Commercial name	Form[b]	Average particle size (μm)	Surface area (m²/g)	Length (cm)	Diameter (mm)	Description
Silica, irregular	BioSil A	B	2–10	400	—	—	Methanol extracted and activated
	Chromegasorb 60 R	C	10	500	30	4.6	Column filled with LiChrosorb
	Chrom Sep SL	B or C	5, 10	400	15, 30	—	Contains LiChrosorb Si60
	HiEff MicroPart	C	5, 10	250	15, 30	—	3 g of silica in 30 cm column
	ICN Silica	B	3–7, 7–12	500–600	—	—	Pore size 60 Å, pH 7
	LiChrosorb Si-60	B or C	5, 10	500	25, 30	2.1–4.6	Pore size 60 Å, 100 Å also
	MicroPak	B or C	5, 10	500	25, 30	2.2, 4.0	Pore size 60 Å contains LiChrosorb Si Prep columns available
	Partisil	B or C	5, 10	400	25	4.6	Pore size 50 Å: 50 cm column available for 10 μm
	μPorasil	C	10	300–350	30	3.9	3000 plates/column at 1.3 cm/sec
	RSL Silica	B or C	5, 10	>200	25	4.6	57 Å pore, ironfree = 1.5 μm distribution
	Sil 60	B or C	5, 10, 20	500	20, 25, 30	4.6	60 Å pore size, 0.75 ml/g volume:

Silica A	B	13 ± 5	400	—	—	60 D has broader distribution
Sil-X-1	B or C	13 ± 5	400	50	2.7	Acid washed, recommended for preparation. Chemically treated surface
Hypersil	B	~7	200	—	—	
Silica, spherical						
LiChrospher Si-100	B or C	5, 10	370	25	2.1–4.6	Pore size 100Å larger available
Nucleosil 50	B or C	5, 10	500	20, 25, 30	4	Pore volume 0.8 ml/g: pore size 50 Å, 100 Å also, 1 ml/g pore volume, 100 V has 1.5 ml/g
Spherisorb SW	B or C	5, 10	220	10, 20, 25	4.6	Packing density 0.6 g/ml: Pore size 80 Å maximum pH 8
Spherosil XOA 600	B	5–8	550	—	—	83 Å average pore diameter: Pore volume 1.2 ml/g: XOA 1000 is 860 m²/g, 35 Å, 0.78 ml/g
Super microbead Si	B	5, 10	380	—	—	95 Å average pore diameter
Vydac TP ads.	B or C	10	100	25	3.2	Pore size 330 Å
Zorbax Sil	C	6–8	300	25	2.4, 4.6	40 Å pore size, pore volume 0.8 ml/g

TABLE XVI (*continued*)

Type	Commercial name	Form[b]	Average particle size (μm)	Surface area (m²/g)	Length (cm)	Diameter (mm)	Description
Alumina, irregular	ALOX 60D		5, 10, 20	60	20, 25, 30	4.6	60 Å pore, basic pH 9.5
	Chroma Sep PAA	B or C	5, 10	70	15, 30	—	Packed with LiChrosorb ALOX T
	HiEff Micropart Al	C	5, 10	—	25	5	Pretested columns
	ICN Al–N	B	3–7, 7–12	200	—	—	Neutral density 0.9 g/ml
	LiChrosorb ALOX T	B or C	5, 10	70	25, 30	2.1–4.6	Pore diameter 150 Å
	MicroPak Al	B or C	5, 10	70	25, 30	2.2, 4.0	Pretested columns; Packed with LiChrosorb ALOX T
Alumina, spherical	Spherisorb AY		5, 10, 20	95	10, 20, 25	4.6	Pore diameter 150 Å, maximum pH 10, pack density 0.9 g/ml

[a] From Majors (1977). (Reprinted by permission of Preston Publication, Inc.)
[b] B = bulk; C = columns.

TABLE XVII

Commercially Available Superficially Porous Packing Materials[a]

Type	Name	Use	Particle size (μm)	Surface area (m²/g)	Loading for LLC			Shape[b]	Description
					Min.	Opt.	Max.		
Silica, active	Actichrom	LSC, LLC	40	25	3	4	6	I	Glass powder, uniform surface activity, higher than PLB
	Corasil I, II	LSC, LLC	37–50	I 7, II 14	0.5	1	1.5	S	Corasil II has a double coating of silica; LLC loadings are twice those of Corasil I
	Pellosil HS, HC	LSC, LLC	37–44	HS 4	0.5	1	1.5	S	HC has a thicker coating than HS: HC means high capacity, HS means high speed
	Perisorb A	LSC, LLC	30–40	14	0.5–0.7	1	3	S	Pore volume 0.05 ml/g
	Vydac	LSC, LLC	30–44	12	1.5	2	3	S	Average pore diameter A: Avoid solvents more polar than methanol
Silica, inactive	Liqua-Chrom	LLC	44–53	10	1.0	3	6.0	I	Silica glass, higher capacity than others
	Zipax	LLC	25–37	1	0.5	1–1.5	2	S	Inactive surface, precoated packings available
Other	Pellidon	LLC	45 (av.)	1	na[c]	na	na	S	Nylon bonded on glass bead, heat before packing

77

TABLE XVII (*continued*)

Type	Name	Use	Particle size (μm)	Surface area (m²/g)	Loading for LLC			Shape[b]	Description
					Min.	Opt.	Max.		
	Perisorb-PA6	LLC	30–40	0.6	na	na	na	S	Polycaprolactam layer, 2-μm thick
	Zipax®-ANH	LLC	37–44	1	na	na	na	S	1% cyanoethylsilicone polymer coated on Zipax®
	Zipax®-HCP	LLC	25–37	1	na	na	na	S	Nonpolar saturated hydrocarbon polymer coated on Zipax®: Avoid temperatures 50°C.
	Zipax® PAM	LLC	25–37	1	na	na	na	S	Nylon coated on Zipax®
Alumina	Pellumina HS, HC	LSC, LLC	37–44	HS 4 HC 8	0.5	1	1.5	S	HC has a thicker coating than HS

[a] From Majors (1974b).
[b] S = spherical; I = irregular.
[c] na means not applicable.

(HETP) to be expected at modest flow rates for the former are about 0.01–0.02 mm, while for the latter they range from 0.1–0.3 mm. Since the square of the pressure drop for a packed column is inversely proportional to the particle diameter [see Eq. (53)], pressure drops for the microparticles (5 to 10 μm) are much larger than for the superficially porous particles. Even when comparing an identical particle size for the two, pressure drops are still larger for the microparticles by factor of about 3. Pressure drops for microparticle columns are decreased by reducing the column length. This is possible because these columns have much higher efficiencies. On the average, commercial columns of the superficially porous particles are 50–100 cm while the commercial microparticle columns are 15–25 cm. In general, the microparticles have been shown to provide a better performance factor (number of effective plates) by about a factor of 5 to 10 when the pressure drop, separation time, and other factors are normalized (Kirkland, 1972).

Sample retention on a polar adsorbent, such as silica, follows the polarity of the most polar functional group in the compound as an approximation. This is shown in Table XVIII. For nonpolar adsorbents the elution order is the opposite as an approximation.

Liquid–solid HPLC on polar adsorbents is generally used for the separation of nonionic compounds which are soluble in organic solvents. Ionic compounds will often tail, while water soluble ones are often easier handled by reversed phase liquid–liquid HPLC or reversed bonded phase HPLC. Recently, liquid–solid HPLC on nonpolar adsorbents has been shown to be suitable for the separation of organic ionic compounds (Chu and Pietrzyk, 1974; Pietrzyk and Chu, 1977; Kroeff and Pietrzyk, 1978a; Cantwell, 1976; Majors and MacDonald 1973).

The number and position of the polar functional groups in the compounds being separated will have a significant effect on their retention on the adsorbents. Also contributing is the number and spatial arrangement of the adsorption sites on the sorbent. Consequently, adsorption is ideal for the separation of polyfunctional and isomeric compounds.

TABLE XVIII

Typical Order of Sorption on a Polar Adsorbent[a]

Fluorocarbons	Nitro compounds
Saturated hydrocarbons	Esters $\approx$ ketones $\approx$ aldehydes
Olefins	Alcohols $\approx$ amines
Aromatics	Amides
Halogenated compounds	Carboxylic acids
Ethers	

[a] k' is increasing as order progresses.

Figures 30–32 illustrate several separations using liquid–solid HPLC. Figures 30 and 31 illustrate the improvement in column efficiency, resolution, and analysis time between using the larger particles and microparticles. Silica type adsorbents were used in Fig. 30, while alumina was used in Fig. 31. It should be noted that part of this improvement is the result of mobile phase selection and recent improvements in the physical form of the adsorbent other than particle size. The ability to resolve isomeric and closely related compounds is illustrated in Fig. 32 where a series of aromatic amines were separated on a microparticle silica column.

Adsorbent activity, which is a measure of the relative number of sites available for chromatographic interactions, can be controlled by adding water to the adsorbent. The water, being highly retained by normal phase adsorbents, selectively covers the most active sites on the surface of the adsorbent. It can be introduced into the unpacked adsorbent or through a packed column by passage of a water modified mobile phase.

Bulk adsorbent is first heated for at least 4 hr to remove residual water ($\sim 125°$C for silica and $\sim 300°$C for alumina). After the adsorbent is cooled in

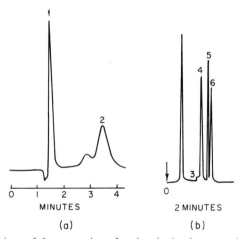

Fig. 30 Comparison of the separation of a vitamin A mixture on (a) large particle and (b) a microparticle silica column. (a) Column conditions: 0.5 m × 2.3 mm, 37–50 μm, Corasil II, using chloroform at 0.75 ml/min and detection by refractive index. Peak identifications: 1, Vitamin A acetate; 2, Vitamin A alcohol; 3, is 13-*cis*-Vitamin A acetate. (Reproduced from K. J. Bombaugh, R. F. Levangie, R. N. King, and L. Abrahams, *J. Chromatogr. Sci.* 8, 657 (1970) by permission of Preston Publications, Inc.) (b) Column conditions: 0.5 m × 4.5 mm, 5 μm, spherisorb silica, using 2% propyl ether in hexane at 2 ml/min and 1100 psi with detection at 254 nm. Peak identifications: 4, 11-*cis*-Vitamin A acetate; 5, 9-*cis*-Vitamin A acetate; 6, All *trans*-vitamin A acetate. (Reprinted by permission from Spectra Physics, Santa Clara, California.)

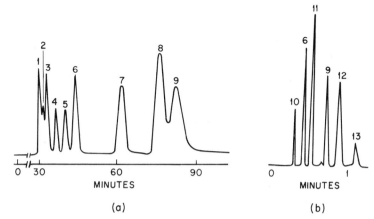

(a)   (b)

Fig. 31 Comparison of the separation of an aromatic hydrocarbon mixture on (a) a large particle and (b) a microparticle alumina column. (a) Column conditions: 2 m × 2 mm, 100–125 μm, Alumina (Woelm), using pentane at 0.19 ml/min and 4 bars with detection at 254 nm. Peak identifications: 1, Benzene; 2, Tetrahydronaphthalene; 3, Styrene; 4, Indene; 5, Naphthalene; 6, Biphenyl; 7, Fluorence; 8, Phenanthrene; 9, Anthracene. (Reprinted with permission from M. Martin, J. Loheac, and G. Guiochon, *Chromatographia* 5, 33 (1972) Copyright (1972), Pergamon Press, Ltd.) (b) Column conditions: 110 mm × 4 mm, 5 μm, Spherisorb Alumina, using hexane at 4 ml/min and 700 psi with detection at 254 nm. Peak identifications: 10, Toluene; 11, o-Terphenyl; 12, Pyrene; 13 p-Terphenyl. (Reprinted by permission from Spectra Physics, Santa Clara, California.)

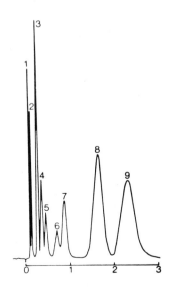

Fig. 32 Separation of an isomeric aromatic amine mixture on a microparticle silica column. Column conditions: 88 mm × 3 mm, 8–12 μm Merckogel Si 150, using 2,2,4-trimethylpentane at 6 ml/min and 130 bar with detection at 270 nm. Peak identifications: 1, Azobenzene $C_{12}H_{10}N_2$; 2, *N,N*-Dimethylaniline $C_8H_{11}N$; 3, Benzo[*h*]quinoline $C_{13}H_9N$; 4, Carbazole $C_{12}H_9N$; 5, *o*-Toluidine C_7H_9N; 6, 1-Naphthylamine $C_{10}H_9N$; 7, 2-Naphthylamine $C_{10}H_9N$; 8, Quinoline C_9H_7N; 9, Isoquinoline C_9H_7N. (Reprinted with permission from H. Oster, S. Van Damme, and E. Ecker, *Chromatographia* **4**, 209 (1971), copyright (1971), Pergamon Press, Ltd.)

a closed container, a known amount of water is added. The amount added depends on whether the adsorbent is totally porous (large or microparticles) or is superficially porous. For the former, this amounts to about 6%–12% water by weight and for the latter about 0.5%–1.5% by weight; this represents about half of a monolayer covering.

If a water modified mobile phase is used, care should be exercised to ensure that the water–organic solvent mixture is homogeneous and that the water is not merely distributed as droplets. Usually a nonpolar organic solvent saturated with water is used. Some workers have suggested that using small amounts (<0.5%) of lower alcohols, such as isopropanol, mixed with the nonpolar organic solvent will serve the same purpose as the water–nonpolar organic solvent mixture.

The water content of the adsorbent must remain constant during use of the column. This is accomplished by having a controlled amount of water in the mobile phase. If the mobile phase is too dry, water is slowly removed from the adsorbent causing its activity to rise slowly; hence, the k' values will increase slowly. If the mobile phase is too wet, the adsorbent's activity slowly decreases due to buildup of water and the k' values slowly decrease. These changes in k' will have a bearing on retention times and on resolution, and may even affect selectivity. Optimizing the adsorbent–mobile phase water activity is accomplished by using a test compound of modest k' value. Variations observed between repetitive injections, are eliminated through modifications of the amount of the water modifier in the mobile phase. In these standardization experiments it is imperative that the chromatographic system be allowed to reach equilibrium anytime changes are made before sample injection.

Usually, directions for activity modification or reconditioning of commercial columns are provided by the manufacturer with the purchase of the column or adsorbent. These standardization procedures are also described elsewhere (Snyder, 1968, 1971; Snyder and Kirkland, 1974; Majors, 1976).

4. *Bonded Phase HPLC*

To overcome several of the problems associated with liquid–liquid chromatography, the stationary liquid layer was chemically bonded to the stationary phase. In general, it was thought that the chemically bonded system would have the physical advantages of an adsorbent stationary phase (stability due to a single component phase) and still provide the column selectivity, efficiencies, and resolution associated with a liquid–liquid system. In practice, both were realized. However, the chromatographic mechanism accounting for sorption on the bonded phase is not necessarily the same as that for conventional liquid-coated phases. For this reason these kinds of supports are treated separately and are part of a broad class of stationary phases known as chemically bonded phases.

Several different types of bonded phase have been prepared·with most being modifications of silica. Progress in synthetic procedures in this field are reviewed elsewhere (Snyder and Kirkland, 1974; Snyder, 1971; Walton, 1978; Grushka, 1974; Leitch and DeStefano, 1973; Majors, 1974a,b, 1976, 1977; Cox, 1977; see also Table I). Typical bonding reactions include formation of siloxanes, silicate esters, and silicon–carbon bonds followed usually by other modifications. Currently the siloxane type bonded phase is widely available through commercial outlets.

The modification of silica involves the reaction of silica with compounds such as $RSiCl_3$ or $RSi(OCH_3)_3$, followed by reaction with water to form oxygen cross-links. By varying the nature of the R group the surface polarity can be altered. Consequently, normal bonded phase particles of differing degrees of polarity and reversed bonded phase particles can be prepared. In these reactions not all of the silica surface is completely covered by the bonded phase and unreacted Si–OH groups will contribute to the chromatographic mechanism. If the bonded phase is hydrophobic or reversed phase, a mixed mechanism is then possible. Treatment with trimethylchlorosilane reduces the number of free Si–OH groups but does not eliminate them completely.

Bonded phase packings were initially prepared with the larger totally porous silica particles ($50 \, \mu m$ average). This was followed by development of the superficially porous bonded phase. At present both of these have been replaced, particularly the former, by chemically bonded microparticles. Table XIX provides a brief list of the types of commercially available bonded phases. Information about these and other bonded phases and a list of commercial suppliers is provided elsewhere (Snyder and Kirkland, 1974; Snyder, 1971; Walton, 1978; Grushka, 1974; Leitch and DeStefano, 1973; Majors, 1974a,b, 1977; Cox, 1977).

At first sight it appears that the same type of bonded phase is available from many different manufacturers. For example, the basic octadecylsilane type bonded phase is commercially available from many sources. Although these all have the same bonded phase, differences are possible and these differences can cause the chromatographic result to vary from manufacturer to manufacturer. Major differences are in particle sizes, the uniformity in size, the amount of coverage by the octadecylsilane group, the ability to be wetted, and the surface area of the silica used in the preparation. It should be noted that polar bonded phases can still be used in the reversed phase mode since the mobile phase used can be of greater polarity than the bonded phase.

TABLE XIX

Type of Commercially Available Reversed and Normal
Phase Microparticles[a]

Reversed phase		Normal phase	
Type	Functionality	Polarity	Functionality
Long chain	Octadecylsilane	Weak	Ester Dimethylamino Diol
Intermediate chain	Octylsilane Cyclohexane	Medium	Fluoroether Nitro Nitrile
Short	Phenyl Short alkyl chain		ε-Aminopolycaprolactam Polyhydroxyethylmethacrylate
	Dimethylsilane	High	Alkylamine Amino

[a] See Majors (1977) for a detailed listing of names of commercial suppliers.

The practical use of bonded phase HPLC has been clearly established. Over 60% of the current publications in HPLC deal with the application of reversed bonded phases in separations. However, even with all this use the actual chromatographic mechanism taking place in bonded phase chromatography is not clearly understood.

Several generalizations are apparent (Walton, 1978; Grushka, 1974; Majors, 1974b; Majors and Hopper, 1974; Brust et al., 1973; Horvath et al., 1976, 1977; Horvath and Melander, 1977) and not only contribute to an understanding of the chromatographic mechanism, but also aid in selecting a column for a particular application. These generalizations, considering a a reversed bonded phase of the type which employs an octadecylsilane layer, are summarized in the following.

(1) Most separations appear to be a function of the hydrocarbon character of the sample.

(2) Solute-bonded phase interactions are generally not strong and the mobile phase may determine the selectivity for a given pair of compounds being separated.

(3) Both the bonded phase and the mobile phase will determine group selectivity.

(4) For a homologous series, particularly those that differ by a methylene group, and a hydrocarbon bonded phase, elution order is often inversely proportional to their solubility in the mobile phase.

(5) A more significant change in elution characteristics for polar compounds compared to nonpolar compounds is observed if the polarity of the bonded phase is increased while still maintaining the reversed mode.

In normal bonded phase HPLC (see Table XIX), the polarity, number, and position of functional groups within the compounds being separated are determining factors. Other generalizations in normal bonded phase HPLC are difficult to make because of the wider differences in types of polar bonded phases. For example, the bonded phases can differ widely in attached functional groups and consequently, polarity. Typical groups such as $-O-$, $-COOH$, $-SO_3H$, $-NH_2$, $-CN$, or NO_2 can be introduced into the bonded layer. In addition, the bonded phase can be monomeric containing the functional groups or polymeric with the groups. Being either monomeric or polymeric will influence whether partitioning or adsorption is the basic chromatographic mechanism. Finally, for some polar bonded phases the level of retention and retention order is only modestly different than that for silica itself, while for others the difference is significant. However, even if the retention is similar, the normal bonded phase is still very useful because of its better and more rapid response to changes in the mobile phase.

Popularity of reversed phase HPLC is the result of many practical considerations.

(1) Chromatography is readily carried out based on partition (adsorption or mixed mechanism) and is strongly influenced by ion suppression and ion-pair partition. The latter two are discussed in Section IV.B.6.

(2) The practical considerations of column resolution, efficiency, and analysis time are very favorable.

(3) Nonionic, ionic, and ionizable compounds are readily separated and generally require only a single mobile phase.

(4) The bonded phase columns are relatively stable to most conditions used in HPLC except mobile phase solutions of pH >9 which will attack the siloxane bonds and those of pH <1, which will attack the silica. Usually, a useful pH range of 2–9 is recommended for bonded phases.

(5) The major component of the mobile phase is water and most frequently used organic modifiers are methanol, ethanol, and acetonitrile, all of which are readily available at modest cost and a high level of purity.

(6) The type of organic modifier and its concentration will have an effect on retention and is generally predictable by the solvent strength parameter. That is, capacity factors will increase as the polarity of the solvent decreases. A small change in concentration of organic modifier will produce a large change in the capacity factor. Thus, the system is ideal for performing gradient elution since only a small change in solvent composition is required.

(7) Elution orders are often predictable from the solubility of the compounds being separated in the mobile phase.

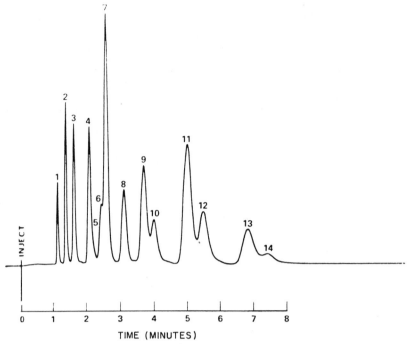

Fig. 33 Separation of aromatic hydrocarbons on a μ bondapak C_{18} column. Column dimensions, 300 mm × 4 mm i.d.; Mobile phase, acetonitrile–water (60:40); Flow rate, 6 cm³ min⁻¹. Peak identifications: 1, Benzene; 2, Toluene; 3, Naphthalene; 4, Biphenyl; 5, Acenaphthene; 6, Phenanthrene; 7, Anthracene; 8, Pyrene; 9, Triphenylene; 10, Chrysene; 11, 1,2-Benzopyrene; 12, 3,4-Benzopyrene; 13, 1,2,3,4-Dibenzathracene; 14, 1,2,5,6-Dibenzanthracene. R. V. Vivilecchia, R. L. Cotter, R. J. Limpert, N. Z. Thimot, and J. N. Little, *J. Chromatogr.* **99**, 407 (1974). (Reprinted by permission from Elsevier Scientific Publishing Company, Amsterdam.)

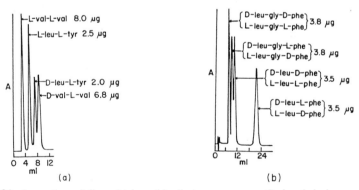

Fig. 34 Separation of di- and tri-peptide diastereomers on a C_8 bonded phase column. (a) Column 250 × 3.2 mm 10 μm LiChrosorb C_8, mobile phase, 10% EtOH–90% H_2O at pH 3.4 (phosphate) and $\mu = 0.1$, flow rate 1.0 ml/min, uv detection at 208 nm (0.16 AUFS). (b) Column, 250 × 3.2 mm 10 μm LiChrosorb C_8, mobile phase, 15% CH_3CN–85% H_2O at pH 3.5 (phosphate) and $\mu = 0.1$, flow rate 1.5 ml/min, uv detection at 208 nm (0.16 AUFS). (Reprinted by permission from E. P. Kroeff and D. J. Pietrzyk, *Anal. Chem.* **50**, 1353 1978. Copyright by the American Chemical Society.)

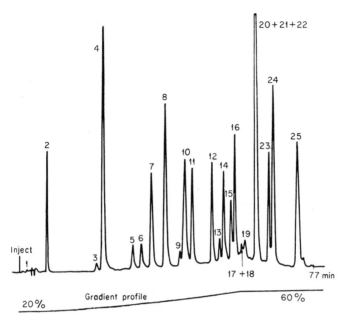

Fig. 35 Separation of 24 carbamate pesticides: Column, μ-Bondapak C_{18}; Mobile phase, 20%–60% MeCN in water, concave gradient (60 min); Flow rate, 1.0 ml/min; uv detector at 220 nm (1.0 AUFS); Inject vol, 1.0 μl: Peaks: (1) Solvent front; (2) Methomyl; (3) Aldicarb; (4) Isolan; (5) Baygon; (6) Carbofuran; (7) Mobam; (8) Carbaryl; (9) Landrin; (10) Propham; (11) Banol; (12) Mesurol; (13) Zectran; (14) Betanal; (15) Chloropropham; (16) Eptam; (17) Bux; (18) Captafol; (19) Barban; (20) Eurex; (21) Vernolate; (22) Pebulate; (23) Butylate; (24) Avadex; (25) Avadex BW. (Reproduced from C. M. Sparacino and J. W. Hines, *J. Chromatogr. Sci.* **14**, 549 (1976) by permission of Preston Publications, Inc.) Company, Niles, Illinois.

Almost all packings used in the reversed phase mode are the octylsilica and octadecylsilane type. Several workers have examined the effect of the hydrocarbon chain length in the reversed bonded phase (Grushka, 1974; Scott and Kucera, 1977; Hemetsberger *et al.*, 1976; Karch *et al.*, 1976; Kitka and Grushka, 1976). It appears that the retention increases with an increase in the chain length, however, a quantitative conclusion is not readily apparent. This is further complicated by the fact that the number of free Si–OH groups vary considerably, not only for different hydrocarbon chain lengths in the bonded phase, but also for the same chain length bonded phase provided by different manufacturers. If a sufficient number of Si–OH groups are not present, the bonded phase surface does not become wetted and hence, mass transfer processes become slow and incomplete (Scott and Kucera, 1977).

Figures 33–35 illustrate three applications of reversed stationary bonded phases. A C_{18} (octadecysilica) microparticle column was used in Fig. 33

where a series of aromatic hydrocarbons were separated. The retention of the aromatic hydrocarbons increases, as previously suggested, as the solubility of the nonpolar aromatic hydrocarbons decreases in the mobile phase. In Fig. 34 a C_8 (octylsilica) microparticle column was used to separate several diastereomeric di- and tri-peptides. The fact that L,L or D,D-dipeptide enantiomers are less retained than the corresponding D,L or L,D-dipeptide enantiomers can be predicted by considering their conformation and poten-

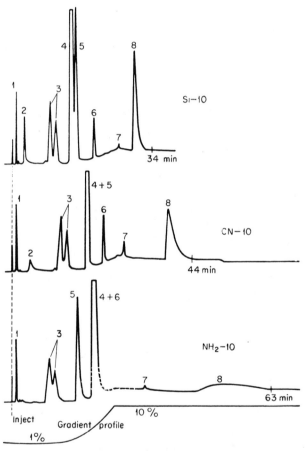

Fig. 36 Comparison of silica microparticles and nitrile and amine bonded phases for the separation of carbamates. Mobile phase, 1%–10% 2-propanol in heptane, concave gradient (20 min.); Flow rate, 1.0 ml/min; uv detector at 220 nm (1.0 AUFS); Inject vol, 2.0 μl. Peaks: (1) Solvent front; (2) Benomyl; (3) Bux; (4) Carbaryl; (5) Carbofuran; (6) Aidicarb; (7) Artifact; (8) Methomyl. (Reproduced from C. M. Sparacino and J. W. Hines, *J. Chromatogr. Sci.* **14**, 549 (1976) by permission of Preston Publications, Inc.)

tial interactions of the side chains with the bonded stationary phase (Kroeff and Pietrzyk, 1978b). Figure 35 illustrates the separation of a series of 24 carbamate pesticides on a C_{18} column using an acetonitrile–water gradient. Again, the elution order correlates with polarity, the more retained being the least polar (Sparacino and Hines, 1976).

Bonded phases of intermediate polarity can be used in both the reversed and the normal phase mode. Figure 36 illustrates the separation of several carbamate pesticides on a nitrile type (intermediate polarity) and an amine type (polar type) bonded phase using a normal phase mode. Also, included is the same separation on the microparticle silica column. Note that the elution orders are similar (compounds 4, 5, and 6 are altered) for the three normal phases and are the opposite of that for the reversed phase (see Fig. 35). Also, note that the two normal bonded phase packings appear to offer little advantage over the microparticle silica. In fact, for the former phases those compounds with larger capacity factors have broader, tailed peaks. Closer examination, however, reveals that advantages in separation can be achieved by using the polar bonded stationary phases (Sparacino and Hines, 1976). For example, compare the elution of compounds 4, 5, and 6 in Fig. 35 for the polar bonded phases. The authors were unable to separate the 24 component carbamate mixture listed in Fig. 35 on the normal bonded phases or on the microparticle adsorbent (Sparacino and Hines, 1976).

The nitrile type bonded phases are particularly useful, because the chromatography on this phase is almost identical to that previously found using a liquid–liquid mode, in which β,β'-oxydipropionitrile was the stationary liquid phase. The bonded phase will provide better efficiency and subsequently reduce analysis time.

The amine type bonded phase has basic properties and is in contrast to silica which has acidic properties. Thus, in aqueous solution the amine can be protonated and the bonded phase will then exhibit properties of a weak anion exchanger. An application of this is shown in Fig. 37; a discussion of ion exchange HPLC is provided in the next section.

Optically active tripeptides have been bonded to silica and these columns were evaluated for the separation of amino acid and dipeptide isomers (Kitka and Grushka, 1977; Fong and Grushka, 1977). The retention orders on these columns do not appear to be exactly the same as those observed for the C_8 column (Kroeff and Pietrzyk, 1978b).

5. Ion Exchange HPLC

In ion exchange HPLC the stationary phase is an ion exchanger and is capable of exchanging its own cations, if it is a cation exchanger, with cations

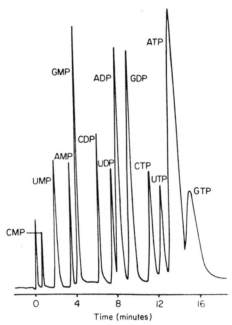

Fig. 37 Separation of a mixture of 5′-nucleotide phosphates on a normal bonded phase. Column MicroPak NH_2. Dimensions, 250 mm × 2.2 mm; Mobile phase, Solvent A, 0.05 M KH_2PO_4 at pH 3.20, Solvent B, 0.5 M KH_2PO_4 at pH 3.95; Gradient, 0%–50% B at 2% min^{-1} for 3 min, 4% min^{-1} for 6 min, 10% min^{-1} for 6 min, hold for 6 min, reset; Detector, uv (254 nm) set at 0.32 AUFS. (Reprinted by permission from R. E. Majors, "Practical High Performance Liquid Chromatography," edited by C. F. Simpson, Heyden and Son Limited, London, England, 1976.)

in the mobile phase on an equivalent basis. If it is an anion exchanger anions are exchanged on an equivalent basis. Cation exchangers can have sites that are strongly or weakly acidic, while anion exchangers can have strongly or weakly basic sites. The stronger sites are ionized throughout the entire pH range, while the weaker sites are ionized only through a limited pH range which is determined by the ionization constant for the exchange site.

The theory of classical ion exchange chromatography has been discussed in detail in many sources (see Table I) and will not be reviewed here. This type of chromatography is ideally suited for the separation of ionic compounds and has been widely used in the separation of inorganic cations and anions. Many organic molecules will ionize or can be made to ionize through pH control, and separations of these compounds are also possible.

To carry out ion exchange HPLC, it is necessary to have ion exchange particles that are of appropriate particle size and physical strength in order

TABLE XX

Comparison of HPLC Ion Exchangers[a]

Property	Superificially porous	Silica	PS–DVB resins
Typical d_p (μm)	30–40	5–10	7–10
Typical ion exchange capacity, (meq/g)	0.01–0.1	0.5–2	3–5
Rigid to pressure deformation	Excellent	Very good	Fair to poor[c]
Shape	Spherical	Spherical or irregular	Spherical
Pressure drop	Low	High	Highest
Efficiency	Moderate	High	Moderate
Packing technique	Dry	Slurry	Slurry
pH range	2–12 (coated) 3–7.5[b] (bonded)	2–7.5[b]	0–12 (anion) 0–14 (cation)
Regeneration rates	Fast	Moderate	Slow

[a] From Majors (1977). (Reprinted by permission of Preston Publications, Inc.)
[b] Some manufacturers claim upper limit of 9.
[c] Depending on degree of cross-linking.

to obtain efficient separations at the flow conditions used in HPLC. Three types of ion exchangers are widely used in HPLC. One type is the conventional or classical ion exchanger based on the polystyrene–divinylbenzene copolymer. The other two types are (1) the modification of porous silica through bonding of a phase which contains the exchange site and (2) the superficially porous type in which the inert core is coated or bonded with a liquid or a copolymer, usually polystyrene–divinylbenzene, which contains the exchange sites. Table XX compares the properties of these three types of ion exchangers. Specific details about the commercially available ion exchangers for HPLC are provided elsewhere (see Walton, 1978; Majors, 1974a,b, 1977; see also Table I).

The ion exchangers are available as cation and anion exchangers containing the strongly acidic $-SO_3H$ and strongly basic $-NR_3{}^+Cl^-$ group, respectively. Weakly acidic and basic ion exchangers, for example ion exchangers containing the $-CO_2H$ and $-NH_2$ group, respectively, are also available.

Each type of ion exchanger offers advantages and limitations. One main advantage of the conventional ion exchanger is that there is a wealth of information already available about the separation of strong and weak organic electrolytes on ion exchangers. These data have been collected, employing classical liquid chromatographic techniques. A second advantage is that the classical ion exchangers are stable throughout the entire pH range and have large capacities. Principal limitations are that the conventional ion exchanger beads are not physically strong; they swell and contract, and they must be slurry packed. Thus, pressure drops across these columns

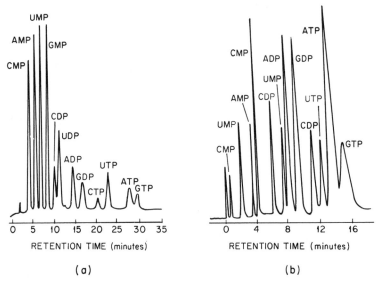

RETENTION TIME (minutes)

(a)   (b)

Fig. 38 Separation of nucleotide mixture on (a) a strong base type exchanger and (b) a weak base type exchanger: CMP = Cytidine-5′-monophosphate; AMP = adenosine-5′-monophosphate; UMP = uridine-5′-monophosphate; GMP = guanosine-5′-monophosphate; CDP = cytidine-5′-diphosphate; UDP = uridine-5′-diphosphate; ADP = adenosine-5′-diphosphate; GDP = guanosine-5′-diphosphate; CTP = cytidine-5′-triphosphate; UTP = uridine-5′-trisphosphate; ATP = adenosine-5′-triphosphate; GTP = guanosine-5′-triphosphate.

(a) Column conditions: Column, 1 m × 2.1 mm i.d.; Packing, Permaphase AAX; Temperature, ambient; Mobile phase, gradient from 0.002 M potassium phosphate, pH 3.3, to 0.5 M potassium phosphate at a gradient rate of 3%/min; Inlet pressure, 67 bars (1000 psi); Flow rate, 1 ml/min; Detector, uv absorbance. (Reproduced from R. A. Henry, J. A. Schmit, and R. C. Williams, J. Chromatogr. Sci. **11**, 358 (1973) by permission of Preston Publications, Inc.)

(b) Column conditions: MicroPak-NH$_2$; Dimentions, 25 cm × 2.2 mm; Mobile phase, concave gradient, Solvent A, 0.005 M KH$_2$ PO$_4$, pH 3.2, Solvent B, 0.5 M KH$_2$PO$_4$, pH 3.95; Slope, 0–50% B at 2% min^{-1} for 3 min 4% min^{-1} for 6 min, 10% min for 6 min, and hold for 6 min; Flow rate, 100 ml/hr; Detector: 2.54-nm uv absorption. R. A. Majors, "Chromatography," Series II, Volume 1, International Scientific Communications, Inc., Fairfield, Connecticut, 1977. (Reprinted by permission from International Scientific Communications, Inc., I, Fairfield, Connecticut.)

tend to be high and will increase as bead fracture occurs. These properties can be modified by increasing the cross-linking in the copolymer matrix.

The silica type ion exchanger is a special type of bonded phase (which contains the exchange site) microparticle. It provides a high efficient column of modest capacity that can be used to separate complex mixtures. The main limitations are its limited useful pH range and the fact that its ionic form is not always clearly identified or readily changed from one form to another.

The superficially porous type ion exchangers are widely used. They are rigid, easily packed into columns, generally provide low column pressure drops, and often provide selectivities different than those observed for other ion exchangers. Efficiencies and capacities are lower than the silica type ion exchangers and if used in the coated mode, there is a risk of a slow bleeding of the coated stationary liquid phase.

Regardless of which type of ion exchanger is used, it is necessary to control the column temperature, ionic strength, pH, type of buffer, and organic solvent concentration, if used. These factors will have a strong influence on the efficiency, selectivity, and resolution. Also, if not controlled, the overall column stability, flow characteristics, and general column performance will deteriorate. Most manufacturers recommend a pretreatment to condition the ion exchanger. This procedure is usually provided by the supplier of the ion exchanger and varies with the type of ion exchanger.

Figure 38 illustrates the separation of a complex mixture of nucleotides. In Fig. 38a the separation is achieved using a superficially porous type packing in which a quaternary ammonium substituted polysiloxane is attached to silica. A microparticle weak base packing where an alkylamine is bound to silica is used in Fig. 38b. This same separation was reported several years earlier to take 2.5 hours when using a superficially porous type anion exchanger containing a strong base type exchange site (Burtis *et al.*, 1970).

6. *Ion Suppression and Ion Pair HPLC*

Reversed phase chromatography employs a hydrophobic stationary phase and a hydrophilic mobile phase. Typical hydrophobic stationary phases are nonpolar adsorbents and inert supports which are coated or bonded with a nonpolar phase, while the mobile phases are typically water mixed with lower alcohols or acetonitrile. Polar substances prefer the mobile phase and elute first. As the hydrophobic character of the solute increases retention also increases.

For organic acids, bases, and ampholytes, their polar character in the mobile phase can be controlled by the pH of the solution. For example, consider the ionization of the weak acid RCOOH or

$$RCOOH + H_2O \rightleftharpoons H_3O^+ + RCOO^- \tag{57}$$

If the pH of the mobile phase is acidic, the equilibrium is shifted to the left, towards the formation of the less polar species, the retention is high on the reversed stationary phase, and the separations are possible based on the differences in retention of the less polar, nonionized form. Alternatively, the pH of the mobile phase can be made basic. Under these conditions the equilibrium is shifted to the right and the acid is converted into the more polar ionized form. Retention is then negligible or modest on a reversed phase depending on the hydrophobic character of the R group. Differences in this hydrophobic character can be utilized to bring about separations on the reversed stationary phase. This kind of chromatography is often referred to as ion suppression chromatography.

Several factors influence the separation. These include (1) the ionization constants for the protonic sites in compounds being separated; (2) the number and type of groups within the molecule which can influence the ionization constants and/or the hydrophobic character of the sample, (3) pH and ionic strength in the mobile phase; (4) the composition of the mobile phase, where eluting power is determined by type and concentration of the organic solvent mixed with water; and (5) the type of reversed stationary phase selected.

The effect of protonic equilibria on sorption on a reversed phase can be quantitatively predicted providing all equilibria are accounted for. Table XXI lists the equations relating capacity factor, k', to pH in terms of the ionization constants, and k_2', k_1', k_0', k_{-1}', k_{-2}', which are the capacity factors for the retention of the doubly positive charged, singly positive charged, neutral, singly negative charged, and doubly negative charged form of the acid, base, or ampholyte, respectively (Chu and Pietrzyk, 1974; Pietrzyk and Chu, 1977; Kroeff and Pietrzyk, 1978a; Horvath et al., 1977; Grieser and Pietrzyk, 1973; Pietrzyk et al., 1978).

If a bonded reversed phase is used, verification of these equations is limited because the pH stability of the bonded phase does not allow measurements at pH values below 2 or above 9 (Horvath et al., 1977). In contrast, verification of each equation was demonstrated using a reversed phase adsorbent that is stable throughout the entire pH range (Chu and Pietrzyk, 1974; Pietrzyk and Chu, 1977; Kroeff and Pietrzyk, 1978a; Grieser and Pietrzyk, 1973; Pietrzyk et al., 1978).

For ampholytes, retention passes through a minimum, if zwitterion formation occurs. If zwitterion formation does not take place, retention passes through a maximum. This is shown in Fig. 39, where k' data for a typical amino acid and peptide (zwitterions) and anthranilic acid (nonzwitterion)

TABLE XXI

Equations Relating k' to pH for Weak Acids, Bases, and Ampholytes[a]

Weak monoprotic acid	Weak diprotic acid
$k' = \dfrac{k_0'}{1 + K_a/[H^+]} + \dfrac{k_{-1}'}{1 + [H^+]/K_a}$	$k' = \dfrac{k_{-1}' + k_0'[H^+]/K_{a_1} + k_{-2}'K_{a_2}/[H^+]}{1 + [H^+]/K_{a_1} + K_{a_2}/[H^+]}$
Weak monoprotic base	Weak diprotic base
$k' = \dfrac{k_0'}{1 + K_b/[OH^-]} + \dfrac{k_1'}{1 + [OH^-]/K_b}$	$k' = \dfrac{k_1' + k_0'[OH^-]/K_{b_1} + k_2'K_{b_2}/[OH^-]}{1 + [OH^-]/K_{b_1} + K_{b_2}/[OH^-]}$

Diprotic ampholyte[b]

$$k' = \frac{k_0' + k_1'[H^+]/K_{a_1} + k_{-1}'K_{a_2}/[H^+]}{1 + [H^+]/K_{a_1} + K_{a_2}/[H^+]}$$

[a] From Pietrzyk et al. (1978).

[b] The ampholyte ionizes according to

$$H_2A^+ + H_2O \xrightleftharpoons{K_{a_1}} H_3O^+ + HA, \qquad HA + H_2O \xrightleftharpoons{K_{a_1}} H_3O^+ + A^-$$

are plotted as a function of pH. Similar results have been found for other ampholytes (Kroeff and Pietrzyk, 1978a; Pietrzyk et al., 1978; Rotsch et al., 1979).

For diprotic acids and bases, which are not shown in Fig. 39, the second protonic ionization appears to have little effect on the change in k' with pH (Pietrzyk et al., 1978). That is, the major drop in k' occurs on passing through the first ionization step to form the singly charged species indicating that k_0' and k_1' (or k_{-1}' for a diprotic base) are significant and k_2' (or k_{-2}') is negligible. Thus, the diprotic acids and bases are similar to the monoprotic systems, where the pH at the break in the k'–pH plot corresponds to the pK_a or pK_b (first ionization constant for the diprotic acids and bases).

Adjustment of pH is a significant variable in carrying out separations on reversed stationary phases. Coupled with choosing a water–organic solvent ratio and type of organic solvent, all of which can be employed in an isocratic and gradient mode, the three provide a powerful set of eluting conditions to effect a separation.

Many examples illustrating this are now in the literature (Walton, 1978; Chu and Pietrzyk, 1974; Pietrzyk and Chu, 1977; Kroeff and Pietrzyk, 1978a; Horvath and Melander, 1977; Grieser and Pietrzyk, 1973; Pietrzyk et al., 1978; Rotsch et al., 1979; Baum and Cantwell, 1978; Mohammed and Cantwell, 1978). Figures 40 and 41 are typical examples. In Fig. 40 improvement in column resolution and efficiency is illustrated by selecting a suitable pH for the elution. In Fig. 41 a complete reversal in elution order is illustrated as a result of a change in pH of the mobile phase.

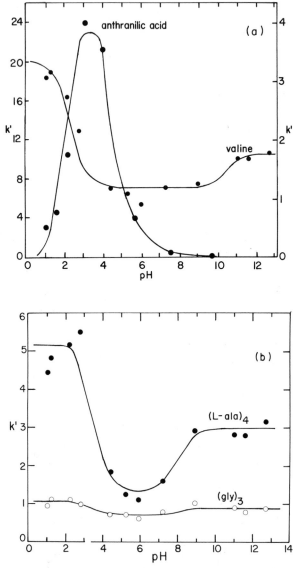

Fig. 39 Capacity factors as a function of pH. (Reprinted by permission from D. J. Pietrzyk, E. P. Kroeff, and T. D. Rotsch, *Anal. Chem.* **50**, 497, 502 (1978). Copyright by the American Chemical Society.)

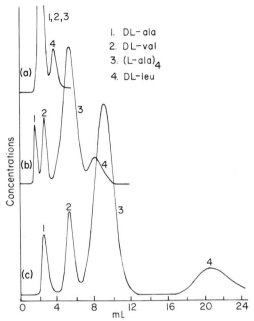

Fig. 40 Illustration of the optimum pH for the separation of an amino acid-peptide mixture. Column conditions 45–65 μm, 0.71 g, 45 × 0.236 cm XAD-2 column at a flow rate of 1.0 ml/min and using 15% CH₃CN–85% H₂O. Sample: (1) 3.6 μg, (2) 3.9 μg, (3) 2.3 μg, (4) 2.3 μg, and (5) 8.3 μg, and detection at 254 nm. (Reprinted by permission from E. P. Kroeff and D. J. Pietrzyk, *Anal. Chem.* **50**, 502 (1978). Copyright by the American Chemical Society.)

Fig. 41 Reversal in elution order as a result of pH change in the eluting condition.

Column conditions, 45–65 μm, 0.31 g, 21 × 0.236 cm XAD-2 column at a flow rate of 0.5 ml/min using detection at 208 nm, a sample of 2.7 μg of each dipeptide, and (a) pH 2.73, phosphate buffer at ionic strength 0.2 *M* in 2% EtOH–98% water, and (b) pH 11.03 phosphate buffer at ionic strength 0.2 *M* in 2% EtOH–98% water. (Reprinted by permission from E. P. Kroeff and D. J. Pietrzyk. *Anal. Chem.* **50**, 502 (1978). Copyright by the American Chemical Society, D.C.)

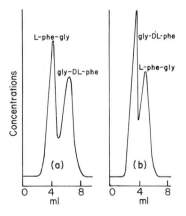

A common technique used in solvent extraction to extract ionizable organic compounds is to increase their solubility in the organic layer via the addition of a counterion. The counterion is usually one containing a hydrophobic group in addition to the charged site.

This technique was originally employed in HPLC with liquid stationary phases coated on inert supports in a liquid–liquid normal phase mode. For example an aqueous buffered tetraalkylammonium salt solution or alkali metal salt solution served as the stationary phase and a hydrophobic organic solvent or mixed solvent was used as the mobile phase. In general, the selectivities and order of elution where predictable from solvent extraction distribution data.

Retention in liquid–liquid ion-paired HPLC was shown to be controlled by several variables. These include the volume ratio of the two phases, temperature, type and concentration of counterion, type and composition of the organic mobile phase, ionic strength of the aqueous phase, and pH of the aqueous stationary phase. Selectivity is altered by changes in one or more of these variables.

A recent development in ion-paired HPLC is the use of reversed bonded stationary phases (Walton, 1978; Majors, 1974a,b, 1977; Gloor and Johnson, 1977; Schill, 1976; Wahlund and Lund, 1976; Fransson *et al.*, 1976). In this case an aqueous solution containing a hydrophobic type counterion is used. Typical countercations and -anions used in eluents in ion paired HPLC are tetraalkylammonium salts and alkylsulfonic acids, alkylphosphonic acids, and alkylcarboxylic acids, respectively. Also, small amounts of organic modifier and buffers for pH control are added to the mobile phase.

The actual chromatographic mechanism accounting for the separation in reversed phase ion-paired chromatography is not fully understood. Whether ion-pair formation between the sample ion and counterion takes place first followed by interaction with the nonpolar bonded phase or whether the counterion is retained by the bonded phase and then participates in an interaction with the solute has not been clearly established. However, the data appear to favor a model based on the latter as being a significant factor in the separation (Walton, 1978; Majors, 1977; Horvath *et al.*, 1976, 1977; Horvath and Melander, 1977; Scott and Kucera, 1977; Gloor and Johnson, 1977; Schill, 1976; Wahlund and Lund, 1976; Fransson *et al.*, 1976; Kissinger, 1977b).

The retention of organic cations such as alkyl- and arylammonium cations on a polystyrene–divinylbenzene reversed phase adsorbent in the presence of small inorganic counteranions can be explained in terms of the Stern–Gouy–Chapman electrical double layer theory. In this model the sample ion is adsorbed onto the surface of the stationary phase as the primary layer and the small inorganic counterions occupy the diffuse part or secondary layer of

TABLE XXII

Variables in Ion Paired Chromatography
on a Reversed Bonded Phase

Variable	Effect
Type of counterion	Retention increases as the ability to ion pair increases.
Size of counterion	Retention increases as the size of the counterion increases.
Concentration of counterion	Retention increases up to the point where ion pair formation is quantitative and then tends to decreases.
pH	Retention, depending on solute, will increase as pH change effects an increase in the ionic form of the solute.
Type of organic modifier	Retention decreases with an increase in lipophilic nature.
Concentration of organic modifier	Retention decreases as modifier's concentration increases.
Temperature	Retention decreases with an increase in temperature.
Stationary phase	Retention increases as the lipophilic character of the stationary phase increases.

the electrical double layer (Cantwell and Puon, 1979). The same model accounts for retention of organic anions such as alkyl- or arylsulfonates on the adsorbent in the presence of inorganic counterions. If lipophilic counterions such as tetraalkylammonium or alkylsulfonate salts are used, the sorption process appears to follow either the ion-pair or ion-exchange sorption mechanism.

The major reversed bonded phases used are the octyl- and octyldecyl–silica type stationary phases. Shorter alkyl chain bonded phases can also be used, however, the exact role and influence of bonded phase chain length is not clearly understood.

The variables in bonded phase ion-paired chromatography are essentially the same as those outlined for liquid–liquid ion-paired chromatography. These are listed in Table XXII accompanied by a brief statement indicating how the variable affects retention. Typical separations are shown in Fig. 42. Note that the elution order is changed as the result of a difference in the counterion in Figure 42a and 42b.

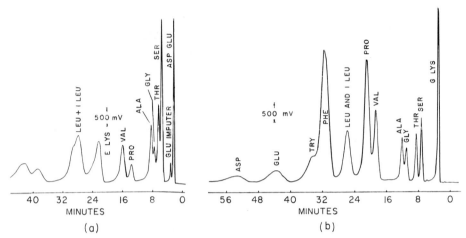

Fig. 42 Comparison of eluting conditions for the ion-paired HPLC separation of a mixture of the Dansyl derivatives of amino acids.

Column conditions, 25 cm × 2.2 mm Micro-Pak-MCH (octadecylsilane bonded phase), 0.5 ml/min using fluorescence detection. (a) 15/85 v/v acetonitrile/H_2O, 0.01 M tetramethylammonium perchlorate, pH = 6.5 with 0.001 M $NH_4H_2PO_4$. (b) 40/60 v/v acetonitrile/H_2O, 0.005 M palmityltrimethylammonium chloride, pH = 6.5 with 0.001 M $NH_4H_2PO_4$. (Reproduced from R. Gloor and E. L. Johnson, *J. Chromatogr. Sci.* **15**, 413 (1977) by permission of Preston Publications, Inc.)

V. Applications

A. Analytical HPLC

1. Scope

The renewed interest in liquid chromatography in the form of HPLC has had a remarkable impact in the scientific laboratory. For example, in the pharmaceutical industry, HPLC has essentially brought about a complete change in the methods of routine quality control. The introduction of HPLC has become one of the major techniques for this kind of analysis and in doing so the technique has led to a substantial decrease in analysis time and an increase in the analytical precision and accuracy.

Clearly, HPLC is applicable to every applied area of science. It is a routine technique in pharmaceutical, biochemical, food and beverage, agricultural, petroleum, environmental, polymer, industrial organic, and inorganic and organometallic analyses. The complexity of the sample does not appear to be a significant handicap. In fact many separations that were at one time impossible to accomplish, or were achieved only after considerable time and effort, have not only been completed by HPLC, but they have been done rapidly and accurately.

HPLC is ideal for trace analysis and the lower limit of detection is generally determined by the type of detector available to the user. Under favorable circumstances current limits of detection are in the order of 1 ppb. Thus, typical routine trace HPLC analysis would include detection of airborne and water pollutants, pesticide residues in plants and foodstuffs, impurities in all sorts of commercial, pharmaceutical, and industrial products, and drugs and their metabolites in body fluids.

Gas chromatography is generally restricted to compounds that have an appreciable vapor pressure or that can be converted into derivatives with appreciable vapor pressure. In contrast, HPLC does not have this restriction and, with proper choice of the stationary phase, it is applicable to the separation of all compounds including those with large molecular weights. The notable exception would be the separation of compounds that are permanent gases; GC not HPLC would be the appropriate choice for their separation. It is clear that the two complement one another in the laboratory and provide a powerful approach to the separation and determination of complex mixtures, particularly those containing organic compounds.

It is beyond the scope of this chapter to survey HPLC applications. However, figures already cited [see Figs. 14, 21, 29–38, 40–42 and Fig. 47 (Section V.B)], although chosen to illustrate specific chromatographic properties, are also examples of typical applications. Many other sources are also available (see Table I; see also (Saunders, 1977; Walton, 1978; Leitch and DeStefano, 1973; Majors, 1974a,b, 1975, 1977; Cox, 1977; Gloor and Johnson, 1977; Kirkland, 1971; Baley, 1976; Marcek *et al.*, 1977; Wheals, 1976; Wheals and Jane, 1977). Another valuable source of applied HPLC information is the technical literature readily available from the manufacturers and suppliers of HPLC instrumentation, columns, and accessories.

2. *Qualitative HPLC*

The method of qualitatively establishing the identity of an eluting component that is most often used is to compare its retention (retention time or volume) to that for standards. For this method to be successful it is necessary that the chromatographic operating conditions for the standard and unknown be the same.

The second method for qualitative identification is to use in-line selective detectors. For example, a photometric detector can be designed to provide a scan of the uv and/or visible spectrum for the elution peak within the detector. Usually, the liquid flow is stopped long enough to obtain the spectrum rather than scanning as the mobile phase carries the elution peak through the cell. Fluorescent and ir spectra can also be obtained with properly designed detectors.

The third method is to collect the eluted peak, remove the mobile phase if necessary, and proceed to characterize the residue with techniques such as melting point, boiling point, ir, nmr, and mass spectroscopy. However, utilization of these methods requires isolation of a sufficient amount of material which is often greater than the sample sizes usually used in analytical HPLC. Techniques for increasing the sample size are briefly described later in this chapter.

As pointed out in Section III.B considerable progress has been made in detector design, which permits passing the eluted sample directly into the mass spectrometer. In this case the compound is isolated from the mobile phase within the design of the detector. This general approach is also possible with ir measurements.

There are pitfalls in qualitative HPLC that can lead to misinterpretation. Usually, these are the result of insufficient attention directed towards the qualitative aspects of HPLC. Some of the more important areas of concern are the following.

(1) The precision of the retention times and volumes must be known.

(2) A single elution peak does not automatically mean that the peak is composed of one compound.

(3) Peaks may appear in the chromatogram as a result of impurities in the components used to prepare the mobile phase.

(4) It is important to verify that all components of the sample are eluted from the column. This can effect not only the first sample examined but also those that follow.

(5) The fact that an unknown sample component and a standard have different retention times under identical chromatographic conditions is evidence that they *are not* the same. In contrast, if the two have identical retention times at one given set of chromatographic conditions, they *may be* the same substance.

(6) Many HPLC detectors are selective and will not respond to the presence of all compounds. Furthermore, each type of detector is characterized by a detection limit.

3. *Quantitative HPLC*

The major application of HPLC is in quantitative analysis. This section briefly deals with the procedures and techniques that are used in the quantitation step in analytical HPLC.

It is important to emphasize that high quality quantitative HPLC data require that the operator pay attention not only to the quantitation step but also to all steps leading to quantitation. These include (1) obtaining a representative sample, (2) preparation of the sample prior to introduction of it into the chromatograph, (3) injection of the sample into the chromatograph, (4) selecting an optimum mobile–stationary phase for the separation, (5) detection, and, finally, (6) quantitation. Errors generated in any or all of the first five steps are automatically carried into the sixth step.

Quantitation procedures are based on the fact that the area under the elution peak is proportional to concentration as in all elution chromatography. Consequently, the procedures used in HPLC are those that have been routinely used in other elution chromatographic techniques, particularly gas chromatography.

Procedures are based on either a direct measurement of the peak area or on measurement of the height of the peak. Both are valid providing the detector response is linear with change in concentration of the component within the eluted peak and that the chromatographic column parameters remain constant. Often the type of analysis being performed will determine which approach is best suited, from an accuracy point of view, for that analysis.

Table XXIII summarizes the basic data handling techniques available for quantitation accompanied by a brief statement describing the procedure. The details for these procedures are readily available elsewhere (see Table I). As pointed out earlier in this chapter in Section III.D, data handling for the quantitation step is readily accomplished as part of a completely automated HPLC instrument.

The quantitation procedures briefly outlined in Table XXIII assume that calibration is achieved by means of an external standard. That is, a series of standards are chromatographed and their peak heights or areas are plotted versus their concentration. The unknown sample is similarly treated and its concentration is determined from the calibration curve. This procedure requires that all the column parameters are held constant and that they are reproduced throughout the standard and unknown measurements. Although these are potentially serious limitations, the major one is not usually one of failure to control one of the column parameters but rather it is a failure to inject the sample uniformly without loss into the liquid chromatograph. Either the sampling injection system or syringe or both can be at fault. (See the III.A.4 for a discussion of the limitations of these components.) Repeated injections are usually desirable to indicate and subsequently minimize these sources of error.

Another approach to minimizing the sample injection errors, as well as the errors due to changes in the column parameters, is to use an internal standard procedure. The peak height or area measurements are still completed in the same fashion, the difference being that a second compound (internal standard) of fixed concentration is added to all solutions used for

TABLE XXIII

Quantitation Procedures for HPLC Data

Peak height

The peak heights for a series of standards are determined and plotted versus concentration of the standard. In general, good accuracy through a peak height calibration curve is possible. This will depend on the reproducibility of the column variables, the number of standard points, and whether the chromatographic peak is a major baseline separated peak or a shoulder. Distorted peaks and overloaded peaks will lead to large errors.

Triangulation

Peak areas can be established through two triangulation procedures. The area is calculated (1) by peak height times the width at half-height and (2) by assuming a triangle for the peak so that the area is given by peak height times 1/2 of its base. The first procedure tends to be more precise since it is less susceptible to errors in judgements. Both require well-defined symmetrical peaks. Calibration is completed by plotting areas of a series of standards versus their concentration. In general precision of around 3% for the former and 4% for the latter are typical.

Counting

Areas under the chromatographic peak can be determined by counting the squares on the chart paper enclosed within the peak. This is tedious and probably offers an advantage only when determining areas for broad peaks with small peak heights. Quantitation is accomplished through a calibration curve of peak area versus concentration of standard.

Planimeter

A planimeter is a mechanical device which accomplishes a counting of squares under the peak. The edge of the peak is traced and its area is read from a scale on the planimeter. A steady hand and good eyesight is required; also areas of irregular shaped peaks can be determined. Quantitation is accomplished through a calibration curve of peak area versus concentration of standard. Precision of about 4% is typical.

Cut and weigh

The area of the peak can be determined in weight units by cutting out the peak with scissors and weighing it on the analytical balance. Either the original or a copy of the original can be used. In both cases it is necessary that the paper weight per unit area be uniform. Quantitation is accomplished through a calibration curve of peak weight versus concentration of standard. Precision is favorable, often approaching 2%.

Disc integration

Area within the peak can be determined by incorporation of a disc integrator, a mechanical device, into the recorder. Under optimum conditions accuracy is limited by the recorder performance and a precision of 1% is often obtained. Errors are large if the baseline drifts or if peaks are not resolved. These are potentially correctible by manual manipulation of the data. Quantitation is accomplished through a calibration curve of peak area versus concentration of the standard.

Electronic integration

Electronic integration ranges widely in sophistication. The basic digital integrator accepts the analog output signal from the detector and provides a digital output of peak

TABLE XXIII (*continued*)

areas accompanied by an analog output signal to drive a strip chart recorder. Depending on design, the basic integration function may include area corrections for baseline drift and for differentiating areas beneath partially resolved peak. A computing integration not only provides a digital output of peak areas and an analog output signal but it also provides memory and computation features. This allows detailed calculations of the data and programming for automated measurements as well as for directing and controlling the operational details for a given analysis.

calibration and to the unknown sample. The calibration curve is established by plotting the ratio of the peak height or area of the standard to the internal standard versus the standard concentration. The unknown sample is then treated in a similar fashion.

The critical factor in employing the internal standard is often choosing the correct one to use. This is particularly difficult if the unknown sample contains many components which will subsequently provide a very complex chromatogram. In general, the internal standard peak should be close to the unknown peak and be completely resolved from the unknown or other peaks in the sample. It should be readily available and be in a highly purified form.

B. Preparative-Scale HPLC

In analytical HPLC the aim is to achieve maximum resolution in a brief period of time while using very small samples. In preparative-scale HPLC speed is less important and the aim is to achieve as high a resolution as possible for purposes of purification with a maximum sample throughout per unit time.

Preparative-scale HPLC can mean different amounts to different users. For example, a few milligrams of a purified peptide or similar type compound might be all that is available for further studies while in typical synthetic work amounts in excess of 1 g are often required. In order to handle these kinds of samples, particularly the larger amounts, specially designed columns (wide diameter), specially designed pumps (high flow rates in order to maintain high linear velocities in the wide diameter column), specially designed injection systems (large sample volumes), and appropriate operating conditions are required. Since large samples are employed, sensitive detectors are not necessary; however, the detector must be stable and provide a rapid response.

Although an analytical HPLC can be used at a preparative scale, it is restricted to small amounts of sample even if recycling is used. Attempts to use large diameter columns ($> \frac{1}{4}$ in.) with an analytical HPLC so that larger samples can be used is limited, because the pumps in the analytical instruments are not designed to produce high flow rates which are required to maintain the high linear velocities in the wider diameter column. Also, the detectors usually used in the analytical HPLC have such high sensitivities that they must be modified or split streaming must be used.

In analytical HPLC, the chromatographic characteristics of a particular chromatographic system can be improved by changes in resolution, capacity, or speed. One or two of these can be improved simultaneously but not all three. Thus, an optimization in analytical HPLC is achieved in one or two of these areas at the expense of the others. Furthermore, because of the limitations briefly described, the parameter most often compromised is that dealing with capacity.

In preparative-scale HPLC, capacity is the main objective. This means speed and resolution are compromised in both elution parameters and instrument design in order to optimize the system. Consequently, the instrumentation, and often the operating conditions, are different for preparative-scale HPLC versus analytical HPLC.

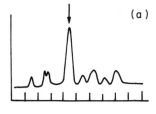

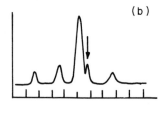

Fig. 43   Typical preparative-scale problems.

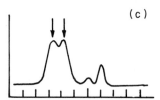

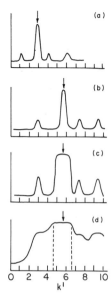

Fig. 44 Steps in scaling up an analytical separation to a preparative-scale for isolation of a single component.

In general, one is interested in using a preparative-scale HPLC either for a crude isolation (or enrichment) of a substance or for a fine separation (or purification) of a substance. Typical examples of the former are the isolation of synthetic intermediates, of natural products, and of trace impurities. Typical examples of the latter include the preparation of pure standards (chemical, pharmaceutical, and biochemical), purification of reagents, recovery of valuable unused reagents, production of fine chemicals, and purification of trace impurities.

Figure 43 shows three typical analytical HPLC chromatograms which represent three of the more common types of problems readily solved by preparative HPLC. If case a in Fig. 43 is selected, where a single component is desired [see curve (a) in Fig. 44], the first step is to improve the resolution in the analytical separation, if possible, by altering the eluting mixture. This is illustrated in curve (b) in Fig. 44. Using these eluting conditions, an analytical HPLC experiment can be used to determine the loading limit [see curve (c) in Fig. 44], and the purification procedure [see curve (d)], can be accomplished at the preparative level using controlled overloading of the prep column.

If a trace component is desired as in case b in Fig. 43, the yield is enhanced by elution at a controlled overload condition and subsequent pooling of collected fractions. This is illustrated in Fig. 45.

For the isolation of two closely related but partially resolved components, as in case c in Fig. 43, recycling is used for the remaining part after removal

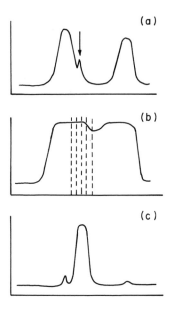

Fig. 45 Steps in scaling up an analytical separation to a preparative-scale for isolation of a minor constituent.

of the leading edge of the first peak and the trailing edge of the second peak. This is shown in Fig. 46.

In each case the analytical HPLC system provides background information for optimizing the preparative-scale separation and for following the progress of the purification. Depending on the column dimensions, column parameters, and available instrumentation, it is often possible to separate over 1 g quantities in a single application.

There are many examples in the literature which illustrate the advantages of preparative-scale HPLC from a column design point of view and from an application point of view (Parris, 1976; Walton, 1978; Fallick, 1977; Pei *et al.*, 1977; DeStefano and Beachell, 1972; Wolf, 1973; DeStefano and Kirkland, 1973a,b; Wehrli, 1975; Karger *et al.*, 1974; Kirkland, 1974; see also Table I). Figure 47 is one such practical application employing a commercially available preparative-scale HPLC (a Waters Prep LC/System 500). Curve (a) in

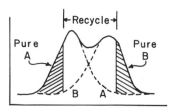

Fig. 46 Scale up for the isolation of two closely related components: pure A and pure B are collected, the rest is recycled.

(a)

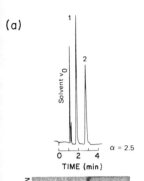

α = 2.5

ΔR$_f$ = 0.15

1. Cholesterol Benzoate

k' 0.82
R$_f$ 0.36

2. Cholesterol Phenylacetate

k' 2.06
R$_f$ 0.21

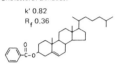

SAMPLE:
Equal mixture (by weight) of Cholesterol Benzoate and Phenylacetate, 10% (w/v) in Benzene/Hexane (50/50).

MOBILE PHASE:
Benzene/Hexane (50/50) for Analytical LC, TLC, *and* PrepLC

ANALYTICAL LC CONDITIONS:
Column: One µPORASIL, 4 mm × 30 cm
Flow Rate: 2.0 ml/min
Detector: Refractive Index @ 8X
Analysis Time: 3.5 min

ANALYTICAL TLC CONDITIONS:
Plate: One precoated, Silica Gel 60, 5 × 20 × .025 cm
Development Time: 40 min
Distance from origin to solvent front: 16.3 cm
Visualization: Phosphomolybdic Acid (10% in ethanol)

(b)

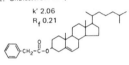

PrepLC CONDITIONS:
Column: One PrepPAK-500 Silica cartridge 5.7 × 30 cm
Flow Rate: 350 ml/min
Detector: Refractive Index @ 5 relative response

QUANTITY:
10 grams of sample in 100 ml volume

SPEED:

Column Preparation:	1.5 min
Wet & Equilibrate Column with Solvent:	5.0 min
Prepare & Load Sample:	2.0 min
Separation & Fraction Collection:	8.0 min
Total Elapsed Time:	16.5 min

(c)

Fraction A

PURITY:
Fraction A - Cholesterol Benzoate - >99.9% by analytical LC

Fraction C

PURITY:
Fraction C - Cholesterol Phenylacetate - >99.4% by analytical LC

Fig. 47 Example of a preparative-scale separation: (a) before Prep LC separation; (b) Prep LC separation; (c) after Prep LC separation.

Sample: Equal mixture by weight of cholesterol benzoate, 1, and cholesterol phenylacetate, 2, in 10% (w/v) in benzene/hexane (1/1); mobile phase: benzene/hexane (1/1) used in analytical and preparative HPLC; analytical conditions: µ Porasil, 4 mm × 30 cm, 2.0 ml/min, refractive index; prep conditions: PrepPak-500 silica cartridge, 5.7 cm × 30 cm, 350 ml/min, refractive index; quantity: 10 g of sample in 100 ml volume. (Reprinted by permission from Waters Associates, Milford, Massachusetts.)

Fig. 47 is the chromatogram for the analytical separation of the two cholesterol derivatives using a 4 mm × 30 cm μPorasil column. In curve (b) of
Fig. 47 a preparative-scale chromatogram is shown for the two cholesterols.
In this case, a silica column of 5.7 × 30 cm and a flow rate of 350 ml/min is
used to separate a 10 g sample of the two cholesterols in 100 ml of solution.
Portions A and C were collected and tested by the analytical column. As
shown in curves (c) in Fig. 47, the two cholesterols are completely free of
each other. It is also important to note the time scales that are involved in
completing this purification.

The large loading capacity of preparative-scale HPLC has potential
commercial applications and cost analysis for this approach has been
reviewed (Rendell, 1975). Applications of this type in the pharmaceutical
industry for the isolation of drugs and/or intermediates from large scale
synthetic reactions are already being done.

References

Arjino, P., Dawkins, B. G., and McLafferty, F. W. (1974). *J. Chromatogr. Sci.* **12**, 574.
Bakalyar, S. R., McIlwrick, R., and Roggendorf, E. (1977). *J. Chromatogr.* **142**, 353.
Baker, D. R., Williams, R. C., and Steichen, J. C. (1974). *J. Chromatogr. Sci.* **12**, 499.
Baley, F. (1976). *J. Chromatogr.* **122**, 73.
Barth, H., Dallmeier, E., and Karger, B. L. (1972). *Anal. Chem.* **44**, 1726.
Baum, R. G., and Cantwell, F. F. (1978). *Anal. Chem.* **50**, 280.
Baumann, W. (1977). *Fresenius' Z. Anal. Chem.* **284**, 31.
Bebris, N. K., Kiselev, A. V., Nikitin, Y. S., Frolov, I. I., Tarasova, L. V., and Yashin, Y. I.
    (1978). *Chromatographia* **11**, 206.
Bollet, C. (1977). *Analusis* **5**, 157.
Bombaugh, K. J. (1977). "Chromatography," Ser. II, Vol. 1, p. 233. Int. Sci. Commun., Fair-
    field, Connecticut.
Brust, O. E., Sebestian, I., and Halasz, I. (1973). *J. Chromatogr.* **83**, 15.
Buchta, R. C., and Papa, L. J. (1976). *J. Chromatogr. Sci.* **14**, 213.
Burtis, A. C., Munk, M. N., and MacDonald, F. R. (1970). *Clin. Chem.* **16**, 667.
Callmer, K., and Nilsson, O. (1974). *Chromatographia* **7**, 644.
Cantwell, F. F. (1976). *Anal. Chem.* **48**, 1854.
Cantwell, F. F., and Puon, S. (1979). *Anal. Chem.* **51**, 823.
Chu, C. H., and Pietrzyk, D. J. (1974). *Anal. Chem.* **46**, 330.
Colin, H., and Guiochon, G. (1976). *J. Chromatogr.* **137**, 19.
Cox, G. B. (1977). *J. Chromatogr. Sci.* **15**, 385.
Davenport, R. J., and Johnson, D. C. (1974). *Anal. Chem.* **46**, 1871.
DeStefano, J. J., and Beachell, H. C. (1972). *J. Chromatogr. Sci.* **10**, 654.
DeStefano, J. J., and Kirkland, J. J. (1973a). *Anal. Chem.* **47** (12), 1193A.
DeStefano, J. J., and Kirkland, J. J. (1973b). *Anal. Chem.* **47** (13), 1103A.
Fallick, G. (1977). "Chromatography," Ser. II, Vol. 1, p. 255. Int. Sci. Commun., Fairfield,
    Connecticut.
Fong, G. W., and Gushka, E. (1977). *J. Chromatogr.* **142**, 299.
Fransson, B., Wahlund, K. G., Johnsson, I. M., and Schill, G. (1976). *J. Chromatogr.* **125**, 327.

Giddings, J. C. (1965). "Dynamics of Chromatography." Dekker, New York.
Gloor, R., and Johnson, E. L. (1977). *J. Chromatogr. Sci.* **15**, 413.
Glueckauf, E. (1955). *Trans. Faraday Soc.* **51**, 34.
Grant, R. A. (1959). *J. Appl. Chem.* **8**, 136.
Grieser, M. D., and Pietrzyk, D. J. (1973). *Anal. Chem.* **45**, 1348.
Grushka, E. (1972). *J. Phys. Chem.* **76**, 2586.
Grushka, E., Yepes-Baraya, M., and Cooke, W. D. (1971). *J. Chromatogr. Sci.* **9**, 253.
Grushka, E., Snyder, L. R., and Knox, J. H. (1975). *J. Chromatogr. Sci.* **13**, 25.
Grushka, R., ed. (1974). "Bonded Stationary Phases in Chromatography." Ann Arbor Sci. Publ., Ann Arbor, Michigan.
Hawkes, S. J. (1972). *J. Chromatogr.* **68**, 1.
Hemetsberger, H., Maasfeld, W., and Ricken, H. (1976). *Chromatographia* **9**, 303.
Horning, E. C., Carroll, D. I., Dzidic, I., Haegele, K. D., Horning, M. G., and Stillwell, R. N. (1973). *J. Chromatogr. Sci.* **11**, 83.
Horvath, C., and Melander, W. (1977). *J. Chromatogr. Sci.* **15**, 393.
Horvath, C., Melander, W., and Molnár, I. (1976). *J. Chromatogr.* **125**, 129.
Horvath, C., Melander, W., and Molnár, I. (1977). *Anal. Chem.* **49**, 142.
Huber, J. F. K. (1969). *J. Chromatogr. Sci.* **7**, 86.
James, A. T., and Martin, A. J. P. (1950). *Biochem. J.* **50**, 679.
Johansson, G., and Karram, K. J. (1958). *Anal. Chem.* **8**, 1397.
Johnson, E., Abu-Shumays, A., and Abbott, S. R. (1977). *J. Chromatogr.* **134**, 107.
Karch, K., Sebastian, I., Halasz, I., and Engelhardt, H. (1976). *J. Chromatogr.* **122**, 171.
Karger, B. L., Martin, M., and Guiochon, G. (1974). *Anal. Chem.* **46**, 1640.
Keller, R. A., and Snyder, L. R., (1971). *J. Chromatogr. Sci.* **9**, 346.
Kikta, E. J., and Grushka, E. (1976). *Anal. Chem.* **48**, 1098.
Kikta, E. J., and Grushka, E. (1977). *J. Chromatogr.* **135**, 367.
Kirkland, J. J. (1971). *Anal. Chem.* **43**, 36A.
Kirkland, J. J. (1972). *J. Chromatogr. Sci.* **10**, 593.
Kirkland, J. J. (1974). *Analyst (London)* **99**, 859.
Kissinger, P. T. (1977a). *Anal. Chem.* **49**, 447A.
Kissinger, P. T. (1977b). *Anal. Chem.* **49**, 142.
Kissinger, P. T., Refshauge, C. F., Dreiling, R., Blank, L., Freeman, R., and Adams, R. N. (1973). *Anal. Lett.* **6**, 465.
Knox, J. H. (1966). *Anal. Chem.* **38**, 253.
Knox, J. H. (1976). *In* "Practical High Performance Liquid Chromatography," (C. F. Simpson, ed.), Chap. 2. Heyden, London.
Knox, J. H. (1977). *J. Chromatogr. Sci.* **15**, 352.
Knox, J. H., and Parcher, J. F. (1969). *Anal. Chem.* **41**, 1599.
Knox, J. H., and Saleem, M. (1972). *J. Chromatogr. Sci.* **10**, 80.
Kroeff, E. P., and Pietrzyk, D. J. (1978a). *Anal. Chem.* **50**, 502.
Kroeff, E. P., and Pietrzyk, D. J. (1978b). *Anal. Chem.* **50**, 1353.
Laird, G. R., Jurand, J., and Knox, J. H. (1974). *Proc. Soc. Anal. Chem.* 310.
Lawrence, J. F., and Frei, R. W. (1976). "Chemical Derivatization in Liquid Chromatography," (Journal of Chromatography Library—Vol. 7). Elsevier, Amsterdam.
Leitch, R. E., and DeStefano, J. J. (1973). *J. Chromatogr. Sci.* **11**, 105.
McFadden, W. M., Schwartz, H. L., and Evans, S. (1976). *J. Chromatogr.* **122**, 389.
McNair, H. M. (1978). *J. Chromatogr. Sci.* **16**, 588.
McNair, H. M., and Chandler, C. D. (1974). *J. Chromatogr. Sci.* **12**, 425.
McNair, H. M., and Chandler, C. D. (1976). *J. Chromatogr. Sci.* **14**, 477.
Magee, R. J., ed. (1970). "Selected Readings in Chromatography." Pergamon, New York.

Majors, R. E. (1974a). "Chromatography," Ser. I, Vol. 1, p. 190. Int. Sci. Commun., Fairfield, Connecticut.
Majors, R. E. (1974b). "Chromatography," Ser. II, Vol. 1, p. 212. Int. Sci. Commun., Fairfield, Connecticut.
Majors, R. E. (1975). *Analusis* **3**, 549.
Majors, R. E. (1976). *In* "Practical High Performance Liquid Chromatography" C. F. Simpson, ed.), Chap. 5. Heyden, London.
Majors, R. E. (1977). *J. Chromatogr. Sci.* **15**, 334.
Majors, R. E., and Hopper, M. J. (1974). *J. Chromatogr. Sci.* **12**, 767.
Majors, R. E., and MacDonald, F. R. (1973). *J. Chromatogr.* **83**, 169.
Marcek, K., Janak, J., and Deyl, Z. (1977). *J. Chromatogr.* **133**, B15.
Martin, A. J. P., and Synge, R. L. M. (1941a). *Biochem. J.* **35**, 1358.
Martin, A. J. P., and Synge, R. L. M. (1941b). *Biochem. J.* **35**, 294.
Martire, D. E., and Locke, D. C. (1971). *Anal. Chem.* **43**, 68.
Maruyama, M., Kakemoto, M., Murakami, K., and Ishii, T. (1977). *Nippon Kagaku Zasshi* 48.
Mohammed, H. Y., and Cantwell, F. F. (1978). *Anal. Chem.* **50**, 491.
Mowery, R. A., and Juvet, R. S., Jr. (1974). *J. Chromatogr. Sci.* **12**, 687.
Parris, N. A. (1976). "Instrumental Liquid Chromatography" (Journal of Chromatography Library—Vol. 5), Elsevier, Amsterdam.
Pecsok, R. L., and Saunders, D. L. (1968). *Anal. Chem.* **40**, 1756.
Pei, P., Ramachandran, S., and Henly, R. S. (1977). "Chromatography," Ser. II, Vol. 1, p. 281. Int. Sci. Commun., Fairfield, Connecticut.
Pietrzyk, D. J., and Chu, C. H. (1977). *Anal. Chem.* **49**, 757, 860.
Pietrzyk, D. J., Kroeff, E. P., and Rotsch, T. D. (1978). *Anal. Chem.* **50**, 497.
Poppe, M., and Kunysten, J. (1972). *J. Chromatogr. Sci.* **10**, 16A.
Rendell, C. M. (1975). *Process Eng.* April, 66.
Rieman, W., III (1961). *J. Chem. Educ.* **38**, 338.
Rotsch, T. D., Sydor, R. J., and Pietrzyk, D. J. (1979). *J. Chromatogr. Sci.* **17**, 339.
Saunders, D. L. (1977). *J. Chromatogr. Sci.* **15**, 372.
Schill, G. (1976). *In* "Assay of Drugs and Other Trace Compounds in Biological Fluids" (E. Reid, ed.), North-Holland Publ., Amsterdam.
Scott, R. P. W. (1976). "Contemporary Liquid Chromatography," (Techniques of Chemistry, A. Weissberger, ed., Vol. XI), pp. 116, 124. Wiley (Interscience), New York.
Scott, R. P. W. (1977a). "Liquid Chromatography Detectors" (Journal of Chromatography Library—Vol. 11). Elsevier, Amsterdam.
Scott, R. P. W. (1977b). *J. Chromatogr.* **142**, 213.
Scott, R. P. W., and Kucera, P. (1977). *J. Chromatogr.* **142**, 213.
Scott, R. P. W., Scott, C. G., Munroe W., and Hess, J., Jr. (1974). *J. Chromatogr.* **99**, 395.
Sherma, J., and Rieman, W., III (1958). *Anal. Chim. Acta* **18**, 214.
Snyder, L. R. (1962). *In* "Advances in Analytical Chemistry and Instrumentation" (C. N. Reilly, ed.), Vol. 3, p. 251. Wiley (Interscience), New York.
Snyder, L. R. (1965). *Chromatogr. Rev.* **7**, 1.
Snyder, L. R. (1968). "Principles of Adsorption Chromatography." Dekker, New York.
Snyder, L. R. (1969). *J. Chromatogr. Sci.* **7**, 352.
Snyder, L. R. (1971). *In* "Modern Practice of Liquid Chromatography" (J. J. Kirkland, ed.), p. 125. Wiley (Interscience), New York.
Snyder, L. R. (1974). *Anal. Chem.* **46**, 1384.
Snyder, L. R., and Kirkland, J. J. (1974). "Introduction to Modern Liquid Chromatography." Wiley (Interscience), New York.
Sparacino, C. M., and Hines, J. W. (1976). *J. Chromatogr. Sci.* **14**, 549.

Sternberg, J. C. (1966). *In* "Advances in Chromatography" (J. C. Giddings and R. Keller, eds.), p. 206. Dekker, New York.

Tjaden, U. R., Lankelma, J., Poppe, H., and Muusze, G. (1976). *J. Chromatogr.* **125**, 275.

Tjssen, R., Billet, H. A. H., and Schoenmakers, P. J. (1976). *J. Chromatogr.* **122**, 185.

van Deemter, J. J., Zuiderweg, F. J., and Klinkenberg, A. (1958). *Chem. Eng. Sci.* **5**, 271.

Vespalec, R. (1975). *J. Chromatogr.* **108**, 243.

Wahlund, K. G., and Lund, U. (1976). *J. Chromatogr.* **122**, 269.

Walton, H. F. (1978). *Anal. Chem.* **50**, 36R.

Waters, J. L., Little, J. N., and Horgan, D. F. (1969). *J. Chromatogr. Sci.* **7**, 293.

Wehrli, A. (1974). *Chimia* **28**, 690.

Wehrli, A. (1975). *Fresenius' Z. Anal. Chem.* **277**, 289.

Wheals, B. B. (1976). *J. Chromatogr.* **122**, 85.

Wheals, B. B., and Jane, I. (1977). *Analyst (London)* **102**, 625.

Wolf, J. P. (1973). *Anal. Chem.* **45**, 1248.

X-Ray Photoelectron Spectroscopy

N. Winograd

Department of Chemistry
The Pennsylvania State University
University Park, Pennsylvania

and

S. W. Gaarenstroom

Analytical Chemistry Department
General Motors Research Laboratories
Warren, Michigan

I. Introduction

A. Historical Perspective

The development of electron spectroscopy has recently progressed to the point where it must be considered a necessary tool in the characterization of a wide variety of materials. The technique involves the irradiation of a sample with a monochromatic beam of photons with energy $h\nu$. Electrons with a binding energy less than $h\nu$ are then ejected into the vacuum, where their energy is carefully measured. The phenomenon of photon induced electron emission, i.e., "photoemission," has a history dating back to Hertz's observation in 1887 of the initation of sparks across a gap by exposure to ultraviolet light (D'Ageostino, 1975). The explanation of the photoelectric effect by Einstein (1905) and the early experimental studies by Robinson and Rutherford (1913) on measuring β-ray spectra set the stage for major developments after the second World War. Although a lot of activity took place up to this period, as noted in a recent historical review of the subject (Jenkin et al., 1977), major progress in studying the electronic structure of materials did not occur until the late sixties, primarily due to experimental difficulties.

The first application of photoemission to chemical analysis probably dates to Steinhardt (1950; Steinhardt and Surfass, 1953), though his group never had the resources to bring the initial ideas to fruition. During the 1950s and 1960s, however, a major effort was undertaken by Professor Kai Siegbahn and his associates at Uppsala University in Sweden to develop electron spectroscopy as a viable tool in electronic structure determination. With the development of strong x-ray sources to irradiate the sample, and the use of a high resolution β-ray spectrometer to measure the kinetic energy of the photoelectrons to 1 part in 10^4 accuracy, these workers were able to map out many of the important implications of the technique. They then published their accumulated results in two definitive works devoted to solid state electronic structure (Siegbahn et al., 1967) and to free molecule spectra (Siegbahn et al., 1969); these publications stimulated most of the recent ideas discussed in this chapter. Siegbahn also suggested in these studies that the acronym ESCA (electron spectroscopy for chemical analysis) refer to the specific application of the technique to problems of chemical interest.

The subsequent literature in ESCA or x-ray photoelectron spectroscopy (XPS) is now so large, with 500–1000 papers appearing annually, that it would be impossible to review it all here. Several key publications include the proceedings of the first two international conferences (Shirley, 1972; Caudano and Verbist, 1974), a textbook (Carlson, 1975), several edited books containing review articles (Brundle and Baker, 1977–1978; Briggs, 1977), as well as many review articles (Hercules, 1970; Shirley, 1973; Jolly, 1974; Jorgensen, 1975; Brundle, 1975; Kowalczyk, 1976; Fadley, 1976). The *Journal of Electron Spectroscopy*, started in 1972 and published monthly, contains important articles in the field. Fortunately, there is also an excellent literature review that appears biannually (Hercules, 1972, 1976; Hercules and Carver, 1974) that helps one keep up with the recent developments.

In this chapter we will be concerned primarily with outlining experimental methods and introducing a necessary theoretical base toward applying XPS or ESCA to the characterization of solids and solid surfaces. At this point we note that the parallel development of electron spectroscopy by Turner *et al.* (1970) and Baker and Betteridge (1972) on gases and Spicer (1972) on solids, using primarily 20 and 40 eV helium radiation to probe the nature of valence molecular orbitals, has opened an equally large scientific discipline. The distinction between XPS and uv photoelectron spectroscopy (UPS) has, in recent years, become muddled with the development of synchrotron radiation (Spicer *et al.*, 1976), but we shall arbitrarily keep that distinction by examining primarily the measurement of core-level photoelectrons. Some overlap with UPS work is, however, necessary and desirable. We have also chosen not to emphasize the XPS spectra of gases, except to illustrate certain aspects, even though these materials are actively being investigated elsewhere. Finally, we shall emphasize the relationship between XPS and Auger spectroscopy (AES) as complementary methods of analysis.

B. Basic Principles

1. Photoemission Process

The basic process of interest in x-ray photoelectron spectroscopy is the absorption of a quantum of energy hv such that

$$A + hv \rightarrow A^+ + e^-$$ (1)

where Einstein's relation predicts that

$$T^V = hv - E_B^V(i)$$ (2)

Here the measured kinetic energy of the ejected photoelectron, referenced to an appropriate zero of energy which we call the vacuum level T^V, is related to the binding energy of the electron in the ith orbital of atom A, $E_B^V(i)$, referenced to the same energy. The equation is rigorous except that the small ($<0.1\,\text{eV}$) recoil energy of the departing electron left with A^+ has been ignored. For most commercial ESCA instruments hv is either 1256.3 eV generated from a Mg K_α anode or 1486.7 eV from an Al K_α source, although other energies are becoming available.

2. XPS Spectrum of Nickel

To illustrate the important features of XPS, we show, in Fig. 1, the spectrum of a solid surface of a clean and oxidized Ni(100) crystal. The spectrum consists of a graph obtained by plotting the intensity of photoelectrons on the ordinate and their E_B on the abscissa. It is conventional to plot E_B increasing to the left (or T increasing to the right), but unfortunately the opposite convention is also used (sometimes both are found in the same paper). The sharp transitions associated with the excitation of the Ni 2p, 3s, 3p, and 3d electrons are clearly visible in the spectrum. The energies of these transitions are readily translated to an energy level diagram as shown in Fig. 2. The relative simplicity of the XPS spectra versus the Auger spectra results from the fact that there is only one photoelectron in XPS, rather than two as in Auger emission, reducing the number of allowed final states. Sometimes Auger transitions are difficult to distinguish from the primary photoemission event. Variation of photon energy will change T^V of the XPS electrons, as is evident from Eq. (2), but will not alter T^V of the Auger electrons.

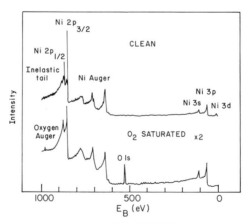

Fig. 1   XPS spectrum of a cleaned and oxidized Ni(100) crystal surface. The oxidized surface was obtained by exposure to 10^{-6} Torr of oxygen for 100 sec.

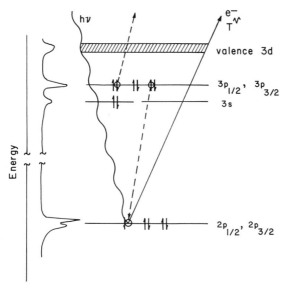

Fig. 2 Electronic structure of Ni obtained from the data given in Fig. 1. The wavy line represents the incident photon to energy $h\nu$. The solid line represents the creation of a 2p core hole by photoexcitation of a 2p electron. Its excess kinetic energy will be $T^V = h\nu - E_B(2p)$. The dotted lines represent a typical Auger process which fills the 2p hole, leaving two holes in the 3p level.

The sharp lines arise from electrons that escape the solid without suffering inelastic collisions with other atoms. When these collisions occur, the electron kinetic energy may be partitioned between many other electrons, giving rise to the background on the high binding energy side of these peaks. The "inelastic tails" are generally absent in gas phase spectra where, at least at low pressure, such inelastic losses are unlikely. Most of the intensity of these electrons can be found near zero kinetic energy, making it difficult to detect any main photoelectron lines which may arise in this region. The weaker valence electrons near zero binding energy are also observable with XPS. The peaks are generally broader than core-level peaks since in solids they originate from the broad valence band. The peak shape of this band is related to the density of electronic states, making this tool a valuable one in determining the electronic structure of metals and semiconductors. Other structures, including plasmon losses, multiplet splittings, and other satellites, can complicate the spectra and will be discussed in detail in subsequent sections.

These spectral bands are uniquely ascribable to Ni. Similar assignments can be made for all the other known elements. A tabulation of these transitions is given in Appendix I which can be used to assign unknown spectra.

The most intense photoemission lines for a specific element are boxed. The most intense Auger features are also indicated, assuming Al K_α excitation.

Upon oxidation, many of the main spectral features for Ni are drastically altered. The intensity of the Ni lines decreases and the peak widths increase due to the appearance of new chemical states. The Ni Auger lines are very different because of the alteration of allowable final states. The presence of oxygen is clearly seen by emission from the O 1s core level and by the appearance of oxygen Auger peaks. It is of interest in this system that the NiO grows to only two atomic layers with the (100) orientation. The fact that the Ni substrate and the NiO(100) overlayer are clearly visible demonstrates that the electron mean free path is only on the order of tens of angstroms at these kinetic energies.

In subsequent sections we will try to cover in more detail the various phenomena that give rise to the features shown in Fig. 1. In Section II we cover experimental aspects and procedures for obtaining reproducible spectra. In Section III the basic ideas behind the chemical shift will be considered, with an emphasis on the role of Auger transitions on the XPS spectra. In Section IV we consider the factors that influence the peak shape, while in Section V the essential elements in using the peak heights in quantitative analysis will be considered. Finally, in Section VI we present a selected series of applications which are admittedly those of special interest to the authors. In general, we will try to keep our discussion to the analysis of x-ray-excited core-level electrons primarily to narrow the scope of this very broad field.

II. Experimental Aspects and Procedures

To carry out the spectroscopic analysis as indicated in Eq. (1), there are several basic required components, as illustrated in Fig. 3. These include a source of monochromatic photons, the sample, placed in an appropriate evacuated housing, an electron energy analyzer, and a detector and data analysis system. The goal is to be able to record binding energy positions to the order of ± 0.1 eV and to resolve photoelectron energies to better than 1 eV. A number of commercial spectrometers have been available since the first Varian instrument was introduced in 1969. A list of the more common ones is given in Appendix II. Currently the price range starts at $80,000, depending upon the number of ancillary techniques required by the user. Most of these are pretty much "turnkey" devices with the conveniences having come a long way from the early days when graduate students had to work only at night to avoid interferences from the local subway (Hercules,

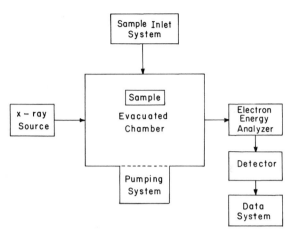

Fig. 3 Block diagram of a typical XPS spectrometer.

1970). In this section we review some of the more important aspects of this instrumentation, with a particular emphasis on how samples can be prepared and confidently analyzed with the XPS method.

A. Photon Sources

1. X-Ray Tubes

Virtually every photon source used to probe atomic core levels is the standard x-ray tube. In this device, a heated filament provides electrons which are accelerated to a potential of between 10 and 20 keV toward a water cooled solid anode. The electrons create core holes in the anode atoms, which are filled by relaxing electrons from higher levels. The relaxation process is then accompanied by x-ray fluorescence. The anode material can be constructed from a wide variety of solids so that the desired x-ray energy can be, in principle, selected by choosing the appropriate material. As we shall see, a number of factors operate to restrict the choice to only a few commonly used metals. The radiation is generally most intense from a single, sharp line providing a reasonably monochromatic x-ray source, although weaker lines and background continuous Bremsstrahlung radiation is also observed. A thin Al window usually serves to separate the x-ray tube from the rest of the spectrometer.

The K_α lines from the first row elements are probably the most suitable sources for XPS studies. As shown in Fig. 4, for the Al x-ray, a doublet arises from the $2p_{1/2}, 2p_{3/2} \rightarrow 1s$ electronic relaxation. Other bothersome spectral

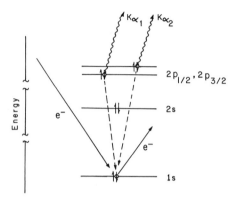

Fig. 4   Origin of the Al K_α x-ray lines.

lines observed on spectrometers that do not provide additional energy selection include the $K_{\alpha_{3,4}}$ lines at $\sim 6\%$ intensity, and a broad K_β line at 45 eV above $K_{\alpha_{1,2}}$ at $\sim 1\%$ intensity (Krause and Ferreira, 1975). The $K_{\alpha_{3,4}}$ lines arise from the 2p $\rightarrow$ 1s transition when the 1s level has been doubly ionized. Its presence can be particularly bothersome; there are still researchers who insist upon assigning its presence to some new "chemical state" of the sample. A few of the other common anode materials are given in Table I.

TABLE I

Some Common Anode Materials

Source	Transition	Energy (eV)	FWHM[a] (eV)
Y	M_ζ	132.3	0.47
Ne	$K\alpha_{1,2}$	849.0	0.30
Na	$K\alpha_{1,2}$	1041.0	0.42
Mg	$K\alpha_{1,2}$	1253.6	0.78
Al	$K\alpha_{1,2}$	1486.6	0.85
Cr	$K\alpha_{1}$	5415.72	2.1

[a] FWHM is the full width at half maximum of a Gaussian peak. The FWHM increases with increasing atomic number due to the 1s hole lifetime and to the increase in spin–orbit splitting of the $2p_{1/2} - 2p_{3/2}$ peaks.

Several important factors limit the choice of anode and explain why Mg and Al are the ones found on most commercial instruments. First, a narrow natural linewidth of the x-ray transition is important to obtain the desired energy resolution of 1 eV since any spread in $h\nu$ will result in an equivalent spread in the observed E_B. Anodes like Cr K_α with a natural width of 2.1 eV would not be effective for high resolution studies. Second, the energy of the characteristic x ray should be sufficient to excite photoelectrons from all the elements. The Y M_ξ anode, for example, has good resolution but does not have sufficient energy to excite the 1s electron from C, a very commonly used transition. The x-ray absorption cross section σ should also be optimized by matching as closely as possible the energy of the excitation to the energy of $h\nu$. This matching produces maximum overlap between the photon field and the electronic orbital involved. That is the reason why the σ_{C1s} for Mg K_α is nearly twice as large as that for Al K_α. A photon energy of 1000–1500 eV seems most suitable as a general probe, although higher energies may be necessary to reach some important core-level transitions like Pt 3d, and lower energies might be desirable to enhance surface sensitivity, as we shall see in a later section. And finally, there are material considerations in fabricating these anodes. It is possible to employ slightly higher electron fluxes for Al than for Mg due to its higher melting temperature, so the resulting x-ray intensity is larger, offsetting the cross-section effects.

2. Monochromators

There are some commercial designs that employ an additional mono-chromator device to reduce background from the Bremsstrahlung, eliminate extraneous spectral lines, and improve resolution (Siegbahn et al., 1972; Siegbahn, 1974). The Al K_α lines can be effectively monochromatized by reflection from a bent quartz crystal although the intensity is significantly reduced. The spectrometer efficiency can be subsequently increased using multichannel detection and a dispersion compensating lens system. Higher x-ray fluxes can also be obtained using a rotating anode (Siegbahn, 1974). Also, since the signal to background ratio is inherently much higher, the resulting spectral response is nearly as strong as without monochromatization. The linewidth improvement is remarkable; the $3d_{5/2}$ line of silver is found to narrow from ~ 1.0 to nearly 0.3 eV.

3. Synchrotron Sources

Complete flexibility in generating continuous soft-x-ray radiation to several kilovolts has recently become available through synchrotron radiation (Perlman et al., 1970). For this source, in principle, the flux is high enough to allow grating monochrometers, tunable over a wide energy range. Only a few studies have been reported for binding energies in the 200–2000-eV energy range, however, due to the low efficiency of these monochromators (Lindau et al., 1974). The facilities are also somewhat inconvenient to use, but the fact that three national centers are currently being constructed should mean easy access for most potential users.

B. Sample Preparation and Pretreatment

Once certain restrictions concerning the nature of the sample are understood, it is possible, in principle, to record the spectrum of virtually any material, be it a solid, a gas, or even a liquid. The characterization of solids is the most straightforward and will be considered here. Several excellent reviews are available for specific studies of gases (Siegbahn *et al.*, 1969) and liquids (Siegbahn and Siegbahn, 1974).

1. Vacuum Requirements

As we have noted, XPS is a surface sensitive method. Impurities can play a major role in the observed spectra and most of the effort that goes into sample handling directly involves this problem. The first criterion then, is that a good vacuum is needed to maintain the integrity of the surface. In In general, 10^{-5} Torr would be sufficient to allow the photoelectron to reach the detector without suffering collisions with other gas molecules. On the other hand, 10^{-9} Torr or lower [ultrahigh vacuum (UHV)] is required to keep an active surface clean for more than several minutes. For most work, 10^{-8}–10^{-9} Torr provides a reasonable pressure where contamination is not too serious, although UHV is essential for well-defined surface characterizations. The conclusion, then, is that the sample characteristics, rather than the requirements on the photoelectron mean free path through space, determine the required instrumental pressure. For the beginner, a review of the available types of pumping would be most helpful (Guthrie, 1963).

2. Sample Introduction Systems

Given these vacuum requirements, sample introduction into the spectrometer becomes an important consideration. Several commercial instruments require venting a rather large chamber each time a sample is changed resulting in a many hour delay. To minimize this inconvenience, most spectrometers are equipped with holders that may contain many samples at once. For example, the carousel holder shown in Fig. 5 can accommodate up to 12 samples. The analysis of a particular sample is then completed by rotating the carousel under vacuum to the appropriate analysis positions. Other systems provide a sample mounted on a rod moveable through polymer seals which are differentially pumped. For this design, samples can be changed in a matter of minutes. The seal system is compatible with UHV, although when the rod is inserted H_2O and N_2 slowly desorb from the rod, increasing the system pressure. Finally, an atmospherically controlled box can be employed about the sample probe for preparing reactive compounds (Kim and Winograd, 1974a). This approach is particularly attractive when catalyst surfaces

Fig. 5 Multisample holder which can accommodate up to 12 samples. The carousel can be rotated to position the appropriate specimen in the x-ray beam.

or other air sensitive samples need to be examined, or for cleaning single crystals outside the vacuum environment.

3. Sample Size Restrictions

The sample size and shape are not critical within limits. The x-ray beam is usually $\sim 1 \text{ mm}^2$ to 1 cm^2, and the sample should be larger than this to maximize sensitivity. Crystals, powders, foils, and odd size chunks have all

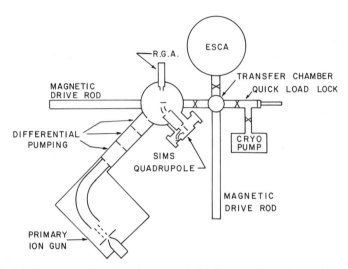

Fig. 6   Block diagram of a multitechnique XPS/SIMS system for surface analysis.

been reported to be examined by appropriate adaptation of the sample holder. Powders are often prepared by pressing them onto double sided adhesive tape.* The C 1s signal from the tape is then used as a reference signal. To avoid unnecessary contamination, however, powders can be pressed into pellets using conventional ir methods. If charging is a problem, the powder can also be pressed into a soft metal like indium (Theriault *et al.*, 1975). For any type of sample that is air sensitive, the preparation should be accomplished in an attached atmospherically controlled box to avoid surface reactions with air.

4. *Multitechnique Capability*

Once introduced into the vacuum environment of the spectrometer, the sample should ideally be characterized using another surface analysis technique. Recently, as indicated in Appendix II, many of the commercial XPS spectrometers also come equipped with Auger spectroscopy, low energy electron diffraction (LEED) for characterizing single crystal surfaces, secondary ion mass spectrometry (SIMS), and UPS. The sample can then be moved in UHV so that it is exposed to the appropriate radiation. An example of how this concept can be carried out is shown in Fig. 6 for a combined

_____

* Recommended only if all else fails!

XPS–SIMS system (Hewitt *et al.*, 1980). The present feeling among most researchers in the field is that information from several techniques is helpful in fully understanding the XPS results.

5. *Pretreatment Chambers*

A pretreatment reaction chamber is also helpful in many aspects of sample preparation and for reacting surfaces *in situ*. Since most samples are covered with an undefined contamination layer, Ar^+ ion sputtering is often employed to physically remove these outermost layers. With this method, a beam of 500–5000 eV Ar^+ ions with current of 1–10 $\mu A/cm^2$ is focused to the sample. Since these ions possess considerable momentum, a significant fraction of the sample atoms can attain enough kinetic energy to be physically ejected from the surface. Thus, the process can be employed to etch away a few of the top layers of impurities, exposing the clean sample surface underneath. Although "sputter cleaning" has been used extensively with both XPS and Auger spectroscopy for pure metals, we are extremely hesitant to recommend it for the preparation of molecular systems. The Ar^+ ion clearly causes a lot of damage to the sample itself (see, e.g., Carter and Colligan, 1968). In many multicomponent systems preferential sputtering can completely change the chemistry at the surface. Most metal oxides, for example, have been found to be reduced to the metal under the influence of 400 eV Ar^+ ion bombardment (Kim and Winograd, 1975a). Since the XPS method is often employed to determine the chemistry, the approach is seemingly self-defeating. We might add, however, that Ar^+ ion bombardment is an excellent way to clean single crystal surfaces if the resulting defects are annealed out by heating the crystal to temperatures below the melting point. The simultaneous use of sputtering with Ar^+ bombardment and an appropriate surface analysis tool can be utilized to obtain a depth profile of the elements in the near surface region. This analysis can be performed with XPS, but it is most conveniently carried out using Auger spectroscopy.

Other methods exist for cleaning surfaces, but the nature of the sample must be well known. Heating in vacuum for a short time often can remove adsorbed water or carbonate impurities (Hammond *et al.*, 1975). An excellent method that works on some foils and crystals simply involves scraping the surface with a tungsten carbide blade (Colton and Rabalais, 1975). It is also possible to cleave crystals in a vacuum, which exposes a clean surface (Chaban and Rowe, 1976).

Other sample handling features that are a desirable part of a reaction chamber include the ability to evaporate metals from a filament onto the sample holder. Provision for the admittance of gases through a precision leak valve is also desirable in many surface chemistry studies.

C.   Reference Level Problem and Sample Charging

1.   Fermi Levels

Measurement of core-level binding energies for gas phase molecules is generally accomplished by measuring the kinetic energy of the photoelectron relative to a standard species with a precisely known ionization potential (Siegbahn *et al.*, 1969). Unfortunately, no such standards are available for solids and the problem becomes immediately messy. A convenient approach for a metal is to reference E_B to its Fermi level. The appropriate energy level diagram is shown in Fig. 7. Note that as in Eq. (1),

$$hv = E_B^V(i) + T_s = E_B^F(i) + \phi_s + T_s \tag{3}$$

where ϕ_s is the sample work function. If the metal sample and the spectrometer are in electrical contact, the Fermi levels must be at equal energies. The potential that the photoelectron experiences, however, will depend on the difference between ϕ_s and the work function of the spectrometer energy analyzer ϕ_{sp}. If these quantities are not equal, the electron will be accelerated or decelerated, depending upon the sign of $\phi_{sp} - \phi_s$, as it enters the field of the energy analyzer. The $\phi_{sp} - \phi_s$ term is often referred to as the "contact potential" between the sample and the analyzer.

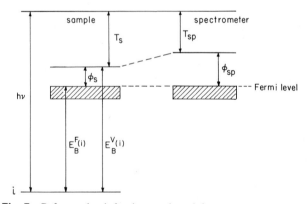

Fig. 7   Reference levels for the sample and the spectrometer system.

Thus from the energy level diagram shown in Fig. 7, it is clear that

$$T_s + \phi_s = T_{sp} + \phi_{sp} = hv - E_B^F \tag{4}$$

The measured E_B^F depends only on ϕ_{sp}, presumably a constant, even when different samples are used. If the E_B were reported with respect to E_B^V, then

$$E_B^V = hv - T_{sp} + (\phi_s - \phi_{sp}) \tag{5}$$

and both the ϕ_s and ϕ_{sp} must be known. Since ϕ_s can vary by several eV, depending on the presence of impurities, the E_B^F reference is clearly the reference of choice.

2. Charging

The situation is only well defined for a metal. For semiconductors and insulators, the Fermi level is not well defined and the sample can exhibit "charging." This ubiquitous problem arises since photoelectrons emitted from the sample are not replaced by electrons from the ground due to poor conductivity. Under these conditions, the Fermi level of the spectrometer and the Fermi level of the sample are no longer held together. The energy level diagram for the sample shown in Fig. 7 may float up or down by some unspecified amount such that

$$E_B^F = h\nu - T_s - C \qquad (6)$$

where C is a correction due to the charging. The kinetic energy of the departing electron is usually altered by the surface charge.

Measurement of E_B^F for these systems obviously presents a very tricky experiment. A long list of ways around the problem have been presented in the literature. The most common one is to try to prepare the insulating sample in as thin a film as possible so the surface charge can be bled off to the spectrometer. To further insure charge equilibrium, the sample can be flooded with very low energy electrons (< 3 eV) from above to externally neutralize the charge. This can be accomplished in some spectrometers more or less automatically due to the presence of secondary electrons created by the Bremsstrahlung radiation of the x-ray source. Where monochromators are used, which eliminate this source of electrons, an electron flood gun can be externally added to the system. We prefer this approach to charge compensation since it provides a controllable means of supplying electrons. In addition, the secondary electrons can be the cause of "x-ray damaged" samples, e.g., those which appear to change their composition under the influence of the x-ray beam.

Other methods developed to measure E_B^F for semiconductors and insulators include the evaporation of a submonolayer film of gold onto the sample surface. If charging does occur, the measured E_B^F for the gold $4f_{7/2}$ level of 83.8 ± 0.2 eV will increase by an equivalent amount since the gold is presumably in equilibrium with the sample and not with the spectrometer. Difficulties include "differential charging" (varying surface potential across the sample), chemical reaction of the gold with the sample (Betteridge et al., 1973), as well as other matrix effects (Kim and Winograd, 1975b). Similar methods include burnishing the sample into a foil of gold or indium to maximize electrical contact or diluting the sample into a graphite powder followed by pelleting. Other approaches which utilize the secondary electron edge (Evans, 1973) or dual photon energies (Johansson et al., 1973) have not been generally applicable.

To sum up this section then, E_B^V provides a convenient reference for gases: E_B^F is most commonly used for metals. Semiconductors and insulators present a difficult problem since a well-defined E^F does not exist and since charging is often a problem. No single satisfactory method currently exists to study these types of compounds and great care must be taken in evaluating spectra.

D. Electron Energy Analysis and Detection

The development of high precision electron energy analyzers has really provided the impetus for the current interest in XPS. Many types of magnetic and electrostatic analyzers are currently in use, although the electrostatic analyzers seem to have gained the most popularity. In general, these devices should have a resolution of $\Delta T/T < 10^{-3}$ (1.0 eV for a 1000-eV photoelectron), high transmission, stability to <1 part in 10^4, and a well-defined electron take off angle from a flat sample. A number of detailed articles have appeared recently which describe the design of most analyzers, so we will try to point out only a few general principles here.

1. Electrostatic Analyzers

The most straightforward method for determining the electron kinetic energy is the retarding grid electron energy analyzer. With this system, as shown in Fig. 8a, the electrons to be analyzed are retarded by a voltage between grids G_1 and G_2. Only those electrons whose kinetic energy is sufficient to pass through G_2 will reach the collector plate C. Note that this configuration produces an integrated spectral response since it responds to all electrons with energy greater than the retarding potential. This aspect of its design generally results in poor signal-to-noise ratio and limited resolution.

The two most commonly used electrostatic analyzers are the hemispherical and cylindrical mirror (CMA) (Palmberg et al., 1969) types, as shown schematically in Figs. 8b and 8c. For the hemispherical design, two-dimensional point-to-point focusing is accomplished by applying a voltage V to the inner and outer spheres such that

$$V_{\text{out}} = T_0(3 - 2R_o/R_{\text{out}}) \quad \text{and} \quad V_{\text{in}} = T_0(3 - 2R_o/R_{\text{in}}) \quad (7)$$

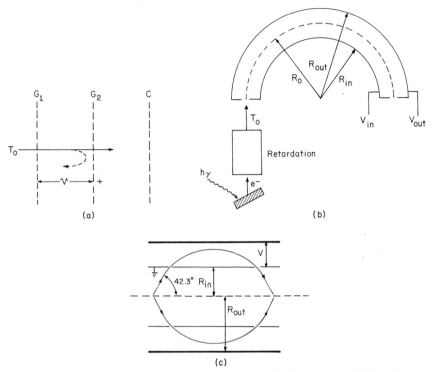

Fig. 8 (a) Retarding grid electron energy analyzer; (b) hemispherical (180°) deflector; (c) cylindrical mirror analyzer.

where T_0 is the pass energy of the analyzer. For the cylindrical design, electrons leave the sample at nearly 45° and are refocused by applying a potential

$$V = 1.3\ T_0 \ln R_{\text{out}}/R_{\text{in}} \tag{8}$$

on the outer cylinder. The CMA is currently very popular due to its high transmission, small size, and excellent resolution. The main disadvantage of the CMA is that it usually has a very short focal length, and electrons of only a specific radial angle are admitted to the analyzer. If a magnetic field is used to deflect the electrons, the shielding problems become considerable. It should be noted, however, that most early XPS measurements were recorded using these magnetic analyzers (Siegbahn *et al.*, 1967, 1969). Currently, one hybrid electrostatic–magnetic system has been marketed commercially, as described in Appendix II. The ultimate choice of analyzers rests on a complicated mix of desired criteria such as resolution, transmission, brightness, angle resolution, size and cost. The factors have been compared in detail elsewhere (Gellender and Baker, 1974).

2. Retardation

The desired resolution can be attained using the hemispherical analyzer although machine tolerances of the sphere radii are critical. A useful approach is to retard the kinetic energy of the photoelectron to a constant value of ~ 100 eV so that the analyzer need only have a resolution of 10^{-2} to obtain the desired spectral response. At lower kinetic energies, the slower moving electrons tend to diverge more readily resulting in a loss of sensitivity, although a focusing lens often compensates for this problem. The electron orbits are susceptible to perturbation by stray magnetic fields, and it is critical to house the entire apparatus in a μ-metal shield. The incorporation of a retardation lens system into the design is illustrated in Fig. 8b.

3. Electron Detection

Once the electron is focused to the back of the analyzer, it is usually counted by a staged electron multiplier or, more commonly, a channeltron detector. These devices are capable of detecting as few as one electron/sec (10^{-19} A) with a background as low as 0.1 electron/sec. Generally the channeltron runs into saturation at a value of about 100 000 Hz, giving it a quite acceptable dynamic range. For spectrometers which employ x-ray monochromators with the concomitant loss of senstivity, it is necessary to employ multichannel counting to gain back sufficient signal. This can be accomplished by coupling a channel-plate detector with the two-dimensional focusing properties of the hemispherical analyzer to measure a band of photoelectrons simultaneously. With commercial channel plates with 1000 or more channels, large improvements in counting rates are clearly possible.

III. Principles of ESCA Chemical Shifts

A. Initial State Effects and the Electrostatic Potential Model

The most important feature of electron spectroscopy to chemists is how $E_B^F(i)$ reflects the chemical environment of a given atom in a molecular environment. The initial experiments indicated that the "chemical shift" of $E_B^F(i)$ was most strongly influenced by the oxidation state of the probed atom. Later, as we shall see, other factors involving mostly final-state effects have muddled this pictured considerably. In this section, then, we will review some of the simpler approaches to understanding the factors that affect the chemical shift, and we will try to provide some insight into what types of molecules are best studied by the ESCA technique.

The quantity measured by an electron spectrometer is the kinetic energy of an electron ejected from a sample. As we have seen, a Fermi level referenced binding energy can be calculated when the spectrometer work function ϕ_{sp} and the exciting radiation energy $h\nu$ are known.

$$E_B{}^F = h\nu - T^V - \phi_{sp} \tag{9}$$

Binding energies and their chemical shifts can be understood at a variety of levels of sophistication. One very simple model, which was first used by Siegbahn to explain chemical shifts, is based purely on classical electrostatics (Siegbahn *et al.*, 1967). An atom is imagined to be a hollow sphere with a surface which is uniformly charged by the electrons of the valence shell. For a charge Q and radius r, the potential energy of a core electron inside the sphere is constant and

$$V = -eQ/r \tag{10}$$

If Q electronic charges are removed from the valence shell, the binding energy shift is

$$\Delta E_B = -\Delta V = -e\,\Delta Q/r = qe^2/r \tag{11}$$

For an atom with a valence shell radius of 1 Å, this is a binding energy shift of about 14.4 eV per electron removed.

In a case where an ion pair AC is formed, the valence electron has not been removed to an infinite distance, as it was in a free ion, but now resides in the valence shell of ion C^-, a distance R away. The binding energy shift compared to the free atom A is now

$$E_B = qe^2/r - qe^2/R \tag{12}$$

For a molecule or lattice, the charge on all the surrounding ions must be considered, and the binding energy shift becomes

$$E_B = \frac{q_A e^2}{r_A} - \sum_{C \neq A} \frac{q_C e^2}{R_{AC}} \tag{13}$$

The chemical shifts for molecules or solids are much less than the free ion shift. Assuming complete charge transfer and all the sphere radii are 1 Å, the chemical shift of ion A^+ will be 7.2 eV for the dimer AC and only 1.8 eV for a rock salt lattice of AC.

There are a couple of other observations that can be noted using this model. First, since the potential is constant inside the sphere, all core electrons, regardless of orbital, should have identical chemical shifts. Experimentally, this has always been found to be true, within a few tenths of an electron-volt. Second, since the chemical shift varies inversely with the valence shell radius, ESCA is more sensitive to small changes in charge for small radii atoms like C, N, O, and F than for large radii atoms in the lower part of the periodic table. Also an atom's radius is not constant with oxidation state, but shrinks as valence electrons are removed. Thus the chemical shift between A^{+1} and A^{+2} should be greater than between A^0 and A^{+1}. In practice, the atomic radius is treated as an adjustable parameter and the chemical shift relative to the binding energy of a reference compound can be predicted from a least squares fit of the chemical shifts from a large training set of compounds using Eq. (14).

$$\Delta E_B = kq_A + \sum_{A \neq C} \frac{q_C}{R_{AC}} + l \qquad (14)$$

where l is an adjustable constant for a series of similar compounds. Equation (14), which is often called the ground state potential model (GPM), has been used successfully to correlate CNDO charges with binding energy shifts for extensive series of carbon, nitrogen, oxygen, sulfur, and other compounds (Siegbahn *et al.*, 1967). An example is shown in Fig. 9 for a series of small carbon containing molecules. The chemical shift parameter k for this figure was 21.9 eV/unit charge (Siegbahn *et al.*, 1969). Although the derivation of

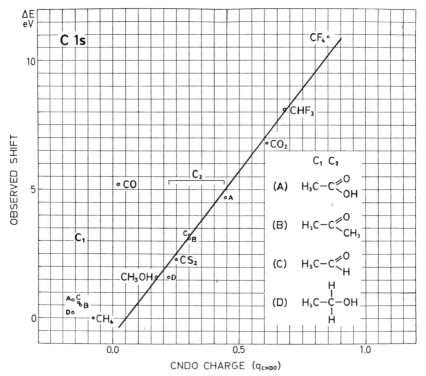

Fig. 9 Measured binding energy shift of carbon 1s electrons versus shift calculated by the CNDO method. (From Siegbahn *et al.*, 1969.)

Eq. (14) has been sketched here from purely electrostatic considerations, Gelius (1974a) has shown that an identical expression is reached using Hartree–Fock orbital energies, with the constant k equal to the Coulomb repulsion integral between a core and a valence electron.

A bothersome aspect with the evaluation of binding energy shifts in terms of theoretical considerations is that seemingly nine-tenths of this work is with gaseous atoms and molecules, while it is the study of solids which makes up well over nine-tenths of all work performed with ESCA spectrometers. There are two problems with Eq. (14) when solids are studied. First, the potential due to the surrounding ions must be replaced with a more complicated expression. For an infinite lattice, Picken and Van Gool (1969) have developed an expression for a "Madelung-type" potential which has been widely used. But since ESCA is a near-surface technique, the use of a bulk lattice potential expression is not completely correct. Several studies have shown that a surface Madelung potential is not drastically different from the bulk value (Slater, 1970; Parry, 1975), but more work is still to be done. The second problem in the study of solids is that binding energies are referenced to the Fermi level and not to the vacuum level. Yet for any equation like Eq. (14) which refers to atomic charge, the ultimate reference must be the free neutral atom, hence a vacuum level measurement is required. The difference between the vacuum level and the Fermi level is the work function ϕ. Thus, Eq. (14) must be adjusted to Eq. (14a) for a solid.

$$\Delta E_B = kq_A + V + l - \Delta\phi \qquad (14a)$$

Unfortunately, except in the case of metals, the work function of a material is almost never known. These problems associated with comparing binding energies of solids to theoretical parameters are not overwhelming, however. There have been numerous studies where good correlations were found between experimental shifts in solids and computed atomic charge. Simply put, there is a little more uncertainty added to the normal assumptions about chemical shifts where solids are involved.

B. Final State Effects

Despite the surprisingly high degree of success of the ground state potential model, there is a flaw which may account for some of the scatter in charge versus binding energy correlations and occasional anomalous binding energy shifts for some molecules, such as CO in Fig. 9. The model presented thus far does not consider perturbations to the remaining electrons in the system caused by the removal of a core electron by photoionization. It is more realistic to consider ESCA just like any other spectroscopy, a measurement of an energy difference between an initial and final state, as

$$E_B = E_f^{tot} - E_i^{tot} \tag{15}$$

Thus, a binding energy can be calculated on an *ab initio* basis with separate variational calculations which minimize the total energies of the ground state and core hole final state of the system. Of course, this has only been done for a few free atoms and simple molecules. The accuracy of hole state calculations varies greatly with the degree of rigor used. The very best calculations agree with experiment to within a few tenths of an electron-volt for light atoms and within a few electron-volts for molecules. Developments in this area are quite rapid, so the accuracy of predictions should continue to improve and calculations will be performed on larger molecules.

1. Relaxation Potential Model

Another way of dealing with final state effects is to consider photoionization as a two step process. In the first step, a core electron is photoionized, but the remaining electrons are "frozen" in their orbit and not allowed to adjust. The binding energy is now simply the one electron orbital energy

$$E_B(i) = -\varepsilon(i) \tag{16}$$

In the second step, the electrons are allowed to relax toward the core hole. This relaxation "screens" the photoelectron from the core hole, thus it will have a slightly higher kinetic energy, equal to the relaxation energy. The binding energy is now the one electron orbital energy minus the relaxation energy

$$E_B(i) = -\varepsilon(i) - E_R \tag{17}$$

2. Equivalent-Core Approximation

To determine the relaxation energy in molecules, it is generally necessary to resort to calculational approximations involving "equivalent cores" (Davis and Shirley, 1974). The relaxation energy is assumed to be half the difference in the potential at the nucleus, V_n, in the ground state and when a nuclear charge of $Z + 1$ is substituted.

$$E_R = (1/2)[V_n(Z + 1) - V_n(Z)] \tag{18}$$

Significant improvement between theory and experiment is achieved by including relaxation effects. One notable study is by Shirley and co-workers on N 1s binding energies of the series NH_3, NH_2CH_3, $NH(CH_3)_2$, and $N(CH_3)_3$ (Davis *et al.*, 1974). The results are summarized in Fig. 10. Good agreement was found between the relaxation potential model (RPM) predictions and experiment, while the ground state potential model actually

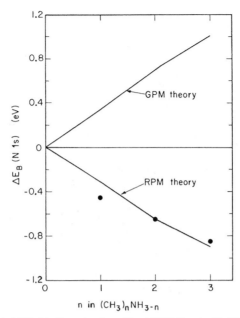

Fig. 10 Calculated N 1s binding energies without (GPM) and with (RPM) relocation terms for NH_3, NH_2CH_3, $NH(CH_3)_2$, and $N(CH_3)_3$. The experimental data are shown as points. (From Davis *et al.*, 1974.)

predicted the shifts to be in the wrong direction. While the RPM method has been employed mainly in conjunction with CNDO charge calculations on organic molecules, a little work has also been done on inorganic solids. Sherwood (1976) has done calculations using both GPM and RPM methods on compounds of chromium and a few other elements. The RPM method using CNDO charges was found to give the best agreement with experimental shifts.

3. Transition Potential Model

Another description of binding energy shifts which includes final state effects has been used extensively by Siegbahn's group (Siegbahn *et al.*, 1976; Howat and Goscinski, 1975; Siegbahn, 1976). It is the transition potential model (TPM) and has the advantage that it resembles the GPM model. Within the framework of the GPM, a binding energy can be given by

$$E_B = kq + V + l \tag{19}$$

In the TPM,

$$E_B = k^T q^T + V^T + l^T \tag{20}$$

The charges and potentials are no longer those of the molecule's ground state, but rather they are calculated by substituting a pseudo-atom with nuclear charge of about $Z + \frac{1}{2}$ in place of the atom of interest. The TPM binding energy can also be written in the form

$$E_B = k^T(q + \Delta q) + (V + \Delta V) + l^T \tag{21}$$

where Δq and ΔV are the differences between the ground state and transition state charges and potentials. Since the GPM usually employs fitted k and l values, there is some automatic consideration of relaxation effects, but only to the extent that Δq and ΔV are linearly dependent on ground state q. Frequently these two dynamic parameters only correlate with ground state charges when compounds which are structurally similar are compared, while dissimilar compounds give much poorer fits of data to theory. Siegbahn et al. (1976) have found much improved agreement between experimental binding energies and CNDO/2 charge densities using TPM compared to GPM for series of boron and carbon compounds.

C. Use of Binding Energy Shifts for Chemical Characterization

In this discussion, it has been shown how binding energy shifts are related to changes in ionic charge, chemical structure, and final state effects. Naturally, for samples with unknown compositions, it cannot be determined absolutely whether the binding energy shift is due to a change in charge, relaxation, or lattice potential. The common practice for determining the chemical state of unknown samples is to compare binding energy shifts, as well as all the other features of the spectra (i.e., satellites, multiplet splitting, peak widths, Auger energies, etc.), to a series of reference compounds. Despite the fact that this procedure does not explicitly employ any of the models mentioned earlier, knowledge of the successes and limitations of these models is still invaluable. For instance, calculation of relaxation energy by the RPM has shown that relaxation energy will increase with the oxidation state of the central atom, the number of nearest neighbors, and the polarizability of the nearest neighbors (Sherwood, 1976). Correlations of E_B versus q and $E_B - V$ versus q have been done for many groups of similar compounds (Siegbahn et al., 1967), and the results give valuable insight into how successfully an unknown sample can be characterized or identified from the ESCA spectrum.

1. Chemical Shifts in Auger Electrons

Since Auger electrons also appear in ESCA spectra and exhibit chemical shifts, it is important to discuss their relation to binding energy shifts. Conventional Auger electron spectroscopy using electron beam excitation usually sacrifices resolution in order to provide high sensitivity and rapid data acquisition. Consequently chemical shift information, for instance to distinguish a metal and its oxide, is not utilized. Using x-ray excitation, however, a number of researchers have systematically compared Auger and core-level shifts for a large number of chemical systems and have discovered that Auger shifts add complementary information to core-level shifts (Wagner and Bileon, 1973; Kowalczyk et al., 1974; Schön, 1973; Wagner, 1975). Table II compares binding energy shifts and Auger shifts between a number of metals and metal oxides. Invariably the Auger shift is larger than the core shift. This is principally because the Auger process is more sensitive to final state effects than the photoelectron process. This phenomenon can be understood by examining the energy relationships of the Auger process. As can be seen from Fig. 1, the Auger energy involves the energy of three different core levels. For the levels j, k, and l the Auger energy (jkl) can be calculated from binding energy data,

$$E(jkl;X) = E_B(j) - E_B(k) - E_B(l^*;X) \tag{22}$$

Here, $E_B(j)$ and $E_B(k)$ are binding energies for the j and k subshells of the neutral atom, and $E_B(l^*;X)$ is the binding energy of an l orbital electron for an ion having a hole in the k subshell. Since the final state of this latter process has two holes, X is the final state term denoting the energy level where the ion finally resides. Kowalczyk et al. (1973) expressed $E_B(l^*;X)$ in terms of $E_B(l)$, together with the interaction energy between the k and l holes $[F(kl;X)$ and the static relaxation $R(kl)]$.

$$E_B(l^*;X) = E_B(l) + F(kl;X) - R(kl) \tag{23}$$

The static relaxation energy is the amount of energy the l orbital has lowered when there is a hole in the k subshell. There are both atomic and extraatomic

TABLE II

Binding Energy and Auger Energy Chemical
Shifts between Metal and Metal Oxide

	Core shift (eV)	Auger shift (eV)
Mg/MgO	1.4	6.3
Al/Al$_2$O$_3$	2.6	6.7
Ti/TiO$_2$	4.6	5.4
Ga/Ga$_2$O$_3$	0.5	5.5
Cu/Cu$_2$O	0.1	1.3
Cu/CuO	0.8	0.7
Zn/ZnO	0.5	4.3
Cd/CdO	−0.8	1.7
Ag/Ag$_2$O	−0.4	0.9

contributions to $R(kl)$. The Auger energy of Eq. (22) can then be given as

$$E(jkl;X) = E_B(j) - E_B(k) - E_B(l) - F(kl;X) + R(kl) \qquad (24)$$

The hole–hole interaction energy is essentially insensitive to chemical environment and the chemical shifts of different atomic levels have experimentally been shown to be closely the same. However, the static relaxation term does change significantly with chemical environment, and this causes the Auger shifts to differ from binding energy shifts. In Table III, the extraatomic static relaxation energy is calculated from experimental Auger and binding energies, together with estimates of interaction energy and atomic static relaxation from standard atomic multiplet coupling theory (Slater, 1960), Slater integrals (Mann, 1967), and optical data (Moore, 1949–1958). Just as was discussed previously for dynamic relaxation in binding energies, the static extra-atomic relaxation energy is large when the nearest neighbors are very polarizable, such as Se or Te (or as in Zn metal where there is significant free electron density) and static relaxation is small when the nearest neighbors are less polarizable, as in F and O.

TABLE III

Calculation of Static Extra-Atomic Relaxation Energy (eV)
for Selected Zn Compounds

	$E^F(L_3M_{45}M_{45}; {}^1G)^a$	$E_B(2p_{3/2})$	$E_B(3d)$	$F(3d3d; {}^1G)$	$R(3d3d)_a$	$R(3d3d)_e{}^b$
ZnF_2	985.6	1023.0	11.5	30.2	10.1	5.7
ZnO	987.6	1022.5	10.7	30.2	10.1	6.6
$ZnCl_2$	986.6	1023.0	11.4	30.2	10.1	6.5
$ZnBr_2$	987.6	1023.1	11.1	30.2	10.1	6.8
ZnI_2	988.7	1022.9	10.7	30.2	10.1	7.3
ZnS	988.2	1022.4	10.4	30.2	10.1	6.7
ZnSe	989.5	1022.4	10.5	30.2	10.1	8.2
ZnTe	991.3	1022.0	10.0	30.2	10.1	9.4
Zn metal	992.3	1021.9	9.6	30.2	10.1	9.7

[a] Experimental Auger and binding energies from Hoogewijs et al. (1976b).
[b] Calculated from Eq. (24).

Recently there have been refinements in the models for Auger shifts. Hoogewijs et al. (1976a) and Kim et al. (1976) have included a relaxation correction term $[-E_R(l) + E_R(l^*)]$, which is added to the static relation term in Eqs. (14) and (15) to account for differences in dynamic relaxation energy between a no-hole and a one-hole initial state. Siegbahn and Goscinski (1976) have analyzed Auger energy shifts using a scheme based on the transition potential model (TPM) that was described earlier.

2. Auger Parameter Shifts

Based on Eq. (24), the Auger chemical shifts can be related to binding energy shifts and changes in static extra-atomic relaxation by Eq. (25).

$$\Delta E_B(jkl) = \Delta E_B(j) - \Delta E_B(k) - \Delta E_B(l) + \Delta R(kl)_e \qquad (25)$$

Since binding energy shifts are nearly the same for all core levels, we can write Eq. (25) as

$$\Delta[E(jkl) + E_B(j)] = \Delta R(kl)_e \qquad (26)$$

Wagner (1975) calls the sum $E(jkl) + E_B(j)$ the "Auger parameter" and has shown that shifts in the Auger parameter have some important advantages over comparisons of E_B. First, the Auger parameter shifts are essentially equal to changes in relaxation energy, which has been shown to be an indicator of an atom's chemical environment. The effects of ground state ionic charge and lattice potential cancel in the Auger parameter, so there are no longer three separate factors which influence the chemical shift as there are with ΔE_B. Table IV illustrates the advantages of using the Auger parameter. The binding energy shifts for Na metal and a series of simple Na salts are quite small and not at all correlated with the ionic charges residing on Na. The Auger parameter, however, shows good agreement with the polarizability of the nearest neighbors. Best of all, since the Auger parameter is a difference measurement, there is no reference level needed, so work function changes and surface charging effects can be ignored.

TABLE IV

Chemical Shifts and the Auger Parameter
for Sodium Compounds (Al K_α Radiation)[a]

Chemical form	Na_{1s} E_B	$KL_{23}L_{23}$ E_K (Auger)	α	ΔE_B (Auger)	ΔE (Auger)	$-\Delta\alpha$
Na(S)	1071.7	994.2	579.3			
NaI	1071.4	991.4	576.2	−0.3	2.8	3.1
NaBr	1071.5	990.8	575.7	−0.2	3.4	3.6
NaCl	1071.4	990.5	575.3	−0.3	3.7	4.0
NaSCN	1071.1	990.7	575.2	−0.6	3.5	4.1
Na_2SO_3	1071.1	990.6	575.1	−0.6	3.6	4.2
$NaNO_2$	1071.4	990.0	574.8	−0.3	4.2	4.5
Na_2SO_4	1071.0	990.0	574.4	−0.7	4.2	4.9
$NaNO_3$	1071.2	989.8	574.4	−0.5	4.4	4.9
NaF	1071.0	988.8	573.2	−0.7	5.4	6.1
Na_2SiF_4	1071.5	987.9	572.8	−0.2	6.3	6.5

[a] From Swingle and Riggs (1975).

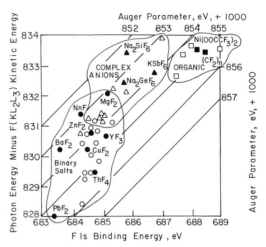

Fig. 11 Chemical state plot for fluorine. A few specific points shown in black are labeled. The number 1000 is added to make the parameter a positive number. (From Wagner, 1977b.)

An approach suggested by C. D. Wagner (1977b) which is even more useful than the Auger parameter alone for chemical state information of unknowns is a two-dimensional plot of Auger energy versus binding energy. A plot of this type for some fluorine compounds is shown in Fig. 11. The diagonal lines are Auger parameter values in 1 eV increments. By using this diagram, the spectroscopist can characterize a fluorine containing unknown as being an organic, a complex fluoro-anion, or a simple fluoride. Wagner *et al.* (1979) have published a comprehensive list of Auger and ESCA chemical shift data in this form. This tabulation is an invaluable reference for comparing unknown chemical shifts to those of most known compounds.

IV. X-Ray Photoelectron Line Shapes and Splittings

The photoemission process is inherently a many-electron event; that is, during photoemission, all of the other electrons in the system must readjust their positions and energy since they are no longer in their ground electronic states. Another way of formalizing this phenomenon in quantum mechanical terms is to note that the remaining electrons no longer possess eigenstates which satisfy the ground state Hamiltonian. Thus, new electronic states become accessible to the final state electrons. These effects give rise to asymmetric peak shapes and various peak widths as well as peak splittings. It is necessary for the electron spectroscopist to be aware of these effects to avoid misinterpretation of spectra.

A. Spin–Orbit Splitting

The simplest type of splitting alluded to previously is the spin–orbit splitting. When the orbital angular momentum l is greater than zero, the electron spin couples with l to yield two levels with designation $j = l \pm \frac{1}{2}$ and degeneracy $2j + 1$. Photoemission from a filled p orbital, for example, gives rise to $^2P_{3/2}$ and $^2P_{1/2}$ states with relative intensities 2:1. This splitting is illustrated nicely for the case of Ni metal shown in Fig. 1. The magnitude of the splitting is generally independent of chemical state for filled shells, but increases systematically with atomic number Z as can be seen in Appendix I.

B. Multiplet Splitting

The next complication, termed "multiplet splitting", is observed when photoemission arises from a closed shell in the presence of an open shell. For example, molecular oxygen has two possible final states when ionizing from the 1s level (Hedman et al., 1969).

$$O_2(1s^2,2s^2,2p^4;\,^3\Sigma) \rightarrow \begin{cases} O_2(1s,2s^2,2p^4;\,^2\Sigma) + e^- \\ O_2(1s,2s^2,2p^4;\,^4\Sigma) + e^- \end{cases}$$

The $^4\Sigma$ state has the 1s electron spin parallel to the π electron spins and the $^2\Sigma$ state has the 1s electron spin antiparallel to the π electron spins. The $^4\Sigma$ configuration is the most stable final state. The magnitude of the splitting, in this case 1.1 eV, and shown in Fig. 12, can be estimated in the Hartree–Fock approximation by the Slater exchange integrals $(2s + 1)G^0(1s,2p)$ with relative intensities 2:1 as given by the ratio of spin degeneracies. The G^0 integrals have been calculated and tabulated in the literature (Mann, 1967). Their value decreases as the binding energy of the photoelectron increases since the exchange probability becomes smaller between deep lying core electrons and valence electrons. These splittings can also be used to indicate the location of the unpaired spin. For the NO molecule, as shown in Fig. 12, the N 1s level is split into a doublet, but the O 1s level is only slightly broadened (Siegbahn et al., 1969). The core hole, then, can only effectively couple with the unpaired electron located primarily on the nitrogen atom. Note also that for the N_2 molecule, only a single peak is observed since the net spin is zero for N_2.

The prediction of multiplet splittings is reasonably straightforward for these cases, but it can get much more complicated when the open shell has $l > 1$. Photoemission from the 3s level of MnF_2 which possesses five parallel spins in the valence 3d shell represents such an example (Fadley and Shirley,

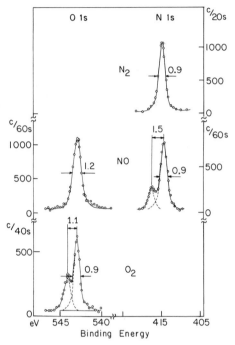

Fig. 12 ESCA spectra from N_2, NO, and O_2 showing spin splitting of the 1s levels in the paramagnetic molecules. (From Siegbahn *et al.*, 1969.)

1970). In this case photoemission from the 3s orbital of Mn^{2+} occurs as follows:

$$3s^2 3p^6 3d^5 ; {}^6s \xrightarrow{h\nu} 3s^1 3p^6 3d^5 ; {}^5s \quad \text{or} \quad {}^7s$$

when the resulting 3s electron spin is either parallel or antiparallel to the five unpaired d electron spins. The resulting energy splittings can again be calculated from the magnitude of the exchange interactions although the calculation is a little more involved than for the simple molecular oxygen case (Bagus *et al.*, 1973). The magnitude of the exchange interaction is related to the charge density around the metal ion, so the splittings have been suggested to be a useful chemical structural parameter, especially for the transition metals. The splittings and intensities are not predictable from simple Hartree–Fock theory, however, due to strong electron correlation effects. The rare-earth valence shells with a $4f^n$ configuration exhibit the most complex multiplet splittings but most of the general features of these spectra have been worked out (Chazalvicl *et al.*, 1976). In fact, a careful analysis of the photoemission spectra from the valence shells themselves represents the most complex example of multiplet splittings.

C. Relaxation

In our discussion of chemical shifts, it was clear that relaxation of the outer electrons in the system had a major influence on E_B. The many-electron nature of the process, however, allows for excitation of additional electrons to bound or unbound continuum states yielding structure in the observed spectra. These excitations result since the remaining electrons suddenly find themselves in a new eigenstate of the atom and can mix with higher levels. When the multielectron excitation results in a bound state, the process is sometimes referred to electron "shake-up," and when electrons are excited into the continuum, the process is referred to as "shake-off." The shake-off process leads to a broad band of photoelectron energies in the solid, since the kinetic energy of the ejected electron can be continuously partitioned between itself and the shake-off electron. In the shake-up process, however, the energy required for the excitation of the second electron is lost from the kinetic energy of the departing electrons. This results in the appearance of a second discrete peak occurring at a higher E_B. The selection rules for the excitation of the second electron are generally believed to be monopole where $\Delta l = \Delta s = \Delta j = 0$ so that the orbital angular momentum and electron spin must remain the same for the process. The phenomenon has also been suggested by Martin and Shirley (1977) to arise from mixing of both initial and final electronic configurations, a concept important in calculating the relative intensities of these satellite lines.

The shake-up phenomenon clearly has implications in the interpretation of photoemission spectra, since the intensities of these peaks depend strongly on electronic structure. For La_2O_3, for example, as shown in Fig. 13, a strong shake-up satellite is observed which is almost as large as the main peak (Burroughs *et al.*, 1976), and one might mistakenly assign it to a second chemical state. For CuS, a strong satellite is also observed (Novakov, 1971), presumably related to the presence of the 3d hole, in Cu^{2+}. For Cu^+ and Cu^0 compounds, however, no satellites are observed suggesting that the structure might be a good diagnostic tool for distinguishing between various oxidation states (Kim, 1975). The explanation for the shake-up phenomenon

Fig. 13   The 3d ESCA spectra of La_2O_3, exhibiting intense shake-up satellites. (From Burroughs *et al.*, 1976.)

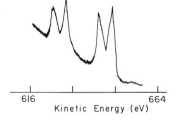

616   664
Kinetic Energy (eV)

in transition metal compounds has been widely debated, but seems to involve a ligand-to-metal charge transfer in the final state. For example, O 2p → Cu 3d would explain why the satellite is intense for Cu^{2+}, but absent for Cu_2O, which has a $3d^{10}$ configuration (Kim, 1975).

A similar mechanism for the shake-up process has been invoked to explain the asymmetric peak shape often observed in metals. Here, the allowed excitations occur to a continuum of states just above the Fermi energy rather than to discrete levels as was found for atoms and molecules. A line shape function derived by Doniach and Sunjic (1970) included the electron density at the Fermi level as a parameter and has been quantitatively used to describe line shapes for a number of simple metals and transition metals (Hufner and Wertheim, 1975). The shape of the spectrum for Ag which has a very low density of states at the Fermi level is much more symmetric than for Pd which has its d band at the Fermi level.

Vibrational effects can also influence the peak shapes. For CH_4 in the gas phase, for example, high resolution studies have shown a 2-peaked structure due to the excitation of the C–H symmetric stretch in the final state (Gelius, 1974b). For solids, the core hole can also couple to its phonon spectrum. The explanation has been confirmed by checking the broadening of the core lines of alkali halides as a function of temperature (Citrin *et al.*, 1974). For insulators, this broadening amounts to ~ 0.5 eV, but is nearly zero for metals.

D. *Plasmons*

Another final state effect which can influence the shape of the photoemission spectra is the occurrence of various inelastic scattering events the electron experiences as it passes through the solid. The most common of these is the discrete energy loss due to plasmon excitation. These excitations have been extensively studied by electron energy loss spectroscopy and are somewhat of a bother in interpreting photoemission spectra. They are typically observed on most solids, although their intensity is strongest on the free electron metals like Na and Al. These excitations can occur anytime a collection of charges (i.e., plasma) experiences a perturbation. For free electrons, the displacement of a single electron initiates a wave of electronic motion throughout the solid, arising from Coulombic electron–electron repulsions. The losses can occur either by intrinsic, extrinsic, or surface processes. A typical example is shown in Fig. 14 for Na where the labels P_1, P_2, . . . refer to the excitation of an integral number of plasmons as the electron moves through the solid (Pardee *et al.*, 1975). These bulk excitations consist of both intrinsic and extrinsic processes since the excitation

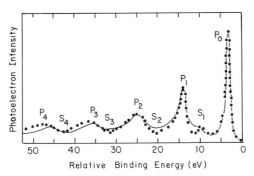

Fig. 14 Magnesium 2s plasmon spectrum: (a) (———) XPS data, (b) (· ·), theoretical spectrum. (From Pardee *et al.* 1975.)

can occur by creation of the core hole, much like the shake-up mechanism, or as the electron is moving through the solid. The current belief is that the intensity is roughly equally distributed between the two processes (Bradshaw *et al.*, 1977). The other feature of note in Fig. 14 is the presence of a surface plasmon loss peak S_1. Simple consideration of boundary conditions shows that $E_s = E_p/\sqrt{2}$, where E_s is the surface plasmon excitation energy. The peaks are very sensitive to contamination and are known to disappear when chemisorption occurs on the surface. The loss features will be observed on all photoemission lines including the Auger peaks, although with differing intensities.

The above discussion covers most of the important features of the photoelectron line shape, which may be of interest to the practicing electron spectroscopist. Of special concern is that these features can be interpreted as arising from different chemical states rather than from the intrinsic nature of a single photoemission event. For the chemist, many of these features may, however, provide additional insight into the electronic structure of the sample. An appreciation for these processes is therefore of double importance.

V.   Quantitative Aspects of Photoelectron Spectroscopy

Electron spectroscopy is increasingly being utilized to make quantitative measurements of chemical composition in the surface region of solids. Normally, electron spectroscopy is considered a semiquantitative technique ($< \pm 50\%$ relative error). This large error is introduced because spectral intensities are strongly influenced by several sample matrix dependent factors (i.e., electron escape depth, energy loss effects, surface roughness, and surface contamination). When better quantitative information is needed ($< \pm 10\%$

relative error), standards can be employed provided they are similar to the unknown. Frequently, however, no standard exists which is satisfactory. Any material exposed to air will have a contaminated surface and its use is suspect. The material may be cleaned by sputter etching, but sputtering may change the composition in the surface region. A pure metal can be cleaned and is sometimes useful as a standard. However, even for closely related sample matrices such as alloys, errors will result from differences in surface roughness and energy loss effects. In this section, each of the factors which influence quantitative electron spectroscopic data will be examined separately, and the best available mathematical expressions for these factors will then be discussed.

For a thick, homogeneous sample, the intensity of photoelectrons detected from a given atomic level can be written as:

$$I = nFk\sigma\lambda \qquad (27)$$

Here n is the number of atoms of the element per cubic centimeter, F is the x-ray photon flux, k is the instrumental response function which is dependent on the kinetic energy of the photoelectrons, σ is the photoionization cross section for the given atomic level, and λ is the mean free path of a photoelectron through the sample.

A. X-Ray Flux

Variations in x-ray flux between the times of data acquisition for two different lines will cause errors of about 5%. This error could be reduced if special precautions were taken to maintain constant x-ray flux. However, no spectrometer manufacturers have made a precision flux x-ray tube available.

B. Spectrometer Response Function

Various electron energy analyzers are used in commercial spectrometers. Their different configurations and characteristics mean that the photoelectron intensities will have an instrumental term. This term involves the electron kinetic energy dependencies of sampling geometry and collection efficiency. The most commonly used analyzers employ preretardation of photoelectrons to a constant analyzer pass energy. The measured photoelectron intensity is then given by

$$I_{\text{meas}} \propto I_{\text{true}}/T_{\text{sp}} \qquad (28)$$

This electron energy dependence is due to the change in sampling area size with the electron kinetic energy selected. For an electron energy analyzer

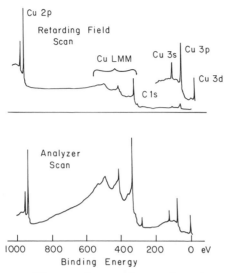

Fig. 15 Comparison to different analyzer mode scans for a clean copper surface. (From Swingle and Riggs, 1975.)

where the pass energy is continually varied (no retardation), the collection efficiency, or window width of the analyzer, will be proportional to the kinetic energy. The intensity is then given by

$$I_{\text{meas}} \propto I_{\text{true}} T_{\text{sp}} \tag{29}$$

These two analyzer modes will produce very different looking spectra, as shown by the copper metal spectra in Fig. 15. It might be noted here that the spectra produced by an instrument employing preretardation is much less influenced by electron energy since the energy dependence due to the spectrometer term is largely cancelled by the energy dependence of the electron mean free path term. For various reasons, the actual electron energy dependence of the spectrometer may differ from the theoretical predictions of Eq. (29), thus the response function for a particular instrument should be determined experimentally.

C. Photoionization Cross Section

The x-ray photoionization cross section is the most significant factor influencing photoelectron intensity data. From Fig. 16, it can be observed that there is a factor of 100 difference in relative cross section between the most sensitive and the least sensitive elements. Subshell cross sections for the elements have been calculated by Scofield (1976) using a relativistic single potential Hartree–Slater atomic model. These values have been

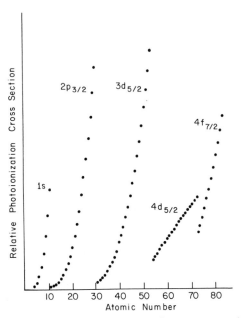

Fig. 16   Relative photoionization cross sections versus atomic number.

widely used by researchers and are employed in our program. Cross sections have also been determined from experimental data by several workers (Leckey, 1976; Nefedov *et al.*, 1975). In general, there is agreement between experiment and theory to within 5% for light elements and 15% for heavy elements.

Two corrections are needed in order to use calculated photoionization cross sections in quantitative work. First, angular effects must be considered. Photoelectrons are not emitted isotropically. Their differential cross section has been given (Huang *et al.*,1975) by

$$dJ/d\Omega = \sigma_{tot}[1 - (1/4)\beta(3\cos^2\theta - 1)]\theta \qquad (30)$$

Here θ is the angle between the photon beam and the ejected electron (the spectrometer angle), and β is a parameter dependent on the x-ray energy. C. D. Wagner (1977a) has calculated the ratio of angular correction factors for the extreme case of Na 2s/Na 2p at a spectrometer angle of 90° to be 1.28.

The second effect that must be considered is a collection of multi-electron processes. Shake-up, shake-off, multiplet splitting, and other many-body effects result in a significant fraction of photoionized electrons having kinetic energies less than that of the principal peak. Figure 17 shows some examples of energy loss regions. Cu^{2+} has a prominent satellite with an

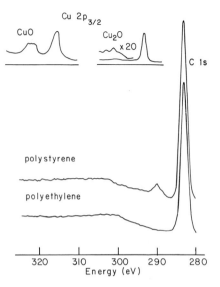

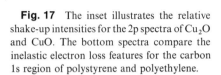

Fig. 17 The inset illustrates the relative shake-up intensities for the 2p spectra of Cu_2O and CuO. The bottom spectra compare the inelastic electron loss features for the carbon 1s region of polystyrene and polyethylene.

area of about 40% of that of the main peak. Cu^{1+} has an energy loss structure that is much less prominent. For both samples, however, the photoionization cross section value that is available is the total cross section, so the intensities of the satellite peaks must be included in the quantitative calculations. In most cases, however, this cannot be done satisfactorily. Comparing the C 1s spectra of polystyrene and polyethylene in Fig. 17, a prominent satellite can be seen for polystyrene at 290-eV binding energy with an intensity of about 7% of the main peak. The broad structure in both spectra which is about 20 eV removed from the main peak is associated with inelastically scattered electrons. This effect should be accounted for in the electron mean free path term. This region of the spectrum, however, probably also includes more satellites which are intrinsic to the photoionization process, but are buried in the high background and cannot be seen. From gas phase spectra where inelastic scattering effects are negligible, such as the Ne 1s spectra in Fig. 18, it can be seen that these satellites extend 100 eV or more from the main peak (see Gelius, 1974c).

Very little quantitative error will result by neglecting satellite intensities when the fraction of photoionized electrons undergoing multielectron processes are nearly the same for all atomic species in a sample. However, the percentage of photoelectrons which undergo intrinsic energy loss processes can vary from 20% in noble gas atoms, to 60% or greater for transition metal atoms (C. D. Wagner, 1977a). Variations in energy loss phenomena represent a significant limitation to the accuracy of quantitative work which can be achieved without employing closely related standards. C. D. Wagner

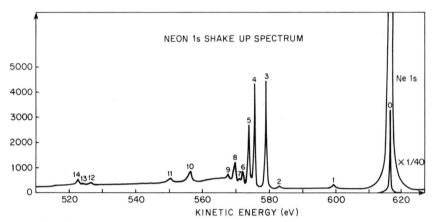

Fig. 18 Neon 1s shake-up spectrum. Note the absence of inelastic scattering of electrons in the background. (From Gelius, 1974c.)

(1977a) has estimated the standard deviation of ground state transition probability ratios for 66 compounds to be 10% when transition elements are excluded.

D. Electron Mean Free Path

Figure 19 shows the variation with kinetic energy of the mean free path of an electron through the specimen based on experimental measurements. These compiled data, perhaps because they are displayed on a log–log plot,

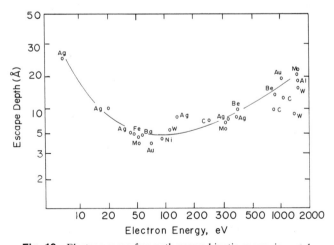

Fig. 19 Electron mean free path versus kinetic energy in metals.

suggest electron mean free path might be material independent. That is, there exists a "universal curve," as shown in Fig. 19, which defines the electron escape depth for a given kinetic energy. This concept has been widely used even when it is not wholeheartedly accepted. The part of the curve of interest for XPS applications, about 200–1500 eV, has been fit mathematically to vary with kinetic energy as about $E^{0.5}$ to $E^{0.82}$.

Recently, theoretical work has made some refinements in the treatment of electron escape depth (C. D. Wagner, 1977a; Penn, 1976). The energy dependence of electron mean free path can be described by

$$\lambda = E/\{a[\ln (E + b)]\} \tag{31}$$

Here a and b are parameters which depend on electron concentration and dielectric function. Penn has calculated mean free paths for most elements using Eq. (31) (Penn, 1976). The shortest mean free path for 1000-eV electrons was 11.3 Å for Ni, while the longest mean free path was calculated to be 49.3 Å for Cs. However, the ratio of mean free paths in the same material is independent of the parameter a and only weakly dependent on b. Penn has found, therefore, that using $b = 2.3$ for all elements and compounds will give at most a 14% error in λ_1/λ_2, and more typically a 5% error (Penn, 1976).

E. Contamination Overlayer

A significant source of error in photoelectron intensity measurements is signal attenuation due to the presence of surface contaminants. Provided the overlayer is uniform and sufficiently thin (< 20 Å), the intensity for a photoelectron line from a homogeneous sample which has been covered by an overlayer is given by

$$I = nFk\sigma\lambda \exp(-d/\lambda_{\text{oV}} \sin \theta). \tag{32}$$

Here, d is the overlayer thickness, λ_{oV} the electron mean free path in the overlayer, and $\sin \theta$ accounts for the "take-off" angle between the plane of the sample and the analyzer entrance. Photoelectron line intensities with widely different kinetic energies will be attenuated to different extents by the overlayer, leading to errors in stoichiometry of 50% or greater if Eq. (27) was used. Any sample which has been exposed to air will have a contamination layer of moisture and carbonaceous material. Many samples will have oxides, hydroxides, and carbonates present as well. Laboratory and industrial environments frequently contribute contaminants containing Si, S, and Cl also. Even samples which have been prepared or cleaned at a good spectrometer vacuum of 1×10^{-9} Torr can accumulate a monolayer of contamination in 15 min., assuming a sticking coefficient of unity. Samples can be cleaned in vacuum by ion sputter etching, but preferential sputtering of one element relative to another will make subsequent quantitative measurements inaccurate.

When the stoichiometry of an overlayer is desired (for instance, to find the composition of the passivated oxide layer on an alloy) Eq. (33) can be used.

$$I = nFk\sigma\lambda_{ov}[1 - \exp(-d/\lambda_a \sin\theta)] \tag{33}$$

The overlayer thickness is calculated from XPS data since

$$d = [-\lambda_i \sin \ln(1 - \eta_{i,\text{homo}}/\eta_{i,\text{ov}})]. \tag{34}$$

Here, $\eta_{i,\text{homo}}$ is the atomic fraction of i calculated assuming a homogeneous sample, and $\eta_{i,\text{ov}}$ is the atomic fraction of i in the overlayer. Since both $\eta_{i,\text{ov}}$ and d are unknown, an iterative procedure is used. First $\eta_{i,\text{ov}}$ is estimated as

$$\eta_{i,\text{ov}} = \eta_{i,\text{homo}} \bigg/ \sum_{i=1}^{m} \eta_{i,\text{homo}} \tag{35}$$

for the m overlayer elements. The overlayer thickness is then taken as the average of the m values for d calculated by Eq. (34). Then $\eta_{i,\text{ov}}$ can be solved using d_{avg}. Treatment of data can easily be extended to include multiple overlayers.

F. Surface Roughness

It has been shown that an increase in surface roughness will cause a decrease in signal intensity (Ebel *et al.*, 1973). However, only sufficient roughness to cause x-ray shadowing of the surface is a factor (roughness on the order of micrometers). Roughness which causes only electron shadowing will not affect absolute photoelectron intensities (Fadley *et al.*, 1974). Furthermore, ratios of photoelectron intensities which are employed in our analysis procedure should not be affected by x-ray shadowing. For these reasons, surface roughness effects can frequently be neglected in ESCA work. This lack of response to surface roughness is in contrast to Auger electron spectroscopy where Holloway (1975) has shown that the use of electron beam excitation will cause surface roughness to influence both absolute intensities and relative intensities.

VI. Some Additional Applications

At this point, the main features of XPS (and ESCA) should be fairly clear. The method is sensitive to virtually all elements. For solids the electron escape depth limits the departing photoelectron to originate from the top few atomic layers. And finally, the measured binding energies and resulting peak shapes, reflect, in a not so straightforward manner, the chemical environment of a given atom in a specific molecular environment. Not surprisingly, then, it is these features which are most often exploited when XPS is used to solve new problems. We next focus on a few of these applications, concentrating entirely on solids in an attempt to illustrate some of these points.

A. Multielement Trace Analysis

When applied to the analysis of bulk compounds XPS does not have an impressive detectability limit, being 0.1 at.% even in the most favorable cases. If the material of interest can be concentrated at the surface, however, the technique can provide analysis of 10^{-8}–10^{-9} at the 5–10% quantitative level. Several workers have exploited this idea by developing unique sample preparation methods which concentrate the elements of interest at the surface. One example is the use of chelating glasses (Hercules *et al.*, 1973) to extract specific metal ions from a solution containing the element. For example, a glass surface with the functional group

$$\text{glass}\!\!\begin{array}{c}|\!\!-\!\!O\\|\!\!-\!\!O\\|\!\!-\!\!O\end{array}\!\!-\!\!Si\!-\!CH_2CH_2CH_2NH\!-\!CH_2CH_2NH\!-\!C\!\!\begin{array}{c}\diagup S\\\diagdown S^{\ominus}\end{array}\ \ Na^{\oplus}$$

reacts very effectively with heavy metals. Detection limits of 10 ppb for lead, calcium, thallium, and mercury could be obtained by flowing 100 ml of solution through a porous chelated glass. A similar method of concentrating material on the surface has been developed by Brinen and McClure (1972) by electrodepositing metals on mercury coated platinum electrodes. In general, then, if the samples can be concentrated to the surface, then XPS can be very sensitive to nanogram quantities of many elements. The approach cannot be considered to yield reliable results for low levels of material in bulk compounds.

B. Chemistry at Surfaces

The chemical shift effect and the surface sensitivity of ESCA provide unique characteristics for the study of surfaces. Of initial interest would be the detection of photoelectrons which arise from the first layer atoms of a pure substrate rather than from bulk atoms. One might expect that these atoms would have a different binding energy since their coordination number is lower and since relaxation or final state effects would be different.

Surprisingly, however, this difference has not been observed until recently (Citrin *et al.*, 1978) since the shift is small (-0.4 eV for Au $4f_{7/2}$) and of fairly low intensity. The effect could be magnified by carrying out the measurements at an electron take-off angle of $80°$, which effectively reduces the electron mean free path to less than 4 Å [see Eq. (32)]. The evidence for this surface peak can be seen in Fig. 20.

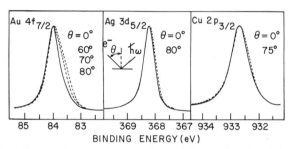

Fig. 20   Normalized XPS spectra from Au, Ag, and Cu as a function of increasing surface sensitivity (increasing θ) (From Citrin *et al.*, 1978.)

The next step is to allow fairly simple molecules to interact with clean single crystal metal surfaces and to use ESCA to identify the chemical nature of the products. These experiments require spectrometer vacuum of $< 1 \times 10^{-9}$ Torr, along with capabilities to heat the metal crystal to 1000 K or more for cleaning and annealing. The first example of this type of experiment involved the chemisorption of CO on a tungsten ribbon (Yates *et al.*, 1974). The presence of CO on the surface could be confirmed from the O 1s and C 1s binding energies and intensities. Of special interest, though, was the direct observation of ESCA peaks which correspond to weakly and strongly held adsorption states that had been previously identified by flash desorption methods. Furthermore, the substrate core levels (i.e., W 3d) were not observed to shift at all. It should be noted, however, that at low electron take-off angles a small shift of $+0.9$ eV has been recently noted for W covered with one monolayer of oxygen (Barrie and Bradshaw, 1975). Thus, it appears that for interactions which only involve the surface layer, most of the chemical information is found in the XPS of the adsorbate.

These phenomena have been further noted by studying the reaction of O_2 with Ni. In this case, the $O\equiv O$ bond is cracked at the metal surface, resulting in adsorbed oxygen atoms. The XPS spectra of this system (Kirshnan *et al.*, 1976; Norton *et al.*, 1977; Fleisch *et al.*, 1978) are shown in Fig. 21 and are typical of oxygen adsorption on many metals with the O 1s binding energy falling in the 531–533-eV range. At low exposures, up to 0.5 monolayer coverage, only a single O 1s peak is found and no shift is observed in the

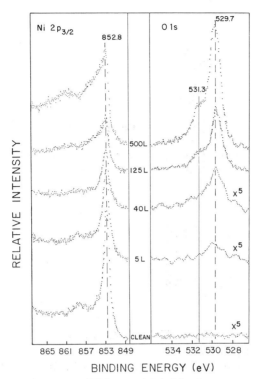

Fig. 21   Representative spectra of Ni $2p_{3/2}$ and O 1s following exposure of a Ni(100) surface to oxygen. (From Fleisch *et al.*, 1978.)

Ni substrate level. At higher oxygen exposures, however, a shoulder appears on the high binding energy side of the Ni $2p_{3/2}$ line, signaling the onset of oxygen incorporation into the lattice and the formation of a Ni^{2+} oxide. This large substrate core-level shift of nearly 2 eV is typical for the appearance of an oxide versus the pure metal. Although it is so far not possible to explain or predict the magnitude of these shifts, it has been speculated that the oxygen interupts the metal–metal bonds that permit the high screening efficiency of the metal conduction electrons (Brundle, 1975).

Note that as the NiO film grows toward approximately two layers thick, a shoulder is observed on the high binding energy side of the O 1s peak. This feature has been assigned to a cation defect structure with stoichiometry of approximately Ni_2O_3 (Kim and Winograd, 1974a) which would give the Ni a +3 oxidation state. This species is only observed on Ni surfaces which are allowed to oxidize slowly. On cleaved NiO crystals and on rapidly oxidized Ni (Kirshnan *et al.*, 1976) surfaces, the defects are not observed. Similar types of mixed oxidation state species have been observed on Ru

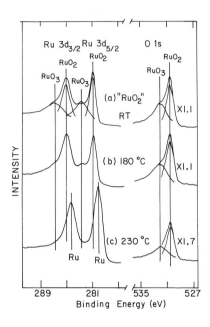

Fig. 22 ESCA spectra of Ru $3d_{5/2,3/2}$ and O 1s levels of RuO_2 powders kept at various temperatures in the spectrometer. The temperature was raised at the rate of about 1°C per min. (From Kim and Winograd, 1974b.)

surfaces (Kim and Winograd, 1974b) as shown in Fig. 22 and for many other systems. The possibility that these defect phases can influence the reactivity of these materials is an interesting one for future study.

C. Energy Loss Features

A final aspect of the use of ESCA to characterize surfaces involves the analysis of the various energy-loss and line-shape features discussed in Section III. A fascinating example involves the chemisorption of molecular CO on metal surfaces and a comparison of the C 1s shake-up structure to bulk metal carbonyls. As shown in Fig. 23, the core ionization from the 1s orbital can pull the unoccupied 2π level below the level of occupied d states in the metal. If this happens, the CO^+ final state is screened by the metal, leaving it as an excited neutral state CO*. Thus two final states are possible involving CO^+ and CO* (Plummer et al., 1978). The degree of screening and hence the fraction of photoemission intensity left in CO* is directly related to the magnitude of the electronic coupling between the CO adsorbate and the metal. In Fig. 23, for example, these two states are shown for a hypothetical variation of the metal–carbonyl bond distance, along with the experimental spectrum for $Cr(CO)_6$. Thus, for CO on Ru, the O 1s spectra show a weak shake-up peak about 7 eV above the main peak, a spectrum

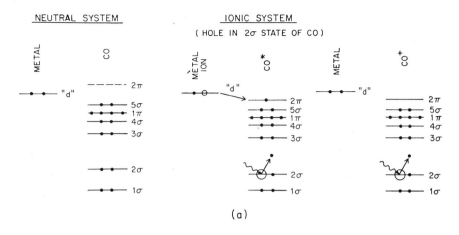

(a)

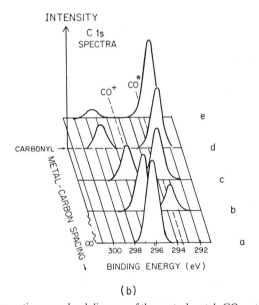

(b)

Fig. 23 (a) Schematic energy level diagram of the neutral metal–CO system (left-hand side) and two ionic states (right-hand side) with a single 2σ hole. (b) Schematic drawing of a C 1s photoelectron spectra from a CO molecule as a function of the metal–carbon distance in a carbonyl. Curve a is a plot for a CO molecule at infinity and curve d is the measured spectrum for Cr(CO)$_6$. (From Plummer *et al.*, 1978.)

virtually the same as condensed $Ru_3(CO)_{12}$. For the case of Cr, however, Norton *et al.* (1980) report that at $77°K$ the shake-up peak is 1.5 times larger than the lowest binding energy O 1s peak and is shifted only 3 eV to higher binding energy. This situation would then correspond to a weak adsorption with a very small amount of screening. This interesting case suggests that the presence of several peaks does not necessarily indicate the presence of multiple adsorption sites of a molecule, but rather may be the result of various electronic relaxation processes.

D.   Polymer Surfaces

Although most of the effort in characterizing surfaces has been directed to inorganic systems (i.e., oxides and other transition metal complexes), a number of workers have focused their attention to the analysis of polymer surfaces. Figure 24 compares the C 1s, N 1s, and O 1s spectra of a sample of acrylonitrile–butadienestyrene (ABS) which has been an injection molded sample. From the peak areas, the elemental composition of the polymer can be estimated, while the measured binding energies give information on the chemical states of the atoms. An obvious difference seen in Fig. 24 is the surface enrichment of acrylonitrile in the injection-molded sample. Thus, from a brief ESCA experiment, information can very quickly be gained on

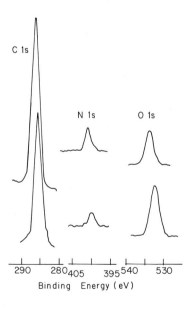

Fig. 24   ESCA spectra of acrylonitrile–butadienestyrene. Upper curve is injection molded and the lower curve is compression molded. (From Wims and Schreiber, unpublished.)

elemental composition, percentage of comonomers, gross structure, and surface modifications. More subtle features of polymer bonding and structure such as degree of branching, structural conformation, molecular weight, and long range order can only be studied with ESCA in special cases. This indicates that ESCA is a good technique for the initial examination of an unknown and the information obtained can aid the analyst in deciding which other techniques need to be employed.

An examination of the use of ESCA to study copolymer compositions was done by Clark and co-workers (1975; Clark and Dilks, 1976), on the alkane–styrene system. The C 1s spectra and general formula of the copolymers are shown in Fig. 25. The binding energies of all carbon atoms

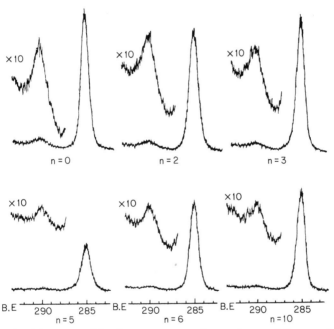

Fig. 25   The C 1s spectra of the alkane–styrene regular copolymers of the general formula shown below. (From Clark and Dilks, *J. Polym. Sci. Polym. Chem. Ed.* **14**, 538 (1976). Reprinted by permission of John Wiley & Sons, Inc.)

in the chain are nearly the same. An aromatic carbon atom can still be distinguished from an aliphatic one on the basis of the $\pi \to \pi^*$ shake-up satellite observed at 6.6-eV higher binding energy from the main peak. From the figure, it is evident that as the alkane chain of the copolymer is lengthened, the shake-up satellite intensity due to the pendant phenyl group is proportionally less. Table V lists the ratio of main photoionization peak intensity to the shake-up peak intensity for a number of copolymers. Interestingly, using the polystyrene shake-up intensity and the repeat units for a given copolymer, the slope calculated with an additive model is 0.90, not 1.0. This suggests there is a specific orientation for the polymers chains in the surface region which would account for the apparent attenuation of the shake-up intensity. One possibility suggested by Clark and Dilks (1976) is a folded chain structure with the alkane component at the surface and the pendant phenyl groups some distance below the surface.

TABLE V

ESCA Data for Alkane–Styrene
Copolymers[a]

| | Binding energies (eV[b]) | | | |
| | | | | Area ratios[c] |
n	C 1s	C³ 1s	Δ	(C 1s/C³ 1s)
0	285.0	291.6	6.6	13.4
1	285.0	291.6	6.6	14.5
3	285.0	291.6	6.6	18.5
5	285.0	291.6	6.6	22.7
6	285.0	291.6	6.6	23.8
10	285.0	291.6	6.6	32.2

[a] From Clark et al. (1975).
[b] Relative to C 1s at 285.0 eV, Δ given with respect to centroid of assymmetric shake-up peak.
[c] Ratio of main photoionization peak to shake-up peak.

The modification of polymer surfaces by plasma treatment has been of considerable research interest recently (Clark, 1975; Clark and Dilks, 1977; Yasuda et al., 1977; Burkstrand, 1977). An important application of this procedure is to improve adhesive bonding. The improved adhesion results because plasma treatment increases the number and type of functional groups present at the surface. Figure 26 shows the C 1s spectra of polypropylene before and after an oxygen plasma treatment. This difference spectrum which is displayed at the bottom of Fig. 26 shows that there is a decrease in the number of carbon–hydrogen bonds and the formation of

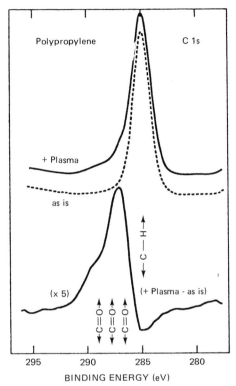

Fig. 26 The C 1s spectra for plasma treated and untreated polypropylene surfaces and the difference between the two. The C 1s curves have been offset from one another for clarity. (From Burkstrand, 1977.)

carbon–oxygen single bonds, carbonyl groups, and carboxyl groups. The mechanism of surface modifications by plasmas is complex and not completely understood. It is believed that reactions are associated both with free radical formation caused by ultraviolet radiation emitted from the plasma and with direct energy transfer from species in the plasma gas. The initial work on fluorinated polymers indicates that the direct energy transfer mechanism dominates (Clark and Dilks, 1977).

Appendix I

The appendix consists of a tabulation, on the following page, of the photoelectron binding energies (in electron volts) of the elements from lithium to uranium.

Binding Energy (eV) Al K X Rays

Strongest lines enclosed in box, second strongest denoted by (xxx), third strongest by ·xxx· (weaker of spin doublets not included).

	1s	2s	2p$_1$[a]	2p$_3$	3s	3p$_1$	3p$_3$	3d$_3$	3d$_5$	4s	4p$_1$	4p$_3$	4d$_3$	4d$_5$		Auger Locations on BE Scale				
3 Li Li$_2$CO$_3$	55	V													KLL	112[b]		122		
4 Be Be^e	111	V																		
5 B Na$_2$B$_4$O$_7$	192		V												B		1305			
6 C (CH$_2$)$_n$	285		V												C	(1226)				
7 N (NH$_4$)$_3$AlF$_6$	399		V												N	(1107)				
8 O CaCO$_3$	531	·23·	V												O	999		(977)		
9 F CaF$_2$	685	·30·	V												F	857		(831)		
10 Ne Ne^e	863	·42·	15												Ne	701		(668)		
11 Na Na$_2$SO$_3$	1071	·63·	30		V										Na	535		(496)		
12 Mg MgO	1305	·90·	51		V										Mg	352		(307)		
13 Al Al$_2$O$_3$		120	(75)																	
14 Si SiO$_2$		(154)	104	103	12	V														
15 P NaPO$_3$		(192)	135	134	15	V														
16 S WS$_2$		·227·	164	163	15	V									LMM	333		334	234	344 244
17 Cl KCl		·269·	199	198	16	V									S	(1336)				
18 Ar Ar^e		·319·	243	241	22	9									Cl	(1305)				
19 K KCl		·377·	295	292	32	16									Ar	(1270)				
20 Ca CaCl$_2$		·639·	351	348	44	25									K	(1236)				
21 Sc Sc$_2$O$_3$[c]		·500·	407	402	54	32									Ca	(1195)				
22 Ti TiO$_2$		570	464	458	62	·37·									Sc	(1152)[d]				
23 V V$_2$O$_5$		632	524	517	70	·42·									Ti	(1105)		1072		
24 Cr K$_3$Cr(CN)$_6$		697	586	577	76	·44·									V	(1059)	1055	1019		
25 Mn K$_3$Mn(CN)$_6$		770	653	642	85	·50·									Cr	1002		(962)		
26 Fe K$_3$Fe(CN)$_6$		846	723	710	94	·56·									Mn	952		(904)		854
27 Co K$_3$Co(CN)$_6$		929	797	782	104	·63·									Fe	900	894	846	(789)	
28 Ni K$_2$Ni(CN)$_4$		1009	872	855	113	·68·									Co	843	837	782	(720)	705
29 Cu Cu		1097	952	932	122	77	·75·								Ni	784	778	718	(646)	629
30 Zn ZnO		1195	1045	1022	138	91	·89·		10		V				Cu	720	712	647	(568)	548
															Zn	664	657	586	(498)	475
31 Ga Ga$_2$O$_3$		1300	1143	1118	160	110	·106·	20			V									
32 Ge GeO$_2$			1252	1221	184	129	125	·33·			V				Ga	519	509	492	(424)	397
33 As As$_2$O$_3$			1363	1327	208	150	143	·46·			V				Ge	451	440	419	(349)	318
34 Se Na$_2$SeO$_3$					234	170	·164·	(58)			V				As	378	367	342	(268)	232
35 Br NaBr					·257·	189	(182)	70	69		V				Se	303	291	261	185	143
36 Kr Kr^e					·287·	217	(209)	89	88	21	8									
37 Rb RbF					·320·	246	(237)	111	110	29	13									
38 Sr SrF$_2$					·359·	280	(270)	135	134	38	21									
39 Y YF$_3$					·396·	314	(302)	161	159	47	27		V							
40 Zr ZrO$_2$					·432·	346	(332)	184	182	53	30		V							

	3s	3p$_1$	3p$_3$	3d$_3$	3d$_5$	4s	4p$_1$	4p$_3$	4d$_3$	4d$_5$	4f$_5$	4f$_7$	5s	5p$_1$	5p$_3$	5d$_3$	5d$_5$	6s	6p$_1$	6p$_3$				
41 Nb Nb^c	469	379	(363)	208	205	58	·34·		V															
42 Mo Na$_2$MoO$_4$	510	415	(398)	235	232	67	·40·		V							MNN	544		444					
43 Tc																								
44 Ru Ru^c	585	483	(461)	284	279	75	·43·		V							Ru		1218						
45 Rh Rh	627	521	(496)	312	307	81	·48·		V							Rh		1189						
46 Pd Pd	670	559	·532·	340	335	86	52		V							Pd	(1159)							
47 Ag Ag	718	603	·573·	374	368	96	59			V						Ag	(1134)		1128					
48 Cd CdO	771	652	·618·	412	405	109	68		11	V						Cd	(1113)		1106					
49 In In$_2$O$_3$	828	704	·667·	453	445	124	79		19		V					In	(1084)		1076					
50 Sn SnO$_2$	885	758	·716·	494	486	138	90		25		V					Sn	(1060)		1053					
51 Sb Sb$_2$O$_3$	946	814	767	539	530	154	101		·34·		V					Sb	(1035)		1027					
52 Te Na$_2$TeO$_4$	1011	875	824	587	577	173	113		·44·		V					Te	(1012)		1002					
53 I KI	1071	930	874	630	619	185	123	50	·49·		V					I	(980)		969					
54 Xe Xe^e	1142	995	934	682	669	206	139	62	·60·		15	4				Xe	(955)		943					
55 Cs Cs^c	1215	1063	997	738	724	229	169	158	77	(75)	21	10				Cs	·932·		918					
56 Ba BaSO$_4$		1137	1063	796	781	253	192	180	93	(90)			40	15		Ba	·905·		890					
57 La La		1206	1125	853	836	274	209	195		(104)	V		36	18		La	·873·		859					
58 Ce Ce^c			1184	900	882	288	221	·206·		(108)	V		37	19		Ce		827						
59 Pr Pr$_2$O$_3$[c]				951	931	305	237	·218·		(114)	V		38	23		Pr		788						
60 Nd Nd$_2$O$_3$				1005	981	319	247	·229·		(122)	V		41	25		Nd		752						
61 Pm																Sm		673						
62 Sm Sm$_2$O$_3$[c]				1108	1082	348	268	·250·		(133)	V		40	22		Eu		635						
63 Eu Eu$_2$O$_3$				1161	1131	360	284	·257·		(134)	V		32	22		Gd		595						
64 Gd Gd$_2$O$_3$[c]				1218	1186	376	289	·271·		(141)	V		36	21										
65 Tb Tb$_2$O$_3$				1276	1242	398	311	·286·		(148)	V		40	26										
66 Dy Dy$_2$O$_3$[c]				1332	1295	416	332	·293·		(154)	V		63	26										
67 Ho Ho$_2$O$_3$[c]				1391	1351	436	343	·306·		(161)	V		51	20										
68 Er Er$_2$O$_3$[c]					1409	449	366	·320·	177	(168)	V		60	29										
69 Tm Tm$_2$O$_3$						472	386	337		180	V		53	32										
70 Yb Yb$_2$O$_3$						487	399	(346)	201	186	V		53	23										
71 Lu Lu$_2$O$_3$[c]						506	410	(359)	205	195	7		28	7		V								
72 Hf HfO$_2$[c]						538	437	·381·	223	(212)	18	17	63	38	32	V								
73 Ta Ta^c						566	465	·405·	242	(230)	27	25	71	45	37	V								
74 W W						593	490	·424·	257	(244)	34	32	75	45	35	V								
75 Re Re^c						625	518	·445·	274	(260)	47	43	83	46	35	V								
76 Os Os^c						655	547	·469·	290	(273)	52	50	84	58	46	V								
77 Ir Ir^c						690	577	·495·	312	(295)	63	60	96	63	51	V								
78 Pt Pt						724	608	·519·	331	(314)	74	71	102	66	51	V								
79 Au Au						760	645	·547·	353	(335)	88	84	108	72	54	V								
80 Hg Hg^c						800	677	·571·	379	(360)	103	99	120	81	58	V								
81 Tl Tl^c						846	722	609	407	(386)	122	118	137	100	76	16	·13·		V					
82 Pb Pb						893	763	644	434	(412)	142	137	147	104	85	20	·18·		V					
83 Bi Bi^c						939	806	679	464	(440)	163	158	160	117	93	27	·25·		V					
90 Th Th^c						1168	968	714	(677)	344	335		290	229	182	95	·88·	60	49	43				
92 U U^c						1045	780	(738)	392	381			324	240	195	105	·96·	71	43	33				

[a] 1 = sub 1/2, 3 = sub 3/2, etc.

[b] 112 = $K_{I}L_{II,III}$; 234 = $L_{II}M_{III}M_{IV,V}$; 544 = $M_{V}N_{IV,V}N_{IV,V}$; etc.

[c] Data from Siegbahn, et al., "ESCA: Atomic, Molecular, and Solid State Structure by Means of Electron Spectroscopy", Nova Acta Regine Soc. Upsaliensis Ser IV, 20 /1967).

[d] Interpolated data point.

[e] Data for rare gases in solid phase.

July, 1974 - Shell Development Company - C. D. Wagner

Appendix II

Commercially Available Electron Spectrometers

Company	Type	Analyzer[a]	Characteristics
Associated Electrical Industries (AEI) Ltd., UK	XPS (UPS and Auger options)	HMA	Solids, surface compatible, monochromator option
DuPont Corporation, USA	XPS		Solids, not suitable for surface work, rapid loading, relatively cheap for an XPS instrument
Hewlett-Packard, USA	XPS	HMA	Monochromator standard, not fully surface compatible, rapid sample loading, expensive
Leybold-Hereaus, Germany	XPS (UPS and Auger options)	HMA	Specialist surface instrument, but gas version available
McPherson Corporation, USA	XPS (UPS option)	HMA	Solids, surface compatible, handles irregular-shaped samples, fully automated
McPherson Corporation, USA	ELS	HMA	Gas-phase instrument
Perkin-Elmer, UK	UPS	127° sector	Gas-phase instrument. Facility for vaporizing solids
Physical Electronics Industries, USA	AES (XPS option)	CMA	Specialist surface physics instrument. Options such as LEED and SIMS available
Varian Associates, USA	XPS	HMA	First commercial XPS instrument. Poor vacuum and sample-handling facilities. No longer in production
Varian Associates, USA	AES	CMA	Specialist surface instrument
Varian Associates, USA	AES/LEED	Retarding grids	Specialist surface instrument
V.G. Scientific, UK	UPS	HMA	Gas-phase or specialist surface instruments available
V.G. Scientific, UK	UPS	HMA	Angular resolved photoemission specialist instrument

Commercially Available Electron Spectrometers (*continued*)

Company	Type	Analyzer[a]	Characteristics
V.G. Scientific, UK	XPS	HMA	Specialist surface instrument. Gas-phase version available
V.G. Scientific, UK	XPS/UPS/AES	HMA	Specialist surface combination instrument. Options such as LEED and SIMS available
V.G. Scientific, UK	AES	HCMA	Specialist surface instrument

[a] HMA—hemispherical mirror analyzer; CMA—cylindrical mirror analyzer; HCMA—hemicyclindrical mirror analyzer. (With permission from C. R. Brundle and A. D. Baker, eds., "Electron Spectroscopy: Theory, Techniques, and Applications," Vol. 1, p. 19. Copyright by Academic Press, Inc. (London) Ltd.)

References

Bagus, P. S., Freeman, A. J., and Sasaki, F. (1973). *Phys. Rev. Lett.* **40**, 850.

Baker, A. D., and Betteridge, D. (1972). "Photoelectron Spectroscopy—Chemical and Analytical Aspects." Pergamon, Oxford.

Barrie, A., and Bradshaw, A. M. (1975). *Phys. Lett. A* **55**, 366.

Betteridge, D., Carver J. C., and Hercules, D. M. (1973). *J. Electron Spectrosc.* **2**, 327.

Bradshaw, A. M., Domcke, W., and Cederbaum, L. S. (1977). *Phys. Rev. B* **16**, 1480.

Briggs, D. (1977). "Handbook of X-ray and UV Photoelectron Spectroscopy." Heyden, London.

Brinen, J. S., and McClure, J. E. (1972). *Anal. Lett.* **5**, 737.

Brundle, C. R. (1975). *Surf. Sci.* **48**, 99.

Brundle, C. R., and Baker, A. D., eds. (1977–1978). "Electron Spectroscopy: Theory, Techniques, and Applications," Vols. 1–3. Academic Press, New York.

Burkstrand, J. M. (1977). *Am. Vac. Soc., Natl. Symp., 24th, Boston, Mass.*

Burroughs, P., Hamnet, A., Orchard, A. F., and Thornton, G. (1976). *J.C.S. Dalton* 1686.

Carlson, T. A. (1975). "Photoelectron and Auger Electron Spectroscopy." Plenum, New York.

Carter, G., and Colligan, J. S. (1968). "Ion Bombardment of Solids." Am. Elsevier, New York.

Caudano, R., and Verbist, J., ed. (1974). "Electron Spectroscopy: Progress in Research and Applications." Elsevier, Amsterdam; *J. Electron Spectrosc.* **5**.

Chaban, E. E., and Rowe, J. E. (1976). *J. Electron Spectrosc.* **9**, 329.

Chazalvicl, J. N., Campagua, M., Wertheim, G. K., and Schmidt, P. H. (1976). *Phys. Rev. B* **10**, 4586.

Citrin, P. H., Eisenberger, P., and Hamann, D. R. (1974). *Phys. Rev. Lett.* **33**, 965.

Citrin, P. H., Wertheim, G. K., and Baer, Y. (1978). *Phys. Rev. Lett.* **41**, 1425.

Clark, D. T. (1975). *In* "Electronic Structure of Polymers and Molecular Crystals" (J. M. Andre and J. Ladik, eds.), p. 259. Plenum, New York.

Clark, D. T., and Dilks, A. (1976). *J. Polym. Sci. Polym. Chem. Ed.* **14**, 533.

Clark, D. T., and Dilks, A. (1977). *J. Polym. Sci. Polym. Chem. Ed.* **15**, 2321.

Clark, D. T., Dilks, A., Peeling, J., and Thomas, H. R. (1975). *Faraday Discuss. Chem. Soc.* **60**, 183.

Colton, R. J., and Rabalais, J. W. (1975). *J. Electron Spectrosc.* **7**, 359.

D'Ageostino, S. (1975). *Hist. Stud. Phys. Sci.* **6**, 261.

Davis, D. W., and Shirley, D. A. (1974). *J. Electron Spectrosc.* **3**, 137.

Davis, D. W., Banna, M. S., and Shirley, D. A. (1974). *J. Chem. Phys.* **60**, 237.

Doniach, S., and Sunjic, M. (1970). *J. Phys. C* **3**, 285.

Ebel, H., Ebel, M. F., and Hillbrand, E. (1973). *J. Electron Spectrosc.* **2**, 277.

Einstein, A. (1905). *Ann. Phys. (Leipzig)* **17**, 132.

Evans, S. (1973). *Chem. Phys. Lett.* **23**, 139.

Fadley, C. S. (1976). *Prog. Solid State Chem.* **11**, 265.

Fadley, C. S., and Shirley, D. A. (1970). *Phys. Rev. A* **2**, 1109.

Fadley, C. S., Baird, R., Siekhaus, W., Novakov, T., and Bergstrom, S. A. L. (1974) *J. Electron Spectrosc.* **4**, 93.

Fleisch, T., Winograd, N., and Delgass, W. N. (1978). *Surf. Sci.* **78**, 141.

Gelius, U. (1974a). *Phys. Scr.* **9**, 133.

Gelius, U. (1974b). *J. Electron Spectrosc.* **5**, 611.

Gelius, U. (1974c). *J. Electron Spectrosc.* **5**, 985.

Gellender, M., and Baker, A. D. (1974). *J. Electron Spectrosc.* **4**, 249.

Guthrie, A. (1963). "Vacuum Technology." Wiley, New York.

Hammond, J. S., Gaarenstroom, S. W., and Winograd, N. (1975). *Anal. Chem.* **47**, 2193.

Hedman, J., Heden, P. F., Nordling, C., and Siegbahn, K. (1969). *Phys. Lett. A* **29**, 178.

Hercules, D. M. (1970). *Anal. Chem.* **42**, 20A.

Hercules, D. M. (1972). *Anal. Chem.* **44**, 106R.

Hercules, D. M. (1976). *Anal. Chem.* **48**, 294R.

Hercules, D. M., and Carver, J. C. (1974). *Anal. Chem.* **46**, 133R.

Hercules, D. M., Cos, L. E., Onisick, S., Nichols, G. D., and Carver, J. C. (1973). *Anal. Chem.* **45**, 1973.

Hewitt, R. W., Shepard, A. T., Baitinger, W. E., Winograd, N., and Delgass, W. N. (1979). *Rev. Sci. Instrum.* **50**, **1386**.

Holloway, P. H. (1975). *J. Electron Spectrosc.* **7**, 215.

Hoogewijs, R., Fiermans, L., and Vennik, J. (1976a). *Chem. Phys. Lett.* **38**, 471.

Hoogewijs, R., Fiermans, L., and Vennik, J. (1976b). *J. Microsc. Spectrosc. Electron.* **1**, 109.

Howat, G., and Goscinski, O. (1975). *Chem. Phys. Lett.* **30**, 87.

Huang, J.-T. J., Rabalais, J. W., and Ellison, F. O. (1975). *J. Electron Spectrosc.* **6**, 85.

Hufner, S., and Wertheim, G. K., (1975). *Phys. Rev. Lett.* **35**, 53.

Jenkin, J. G., Leckey, R. C. G., and Liesegang, J. (1977). *J. Electron Spectrosc.* **12**, 1.

Johansson, G., Hedman, J., Benndtsson, A., Klasson, M., and Nilsson, R. (1973). *J. Electron Spectrosc.* **2**, 295.

Jolly, W. L. (1974). *Coord. Rev.* **13**, 47.

Jorgensen, C. K. (1975). *Struct. Bonding (Berlin)* **22**, 49.

Kim, K. S. (1975). *Phys. Rev. B* **11**, 2177.

Kim, K. S., and Winograd, N. (1974a). *Surf. Sci.* **43**, 625.

Kim, K. S., and Winograd, N. (1974b) *J. Catal.* **35**, 66.

Kim, K. S., and Winograd, N. (1975a). *Surf. Sci.* **52**, 285.

Kim, K. S., and Winograd, N. (1975b). *Chem. Phys. Lett.* **30**, 91.

Kim, K. S., Gaarenstroom, S. W., and Winograd, N. (1976). *Phys. Rev. B* **14**, 2281.

Kirshnan, N. G., Delgass, W. N., and Robertson, W. D. (1976). *Surf. Sci.* **57**, 1.

Kowalczyk, S. P. (1976). Photoelectron spectroscopy and Auger electron spectroscopy of solids and surfaces. Ph.D. Thesis, Lawrence Berkeley Lab., Rep. No. LBL-4319.

Kowalczyk, S. P., Pollak, R. A., McFeely, F. R., Ley, L., and Shirley, D. A. (1973). *Phys. Rev. B* **8**, 2387.

Kowalczyk, S. P., Ley, L., McFeely, R. F., Pollak, R. A., and Shirley, D. A. (1974). *Phys. Rev. B* **9**, 381.

Krause, M. O., and Ferreira, J. G. (1975). *J. Phys. B* **8**, 2007.

Leckey, R. C. G. (1976). *Phys. Rev. A* **13**, 1043.

Lindau, I., Pianetta, P., Doniach, S., and Spicer, W. E. (1974). *Nature (London)* **250**, 214.

Mann, J. B. (1967). "Atomic Structure Calculations I. Hartree–Fock Energy Results for the Elements Hydrogen to Lawrencium," Los Alamos Sci. Lab., Rep. No. LASL-3690 (unpublished).

Martin, R. L., and Shirley, D. A. (1977). *In* "Electron Spectroscopy: Theory, Techniques, and Applications" (C. R. Brundle and A. D. Baker, eds.), Vol. 1, p. 76. Academic Press, New York.

Moore, C. E. (1949–1958). "Atomic Energy Levels," Natl. Bur. Std., Circ. No. 467, Vols. 1–3. U.S. Gov. Print. Off., Washington, D.C.

Nefedov, V. I., Sorgashin, N. P., Salyn, Y. V., Band, I. M., and Trzhaskovkaya, M. B. (1975). *J. Electron Spectrosc.* **7**, 175.

Norton, P. R., Tapping, R. L., and Goodale, J. W. (1977). *Surf. Sci.* **65**, 13.

Norton, P. R., Tapping, R. L., and Goodale, J. W. (1980). *Surf. Sci.* to be published.

Novakov, T. (1971). *Phys. Rev. B* **3**, 2693.

Palmberg, P. W., Bohn, G. K., and Tracy J. C. (1969). *Appl. Phys. Lett.* **15**, 254.

Pardee, W. J., Mahan, G. D., Eastman, D. E., Pollak, R. A., Ley, L., McFeely, F. R., Kowalczyk, S. P., and Shirley, D. A. (1975). *Phys. Rev. B* **11**, 3614.

Parry, D. E. (1975). *Surf. Sci.* **49**, 433.

Penn, D. R. (1976). *J. Electron Spectrosc.* **9**, 29.

Perlman, M. L., Rowe, E. M., and Watson, R. E. (1970). *Phys. Today* **27**, 30.

Picken, A. G., and Van Gool, W. (1969). *J. Mater. Sci.* **4**, 95.

Plummer, E. W., Salaneck, W. R., and Miller, J. S. (1978). *Phys. Rev. B* **18**, 1673.

Rutherford, E., and Robinson, H. (1913). *Philos. Mag.* **26**, 717.

Schön, G. (1973). *J. Electron Spectrosc.* **2**, 75.

Scofield, J. H. (1976). *J. Electron Spectrosc.* **8**, 129.

Sherwood, P. M. A. (1976). *J.C.S. Faraday II* **72**, 1791.

Shirley, D. A., ed. (1972). "Electron Spectroscopy." North-Holland Publ., Amsterdam.

Shirley, D. A., (1973). *Adv. Chem. Phys.* **23**, 85.

Siegbahn, H., and Goscinski, O. (1976). *Phys. Scr.* **13**, 225.

Siegbahn, H., and Siegbahn, K. (1974). *J. Electron Spectrosc.* **5**, 1059.

Siegbahn, H., Medeiros, R., and Goscinski, O. (1976). *J. Electron Spectrosc.* **8**, 149.

Siegbahn, K. (1974). *J. Electron Spectrosc.* **5**, 3.

Siegbahn, K. (1976). *Uppsala Univ. Inst. Phys., Rep.* UUIP- 940.

Siegbahn, K., Nordling, C., Fahlman, A., Nordberg, R., Hamrin, K., Hedman, J., Johansson, G., Bergmark, T., Karlsson, S.-E., Lindgren, I., and Lindberg, B. (1967). "ESCA: Atomic, Molecular, and Solid State Structure Studies by Means of Electron Spectroscopy," Nova Acta Regiae Soc. Sci. Ups., Ser. IV, Vol. 20. Almqvist & Wiksell, Stockholm. (Also available as Natl. Inf. Serv. Tech. Rep. No. AD 844315, Dep. Commerce, Springfield, Virginia, 1968.)

Siegbahn, K., Nordling, C., Johansson, G., Hedman, J., Hedèn, P.-F., Hamrin, K., Gelius, U., Bergmark, T., Werme, L. O., Manne, R., and Baer, Y. (1969). "ESCA Applied to Free Molecules." North-Holland Publ., Amsterdam.

Siegbahn, K., Hammond, D., Fellner-Felldogg, H., and Burnett, E. F. (1972). *Science* **176**, 245.

Slater, J. C. (1960). "Quantum Theory of Atomic Structure," Vol. 2. McGraw-Hill, New York.

Slater, R. R. (1970). *Surf. Sci.* **23**, 403.

Spicer, W. E. (1972). *In* "Optical Properties of Solids" (F. Abelés, ed.), p. 755. North-Holland Publ., Amsterdam.

Spicer, W. E., Yu, K. Y., Lindau, I., Pianetta, R., and Collins, D. M. (1976). *In* "Surface and Defect Properties of Solids" (J. M. Thomas and M. W. Roberts, eds.), Vol. V, p. 103. Chem. Soc., London.

Steinhardt, R. G., Jr. (1950). An X-ray photoelectron spectrometer for chemical analysis. Ph.D. Thesis, Lehigh Univ., Bethlehem, Pennsylvania.

Steinhardt, R. G., and Surfass, E. J. (1953). *Anal. Chem.* **25**, 697.

Swingle, R. S., and Riggs, W. M. (1975). *CRC Crit. Rev. Anal. Chem.* **5**, 267.

Theriault, G. A., Barry, T. L., and Thomas, M. J. B. (1975). *Anal. Chem.* **47**, 1492.

Turner, D. W., Baker, C., Baker, A. D., and Brundle, C. R. (1970). "Molecular Photoelectron Spectroscopy." Wiley (Interscience), New York.

Wagner, C. D. (1975). *Faraday Discuss. Chem. Soc.* **60**, 291.

Wagner, C. D. (1977a). *Anal. Chem.* **49**, 1282.

Wagner, C. D. (1977b). *J. Electron Spectrosc.* **10**, 305.

Wagner, C. D., and Bileon, P. (1973). *Surf. Sci.* **35**, 82.

Wagner, C. D., Gale, L., and Raymond, R. (1979). *Anal. Chem.* **51**, 466.

Wims, A. M., and Schreiber, T. P. (unpublished).

Yasuda, H., Marsh, H. C., Brandt, E. S., and Reilley, C. N. (1977). *J. Polym. Sci. Polym. Chem. Ed.* **15**, 991.

Yates, J. T., Jr., Madey T. E., and Erickson, N. E. (1974). *Surf. Sci.* **43**, 257.

PHYSICAL METHODS IN MODERN CHEMICAL ANALYSIS, VOL. 2

X-Ray Diffraction Methods
Applied to Powders and Metals

William L. Davidson

Dunedin, Florida

171

I. Introduction

X rays can assist the chemical analyst in three major ways. The first involves beam absorption, whose utility depends on the finding that such rays are absorbed exponentionally while traversing matter and the fact that absorption coefficients vary widely with element and wavelength. Second, fluorescent (characteristic) x rays emitted by an irradiated sample can be measured to identify the nature and amount of constituent *elements* present. Finally, x rays diffracted by crystalline materials produce unique "fingerprints" (diffraction patterns) which may be analyzed to deduce both qualitatively and quantitatively the *structures* present. This chapter is dedicated to describing techniques and applications which employ the x-ray diffraction phenomenon in analyzing powders and metals.

When this material appeared first (Davidson, 1950), practically all powder diffraction patterns were recorded photographically. The diffractometer had been invented, but its name had not. Until 1952 it was known simply as an x-ray spectrometer. By the time the second edition came out (Davidson, 1960), real competition had developed between camera techniques and the diffractometer. Each had, and still possesses, certain merits as well as admitted disadvantages. And today, while the diffractometer is clearly the most popular, we have yet a third contender vying for honors, the energy-dispersive diffraction technique. This latter procedure is an outgrowth of the remarkable advances made in solid state and nuclear physics during the past decade.

No discussion of x-ray diffraction as applied to powders and metals can ignore any of these three major tools, since each has carved out its own special niche in the analytical arena. But it is patently impossible to do full justice to all three in space originally devoted to only one. Thus, while the author will endeavor to cover all three adequately, conciseness must be emphasized and liberal use will be made of appropriate references, which may be studied to provide the reader more complete information on the specific topic under discussion.

A. *Advantages of the Powder Method in Analysis*

X-ray diffraction can be a powerful aid in chemical analysis. It does not alter the sample and thus is nondestructive; it reveals the chemical structures present and, with some additional effort, the concentration of each; it can disclose a wealth of information on crystallite orientation, overall sample texture, and grain size; it can serve to evaluate residual stress in macrostructures, again nondestructively, and frequently as little as 1 mg of material can suffice as an adequate sample. Of equal importance, the interpretation of results is for the most part reasonably straightforward. Though more expensive than a camera setup, the diffractometer, and to a lesser degree energy-dispersive apparatus, have expanded the versatility and productivity of the diffraction technique enormously. Multiple sample holders and the addition of automating software make 24 hour a day operation possible with minimum operator attention. In raising the automation feature, two distinct modes of operation should be distinguished. In a so-called "hard-wired" setup, the desired program is essentially fixed, and once initiated the machine follows these orders as directed to completion. In a more sophisticated operational mode, a minicomputer becomes an integral part of the system. Such a computer is termed "on-line" to distinguish it from "off-line" use of computers for the analysis of stored data. An on-line system is under the command of the diffractionist. Changes in the program can be ordered while the experiment is in progress and the data being generated may be studied visually on a cathode ray tube as it develops; this data may be added to or subtracted from other data in the memory; data may be shifted from one section of the storage unit to another and generally manipulated in sundry ways to provide quick results. Many laboratories have found that the added cost of this complex equipment represents a worthwhile tradeoff in terms of the gain in efficiency and increased productivity balanced against manpower costs. For the modern chemical laboratory, x-ray diffraction can prove an almost indispensable tool.

B. Production of X Rays

X rays are electromagnetic waves similar in character to light but of much shorter wavelength. They are generated when high velocity electrons impinge on a target. Standard commercial diffraction tubes can be obtained with the following eight targets: tungsten (W), silver (Ag), molybdenum (Mo), copper (Cu), nickel (Ni), cobalt (Co), iron (Fe), and chromium (Cr). Figure 1a illustrates the basic construction of a sealed-off tube suitable for x-ray diffraction and Fig. 1b is a photograph of such a tube.

If one were to examine the x rays emanating from the Mo target of Fig. 1a he would discover that the intensity varied with wavelength as shown in Fig. 2, curve (a).

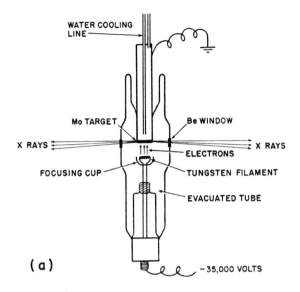

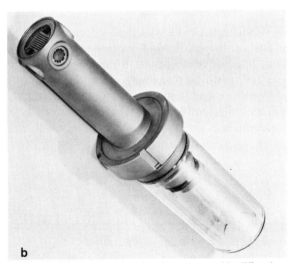

Fig. 1   (a) Sketch of molybdenum x-ray tube of the type used in diffraction work. Electrons from heated tungsten filament bombard the Mo target and generate x rays. The latter emerge through low absorption beryllium windows. Table I lists other metal targets available. (b) Photograph of a typical commercial diffraction tube. Photo courtesy of Diano Corp.

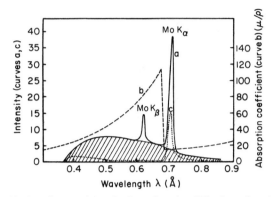

Fig. 2   Curve (a) plots the x-ray intensity from the tube of Fig. 1 as a function of wavelength. Curve (b) shows how the mass absorption coefficient of zirconium varies with wavelength. Curve (c) illustrates the filtering effect of 0.008 cm of zirconium on the radiation. The Mo K_β and white radiation are substantially removed from the beam, whereas almost half of the Mo K_α intensity remains. This provides a beam sufficiently monochromatic for most powder work.

C.   Types of X Radiation

Two kinds of x radiation are distinguished, depending on the mechanism of production. The shaded region under curve (a) comprises the *white* or *general* radiation. This radiation results from interaction of the electrons with target nuclei and, as will be seen later, is important when one is employing the energy-dispersive technique. It is present irrespective of applied voltage or target materials, although the shape of the curve changes when either of these quantities is varied. In fact, there is always a sharp cutoff at the wavelength λ_{min} given by the equation

$$\lambda_{min} = \frac{12,396}{V_0} \tag{1}$$

where V_0 is the potential difference across the tube in volts and λ_{min} will be expressed in units of 10^{-8} cm or angstroms (Å).

If the voltage is increased beyond a certain critical value, sharp radiation peaks begin to appear, superposed on the white spectrum. These peaks constitute the *characteristic* x rays, since they are associated with the target element. From the Bohr picture of an atom, the critical voltage to generate these rays is that which gives a bombarding particle (electron here) sufficient energy to eject one of the shell electrons from the target atom. When this happens, an electron from one of the outer shells immediately negotiates a jump, replacing the dislodged electron and emitting a quantum of radiation.

TABLE 1

Properties of Radiation Most Used in Diffraction Work

Target	Minimum excitation potential (V)	Characteristic wavelengths (Å)			Filter to remove K_β	$\bar{K}_\beta$	Approximate filter thickness (cm)
		K_{α_1}	K_{α_2}	$\bar{K}_\alpha = \frac{1}{3}(2K_{\alpha_1} + K_{\alpha_2})$			
Ag	25 500	0.5583	0.5627	0.5597	Pd	0.4960	0.0013
W	59 000	0.2086	0.2135	0.2109	Hf	0.1842	0.0013
Mo	20 000	0.7093	0.7135	0.7107	Zr	0.6323	0.0080
Cu	8100	1.5405	1.5443	1.5418	Ni	1.3922	0.0020
Ni	7500	1.6578	1.6617	1.6591	Co	1.5001	0.0020
Co	6900	1.7889	1.7928	1.7902	Fe	1.6207	0.0015
Fe	6400	1.9360	1.9399	1.9373	Mn	1.7565	0.0013
Cr	5400	2.2896	2.2935	2.2909	V	2.0848	0.0013

If the initial electron is ejected from the K shell and the "jumping" electron comes from the L shell, the radiation is termed K_α; if the jumping electron arises from the M shell, the radiation is termed K_β, etc.

There are two L electrons in slightly different energy states which may fill a vacancy in the K shell. Consequently, the K_α radiation is really a close doublet, $K_{\alpha 1}$ and $K_{\alpha 2}$. For most powder diffraction work this doublet is not resolved and a weighted average wavelength of the two used. However at the larger diffraction angles and with high resolution techniques, this pair may be resolved and show up as a doublet.

Table I records the $K_{\alpha 1, \alpha 2}$ and $\bar{K}_\alpha$ (average) and $\bar{K}_\beta$ wavelengths plus other pertinent data for the eight target elements most used in diffusion work.

Now, as the voltage is increased above the critical value V_0, the characteristic x-ray intensity increases approximately as $(V - V_0)^{1.7}$. Thus, although $V_0 \approx 20\,000$ V for Mo K, one normally uses considerably higher voltages in order to increase the characteristic output. For most photographic and diffractometer powder work a monochromatic beam is needed. The simplest way of achieving this is to introduce a selective filter into the beam where it emerges from the tube. This is illustrated in Section VI.A (Figs. 16 through 19).

D. How a Selective Filter Produces a Monochromatic Beam

When x rays travel through matter, the intensity is reduced, partly by scattering but mainly by a transformation of x-ray energy into that of moving electrons (photoelectric effect). The fraction I/I_0 of the original beam penetrating any thickness x of material is given by the well-known expression

$$I/I_0 = e^{-\mu x} \equiv e^{-(\mu/\rho)(\rho x)} \tag{2}$$

where μ is the linear absorption coefficient and ρ is the density of the absorber. The *mass* absorption coefficient μ/ρ is often preferred, since it is independent of the physical (solid, liquid, gas) state of the material. Curve (b) (dashed line in Fig. 2) shows how μ/ρ for Zr varies with x-ray wavelength. The important feature of this plot is the sharp break at around 0.69 Å. This is known as the K absorption edge for Zr. X rays with a wavelength shorter than this limit have energy ($E_{X\,ray} \propto 1/\lambda$) sufficient to knock electrons out of the Zr K shell. In this way the initial beam is effectively attenuated since the resulting secondary Zr K x rays are emitted equally in all directions. Owing to the phenomenon of resonance, the chance of ejecting a K electron decreases slowly as the x-ray energy increases beyond the threshold value. Conversely, an x-ray quantum with $\lambda > 0.69$ Å has insufficient energy to remove Zr K electrons and the absorption for such quanta plummets to a low value. Since the Zr K absorption edge falls between Mo K_α and Mo K_β, and inasmuch as μ/ρ appears as an exponent, it is clear that a thin Zr foil will pass a goodly fraction of Mo K_α radiation while absorbing most of the Mo K_β and white radiation. Curve (c) (dotted line) in Fig. 2 depicts what remains of the curve (a) radiation after passing through a 0.008-cm Zr foil. The white and Mo K_β are scarcely visible whereas almost half of the Mo K_α remains. For most powder work this beam is adequately monochromatic. Crystal monochromators (for precise camera work) and a pulse height discriminator or selector (diffractometer) may be resorted to in instances demanding the ultimate in wavelength purity.

II. X-Ray Diffraction

A. *Why Matter Diffracts X Rays*

(a) X-ray wavelengths are the same order of magnitude as the spacing of atom centers in matter. This is a necessary condition for interference among electromagnetic waves.

(b) Many solids are constructed from a relatively simple assemblage of components (atoms or atomic groupings) repeated over and over again at regular intervals in three dimensions. This is equivalent to stating many solids are crystalline.

Both of these concepts were known long before 1912. But in that year the brilliant German physicist von Laue merged these two ideas and predicted from mathematical considerations that when a beam of x rays traversed crystalline matter constructive interference should occur, splitting the beam into a number of secondary components, each component being emitted in a direction determined by the incident beam and structure of the crystal. In brief, this states that x rays should be diffracted by crystalline materials. This prediction was verified experimentally a year later and thus was born a new means for probing and elucidating the structure of matter. Von Laue's mathematical arguments were quite complex; so much so that the full significance of his analysis was not appreciated immediately. It remained for the Braggs in England to show that equivalent results could be expected by assuming that x rays are *reflected* by sets of parallel planes slicing through the crystal being irradiated. The planes are always chosen so they contain a heavy population of atom centers.

B.   The Bragg Equation

Consider a two-dimensional array of atom centers as depicted in Fig. 3 with x radiation incident at angle θ to the plane PP' and all points on the wavefront AB are in phase. On the Bragg picture I_1 and I_2 will be reflected as shown. In order that points A and D be in phase after reflection, it is evident that BC + CD must equal some multiple of the x-ray wavelength; i.e., BC + CD = $n\lambda$. Now AB and AD are perpendicular to I_2C and CI_2', respectively, and AC is perpendicular to PP'. Therefore $\angle$ BAC = $\angle$ CAD = θ; thus BC = CD = $d \sin \theta$. Hence, the condition for constructive reflection is

$$n\lambda = 2d \sin \theta \qquad (3)$$

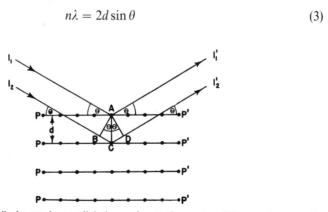

Fig. 3 "Reflection" of x rays by parallel planes of scattering centers. When electromagnetic radiation traverses an atomic array, each electron present becomes a source of secondary radiation. The result may be pictured as a "reflection" of the x rays by the atomic planes. Unlike an ordinary mirror, these planes reflect rays only for particular angles of incidence.

This is the famed Bragg equation and is the fundamental relation underlying all x-ray diffraction measurements. It is important to remember that only for an angle of incidence such that $\sin \theta = n\lambda/2d$ will x rays be reflected. At all other angles destructive interference will occur. Also, since $\sin \theta \leq 1$, it is clear that $n\lambda$ can never be greater than $2d$. For example, if we are dealing with atomic planes spaced 1 Å apart ($d = 1$ Å), no reflection of higher order than the second ($n = 2$) can be obtained with Mo K_α radiation.

In Eq. (3) to calculate interplanar spacings, one usually knows n and λ and has only to measure θ. In powder work the diffracted rays normally are allowed to fall on a photographic plate. From the resulting pattern and camera geometry θ can be readily determined for every reflection.

III. Elements of Crystal Structure

It would be quite erroneous to think that our knowledge of crystal structure dates only from the discovery of x-ray diffraction. Crystallographers had made remarkable strides in this direction long before the discovery of x rays, even though their observation were limited to the external, optical, and cleavage features of crystals. X rays in essence permitted one to look inside the crystal and thus provided a potent tool for studying ultimate structures. Before x rays can be used for this purpose, some knowledge of crystallography is necessary. To attain this end, several basic concepts must be mastered. These include the following: (1) space lattice, (2) unit cell, (3) crystal systems, (4) elements of symmetry, (5) point groups, (6) space groups, and (7) Miller indices. In the following discussion we propose to consider these concepts in the order listed.

A. Space Lattice

Were we able to see the ultimate structure of a simple crystal composed of like atoms, a small fragment might appear somewhat as pictured in Fig. 4a. If we should now plot the position in space of each atom center, a series of points would be obtained as shown in Fig. 4b. This array of points constitutes the *space lattice* for our crystal. A space lattice is defined as a series of points, formed by the intersection of three sets of parallel planes, which locates equivalent positions in the crystal. When more than one kind of atom is present, e.g., sodium chloride, each type would possess its own space lattice. Sometimes even for crystals containing but one atom type, certain of the atoms fit one lattice whereas others require a different one. In either event the combination of separate lattices will then determine the symmetry properties

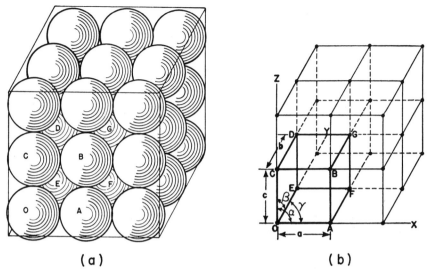

(a)   (b)

Fig. 4 (a) Small section of a hypothetical crystal of like atoms, magnified about one hundred million times. The sketch is purely for illustrative purposes; no *element* actually crystallizes in this fashion, although it could be thought of as representing the structure of KCl, since K^+ and Cl^- have the same scattering power for x rays. (b) A plot showing the location of the centers for the atoms in (a). These points (ignore the connecting lines) comprise the space lattice for the crystal. The parallelepiped set off by heavy lines constitutes the unit cell.

of the crystal. There are many ways in which planes may be erected to form our particular lattice, but the simplest is as shown in Fig 4b, i.e., planes parallel to OCDE, OABC, and OAFE. In the case we are considering, each lattice point corresponds to an atom center. This is not essential. Sometimes we find many atoms grouped around a lattice point. This can be seen in Fig. 5, which depicts a projection of the unit cell for crystalline benzene.

Fig. 5 Projection of a unit cell for crystalline benzene. The plane of all benzene molecules is perpendicular to the plane of the paper, so only three carbons are visible. For simplicity the hydrogen atoms are not shown. The shaded molecules are displaced into the paper one half the length of a unit cell. Note that a whole benzene molecule surrounds each lattice point.

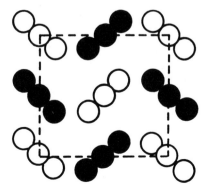

The only requirement is that each point represents an equivalent position in the crystal.

B.   Unit Cell

The volume bounded by the labeled points in Fig. 4b is known as the *unit cell*, since it represents the smallest "brick" which, by repetition of itself in all directions, will build the crystal.

C.   Crystal Systems

It is clear that the lattice pictured can be referred to lines OX, OY, and OZ. These are termed the *crystallographic axes*. The system to which a crystal belongs is determined by the angles, α, β, and γ, formed by these axes and the relative lengths of the *primitive translations a, b*, and *c*. It has been found that all crystals can be fitted to one of just seven systems of coordinates. These represent the seven major divisions into which crystals are classified. Table II records these systems and the properties of each. Some writers recognize only six systems, since by a transformation of axes rhombohedral crystals can be referred to hexagonal coordinates.

TABLE II

Crystal Systems

System	Axes	Angles
Triclinic	$a \neq b \neq c$	$\alpha \neq \beta \neq \gamma \neq 90°$
Monoclinic	$a \neq b \neq c$	$\alpha = \gamma = 90°; \beta \neq 90°$
Orthorhombic	$a \neq b \neq c$	$\alpha = \beta = \gamma = 90°$
Tetragonal	$a = b \neq c$	$\alpha = \beta = \gamma = 90°$
Hexagonal	$a = b \neq c$	$\alpha = \beta = 90°; \gamma = 120°$
Rhombohedral	$a = b = c$	$\alpha = \beta = \gamma \neq 90°$
Cubic	$a = b = c$	$\alpha = \beta = \gamma = 90°$

It is seen that the crystal considered above belongs to the cubic system, since $\alpha = \beta = \gamma = 90°$ and $a = b = c$. The arrangement of points which represents our crystal is termed a *simple* cubic lattice to distinguish it from two other possible space lattices having cubic symmetry. All three of the cubic space lattices are depicted in Fig. 6. In addition to the simple cubic arrangement the body-centered and face-centered lattices exist, so called because of additional lattice points appearing at the cube center and each face center, respectively. Note that, although the simple cubic cell contains eight lattice points, each must be shared equally with seven other cells

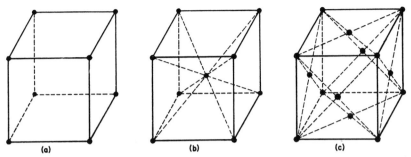

Fig. 6 A unit cell for each of the three cubic space lattices: (a) simple cubic; (b) body-centered cubic (b.c.c.); (c) face-centered cubic (f.c.c.).

meeting at the point. Thus the simple cube possesses but one complete point per unit cell. Similarly we notice that the body-centered cubic cell has two points (the body-centered point is not shared), whereas the face-centered cube has four points per unit cell (each of the six face-centered points is shared equally with one other cell). This feature will prove important later for checking unit cell determinations. All told, fourteen space lattices are geometrically possible. Each can be referred to one of the seven coordinate systems.

D. Crystal Symmetry

The symmetry of any object is an expression of the fact that the object has identical properties in different directions. For example, the cross-hatched areas in Fig. 7 resume their initial appearance after a rotation around O of 90°, 180°, 270°, or 360°. In crystal terminology O would be called an axis of symmetry. When we consider figures in space, there are three types of symmetry operation which may be performed to bring equivalent points into

Fig. 7 Diagram to illustrate the concept of symmetry.

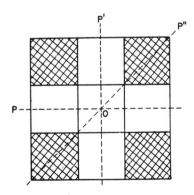

coincidence. These are termed symmetry elements and include:

(1)   Axes of symmetry. Points in crystals may have two-, three-, four-, or sixfold axes. This implies rotation of 180°, 120°, 90°, or 60° to bring equivalent points into coincidence. As previously noted, our two-dimensional diagram, Fig. 7, has a fourfold axis through O.

(2)   Plane of symmetry. Here points on one side of the plane are mirror images of points on the other side. P, P′, and P″ represent planes of symmetry for the cross-hatched areas.

(3)   Center of symmetry. For this to hold true every point in the crystal must be matched by a corresponding point such that the line joining the two is bisected by the center of symmetry; O is such a center for Fig. 7.

E.   Point Groups

These operations can be used to describe the symmetry properties of crystals. Imagine a perfect cube. It has the following symmetry elements (these can be demonstrated quite readily by reference to an actual cube): (1) three fourfold axes, (2) four threefold axes, (3) six twofold axes, (4) three reflection planes (perpendicular to the fourfold axes), (5) six reflection planes (perpendicular to the twofold axes), and (6) center of symmetry.

These elements collectively are termed the *point group* for a cube. It will also be found that each of the cubic lattices in Fig. 6 possesses this high symmetry. It is not essential, however, that *every* cubic crystal have *all* these symmetry elements. This fact can be ascertained by analogy, from consideration of the two-dimensional plots in Fig. 8. Here we are in effect viewing the projection on a plane of a simple cubic lattice [Fig. (33a)] having atom clusters grouped around the lattice points. Figure 8b has all the symmetry properties associated with a square. This would correspond to the perfect cube in three dimensions. Figure 8c lacks several of the symmetry elements found in Fig. 8b (the planes of symmetry have disappeared). Yet this crystal has identical properties along X and Y and can be referred to a square lattice

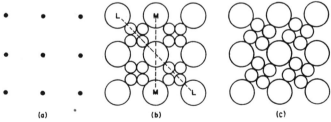

Fig. 8   Both (b) and (c) belong to the square lattice of (a). The planes of symmetry LL and MM in (b) are absent in (c), however. (From Bragg and Bragg, 1934.)

just as well as the grouping in Fig. 8b. If a crystal has the full symmetry of the system to which it belongs, it is called *holohedral*; if it possesses fewer symmetry elements, it is called *hemihedral*.

All seven crystal systems have been analyzed with regard to the possible symmetry elements they can accommodate. Such an analysis leads to a total of 32 different point groups, and every known crystal can be assigned to one of these 32 *crystal classes*, as they are called. These are divided among the crystal systems so that each system has at least the symmetry elements shown or their equivalent. (See the accompanying tabulation. The number of classes associated with each system is also included.)

System	Minimum symmetry	No. of classes
Triclinic	None	2
Monoclinic	A twofold rotation axis	3
Orthorhombic	Two reflection planes perpendicular to each other	3
Tetragonal	A fourfold rotation axis	7
Rhombohedral	A threefold rotation axis	5
Hexagonal	A sixfold rotation axis	7
Cubic	Four threefold rotation axes (along the cube diagonal)	5

Crystallographers are able to make class assignments through studies on face angles and certain physical properties of the crystals. It must be emphasized that the point group assignment is basically a macroclassification. The ideas of space lattices and atoms as ultimate units do not enter here, even though we have used them to illustrate the consequences of symmetry operations. Real progress in deducing ultimate crystal structure was delayed until it was realized that *atomic arrangement* is the *sine qua non* of a crystal, its outward appearance being an unimportant consequence of this. It is a fact that the atomic arrangement bears little relation to the symmetry. Crystals closely related in their atomic groupings and bonding may differ widely in symmetry. As an example of this, it is found that benzene crystallizes in the orthorhombic system whereas hexamethylbenzene fits the triclinic system of much lower symmetry. It is by contributing information on the true arrangement of atoms in matter that x rays have proved invaluable.

F. Space Groups

By its very nature the point group idea has a limited utility. By definition all the symmetry elements of each point group are associated with a central point, and translation is impossible. In order to describe the true atomic structural relationships in crystals a further concept is necessary. It remained for Federov, Schoenflies, and Barlow, working independently (circa 1890), to supply the missing link. Their contribution, incorporating the theory of *space groups*, combined the 32 classes of symmetry around a point with translation in three directions to other equivalent points. This successfully merged the point group and space lattice. In so doing, new symmetry operations occur. These are the *screw axis* and the *glide plane*. The former rotates a structure $360°/n$, and at the same time translates it parallel to the axis. This operation, repeated n times, is equivalent to a translation of the space lattice. The glide plane operation entails a reflection across the plane plus a translation parallel to the plane. These new operations are possible because each point of the structural unit need never return to its original position (as is required of a point group) but only to a similar position in another cell of the space lattice. The space group may be defined as a set of self-consistent symmetry operations extending in space and associated with a space lattice. Any symmetry operation or lattice translation brings all the remaining symmetry operations into self-coincidence. In all, 230 different space groups exist. Several notation schemes have been proposed to identify the various space groups. These may be found in any standard work on crystallography (e.g., Davey, 1934).

Since the translations associated with a screw axis are of atomic dimensions, it is a fact that the finished crystal appearance does not distinguish between a mere rotation axis and a screw axis. The same situation obtains for a glide plane versus a straight reflection. To illustrate the importance of this, let us focus our attention on a diamond crystal. It shows externally the highest cubic symmetry. Therefore, one might naively demand a unit cell showing such symmetry. The actual situation is quite different, however. Looking down on a unit cell (Fig. 9), we see that the A carbon atoms occupy the top (and bottom) layers, B atoms comprise the second level, C atoms the

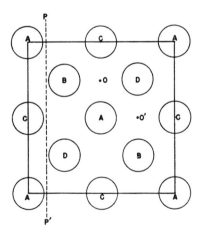

Fig. 9 Projection of the diamond unit cell to illustrate the function of screw axes O (clockwise), O′ (counterclockwise), and glide plane PP′.

middle layer, etc. Obviously this structure has no fourfold rotation axis, but O and O′ represent fourfold screw axes, clockwise and counterclockwise, respectively. These turn A into B, B into C, C into D, in a spiral manner. Likewise, there is no reflection plane parallel to a cube face, but PP′ is a glide plane. A reflection across PP′ together with motion downward a quarter cell length, parallel to PP′ and at 45° to the cell edge, makes A coincident with B, B with C, C with D, etc. Therefore, it is only by means of the space group concept combined with information obtained from x-ray diffraction that the basic structures of diamond and hundreds of other crystals, organic and inorganic, have been elucidated.

An ultimate structure determination normally requires data secured from single crystals by optical, goniometric, and x-ray diffraction methods. Since we are confining ourselves substantially to the powder method of analysis, the possibility of finding a complete structure will not usually arise. Even so, we shall see that powder patterns can provide us with a wealth of information having unique value to the fields of chemical and metallurgical analysis.

G. Miller Indices

Consider the unit cell for a face-centered lattice depicted in Fig. 10. In addition to the faces there are numerous other planes, e.g., ABYX, CDEF, XYZ, which may be passed through this cell, cutting lattice points. When extended through neighboring cells, each plane will cut corresponding lattice

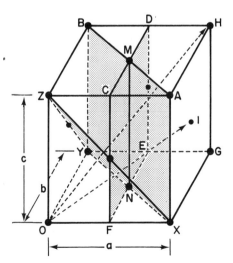

Fig. 10 Plot of unit cell for a face-centered structure to provide an understanding of nomenclature for planes and directions within a crystal.

points. In addition, each plane will have associated with it a family of equally spaced parallel planes. It should be reemphasized that the points contained in these planes are not the atoms themselves but represent equivalent points around which atoms are grouped. Nevertheless, we can assume that these planes will reflect x rays, since the atomic arrangement around each point is symmetrical. For this reason it is important that we possess a notation that will allow us to index these planes in a logical manner. The scheme actually employed involves the use of so-called *Miller indices*. The planes are designated by the reciprocals of the fractional intercepts on the axes XYZ of the unit cell. For example, CDEF cuts X at $\frac{1}{2}a$ and is parallel to Y and Z. Hence the intercepts are $\frac{1}{2}$, ∞, ∞, and the reciprocals 200. Thus we identify CDEF as a two-zero-zero (200) plane, the figures being enclosed by parentheses. Similarly ABYX is seen to be a (110) plane, and XYZ a (111) plane. It is evident that the cell faces have indices (100), (010), and (001). The general expression is written (hkl), h being the reciprocal of the fractional intercept on the X axis, k and l being similar quantities for the Y and Z axes. If any of the reciprocals should be fractional, (hkl) is taken as the smallest set of integers which leaves the relative values unchanged. For example, if the reciprocals are $\frac{3}{2}$, $\frac{4}{3}$, $\frac{1}{2}$, the plane is written (983). If the intercepts of a plane on the axes are negative, then the indices of that plane are negative. Convention requires that the minus sign be placed above the number. For example, a plane striking the axes at $-\frac{1}{3}a$, $\frac{1}{2}b$, $1c$, would be written ($\overline{3}21$).

If a particular line has components on the XYZ axes of ua, vb, wc, respectively, its *direction* is written [uvw], consistent with the requirement that u, v, w have minimum integral values. For example, in Fig. 10, OI has components $1a$, $\frac{1}{2}b$, $\frac{1}{2}c$. Hence direction OI is $[1, \frac{1}{2}, \frac{1}{2}]$ or more properly [211]. Similarly OH = [111], HO = [$\overline{111}$]. In the cubic system a direction is always normal to the set of planes having identical indices; e.g., [111] is normal to (111), etc. Certain sets of crystal planes will intersect along a line or parallel lines. This line is termed a *zone axis* (such as MN of Fig. 10), and all planes which pass through this or parallel lines constitute a *zone*. Like a crystal direction, the zone axis is designated by indices in square brackets. MN (and parallel lines such as OZ, BY, DE) form the [001] zone axis, and planes ABYX, CDEF, AZOX, AHGX, etc., all are a part of the [001] zone. The essential distinction between a direction and a zone axis is that the former passes through the origin, whereas the latter may not.

Special mention should be made regarding the notation for planes in the hexagonal system. By definition this system has two axes, X and Y, at an angle of 120°, with a third axis perpendicular to the plane of the first two. To take full advantage of the symmetry inherent in the hexagonal system, a fourth axis, W, is often added. W lies in the XY plane and makes an angle of 120° to both X and Y. The (hkl) indices for a plane are now replaced by

($hkil$), called Miller–Bravais indices; i refers to the intercept on W and is always equal to $-(h + k)$. To illustrate this point, the conventional (010) plane becomes ($01\bar{1}0$) in the new notation. Since i never enters into the calculation of interplanar spacings, it is often replaced by a dot; i.e., ($01\bar{1}0$) becomes ($01 \cdot 0$).

IV. Formulas for Interplanar Spacings

The spacing between adjacent planes of a set having indices (hkl) is usually written d_{hkl}. It is this d that occurs in the Bragg equation $n\lambda = 2d \sin \theta$. A certain simplification results if one includes n in the d. In this way every diffraction line can be thought of as a first-order reflection. For instance, one would interpret a second-order reflection from the (111) set of planes as a first-order 222 reflection. Note that no parentheses surround the 222. This enables true planes to be distinguished from reflections. Obviously the above point of view requires that $d_{222} = \frac{1}{2}d_{111}$.

Now d_{hkl} is a function of the unit cell parameters $a, b, c, \alpha, \beta, \gamma$. Let us derive an expression for d, limiting ourselves to the case where $\alpha = \beta = \gamma = 90°$. This expression will be adequate for the cubic, tetragonal, and orthorhombic systems.

In Fig. 11 we choose a plane (hkl) such that the neighboring plane of this set cuts the origin. Then OA, a normal to the plane represents the interplanar

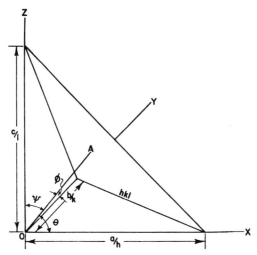

Fig. 11   OA represents the interplanar spacing d for the set (hkl). θ, ϕ, and ψ are the angles OA makes with X, Y, and Z, respectively. If the axes are mutually perpendicular, d can be readily evaluated in terms of a, b, c, h, k, l.

spacing d_{hkl}. It can be seen that

$$d_{hkl} = \frac{a}{h} \cos \theta = \frac{b}{k} \cos \phi = \frac{c}{l} \cos \psi \qquad (4)$$

Now if $\cos \theta$, $\cos \phi$, and $\cos \psi$ are the direction cosines of the plane normal referred to the coordinate axes, it is true that

$$\cos^2 \theta + \cos^2 \phi + \cos^2 \psi = 1 \qquad (5)$$

Squaring the expressions in Eq. (4) and putting them in Eq. (5) gives

$$d^2 \left(\frac{h^2}{a^2} + \frac{k^2}{b^2} + \frac{l^2}{c^2} \right) = 1 \qquad (6)$$

or

$$d = 1 \bigg/ \left(\frac{h^2}{a^2} + \frac{k^2}{b^2} + \frac{l^2}{c^2} \right)^{1/2} \qquad (7)$$

for the cubic system $a = b = c$, and Eq. (7) reduces to

$$d = a/(h^2 + k^2 + l^2)^{1/2} \qquad (8)$$

In tetragonal crystals $a = b$, and we find

$$d = a \bigg/ \left(h^2 + k^2 + \frac{l^2 a^2}{c^2} \right)^{1/2} \qquad (9)$$

The expression for d in the hexagonal system is

$$d = a \bigg/ \left(\frac{4}{3} (h^2 + hk + k^2) + \frac{l^2 a^2}{c^2} \right)^{1/2} \qquad (10)$$

In the other crystal systems the equations for d are much more involved. It is fortunate that most solids crystallize in systems possessing high symmetry, for these are the ones most amenable to study by the powder method.

V. X-Ray Diffraction Technique

A. Major Methods and Equipment

Selecting the most appropriate x-ray diffraction apparatus is much like buying an automobile. In both cases there are numerous makes, models, and prices from which to choose. And just as one can purchase an automobile powered by a piston engine (gasoline or diesel), rotary engine, or even rechargable batteries, so can one acquire an x-ray diffraction unit employing film cameras, a diffractometer utilizing various kinds of detectors and data storage and output devices, or a no-moving-parts energy-dispersive machine involving a solid-state detector-multichannel analyzer, the latter boasting a choice of display, storage, and recording devices. Carrying the automobile analogy one step further, optional accessories for diffraction units are equally numerous as for motor vehicles. Thus there can be no one "best" choice in either case. It will depend on the user's needs and available finances. In the following pages we will describe all three major techniques, emphasizing the strengths and shortcomings of each. For the reader who would wish to go beyond our necessarily abbreviated descriptions, we strongly recommend making use of the following source material.

Each year the American Chemical Society journal *Analytical Chemistry* publishes a special "Laboratory Guide" issue which carries a comprehensive alphabetical listing of chemical laboratory equipment and supplies. Under each heading is listed the manufacturer and supplier. In a separate section the names and addresses (also in alphabetical order) for all these firms appear. The American Association for the Advancement of Science magazine *Science* renders a similar service annually under the name: "Guide to Scientific Instruments." For example, reference to a recent guide revealed eight firms offering x-ray film, six under the heading "x-ray film processing chemicals", ten under "x-ray diffraction cameras," ten under the heading "x-ray diffractometers," and 17 producers of x-ray tubes (not all of these firms offer *diffraction* tubes however). The wealth of literature available from these sources will be most useful in helping the prospective user find equipment best suited to his requirements and purse.

B. X-Ray Generation

The generation of x rays for diffraction work is a step common to all three techniques noted previously. The high voltage supply is normally an oil-immersed transformer having an output around 60 kV. The primary is usually fed through an autotransformer, permitting a wide choice of secondary voltage. The transformer voltage may be applied to the tube in various ways. The four most common are depicted in Fig. 12. With arrangement (a) or (b), one pulse of x rays is produced per ac cycle. Methods (c) and (d) require additional rectifier tubes but yield two x-ray pulses per cycle. This effectively doubles the x-ray output compared to schemes (a) and (b), or alternatively provides the same output with reduced filament emission in the x-ray tube. Under the latter condition tube life will be greatly increased.

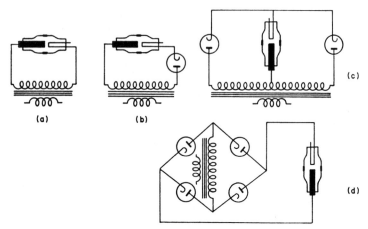

Fig. 12 Common rectification schemes used in diffraction apparatus: (a) self-rectification, (b) half-wave rectification employing one rectifier tube, (c) full-wave rectification using two tubes, and (d) full-wave bridge rectifier. For the same output voltage the transformer used in (c) must produce twice as much voltage across the terminals as that given by (a), (b), or (d).

Another way of enhancing the life and reliability of the high voltage system is to utilize stacks of solid-state diodes in lieu of tube rectifiers.

With photographic or energy-dispersive recording, fluctuations in x-ray output are not serious since the whole pattern is being recorded simultaneously and all portions of the pattern are affected equally. Fluctuating x-ray output is much more critical in the case of a diffractometer, however, because each line is scanned sequentially in time. Diffractometer manufacturers have solved this difficulty in one of two ways. The usual approach involves the addition of a line voltage stabilizer unit whose output supplies all the diffractometer electric requirements, including the x-ray tube. Such units will maintain the x-ray output to within $\pm 0.1\%$ for long periods of time. The other solution, rarely used, involves the addition of a second radiation detector which continuously monitors a portion of the main beam (usually picking up the rays scattered from a thin film in the main beam). The electrical output from the monitor and line detector may be fed to a ratio recorder so that the chart record is a true measure of the intensity of the diffraction pattern at all times.

In conventional installations (a), (b), and (c), the x-ray target (anode) is grounded while the filament floats at the required negative potential. This enables one to carry cooling water to the target support without shock or insulation worries. Cooling is essential, since of the ≈ 1500 W in the electron beam, only 1% to 2% appears as x-ray power, the remainder being dissipated as heat.

C. X-Ray Tube

The hot filament, water-cooled, evacuated, sealed-off x-ray tube with Be windows, as shown in Fig. 1 is standard for essentially all commercial diffraction units today even though low output field emission tubes are finding favor for a few special applications, particularly when portability is a factor. Targets comprising any of the eight metals listed in Table I are offered by most tube manufacturers. These tubes are rated to handle voltages in the 60 kV range and electron emission currents up to 30–40 mA. The electron beam produced by the heated filament is focused on a slender rectangular target area (normally about 1 mm × 12 mm) as shown in Fig. 13 and the x rays are emitted from this area. Normally four circular Be windows 90° apart are built into the tube to provide two spot focus beams and two line focus beams. Diano Corp. tubes have three ports, two spot focus and one line focus, but on the other hand, this company offers a double tube basic diffraction unit which one can use to operate six film cameras simultaneously. As will be noted later the line focus beam is ideal for slit system applications such as the diffractometer, Seeman–Bohlin and Guinier cameras, whereas the spot focus beam provides maximum intensity for pinhole geometry systems including some Debye–Scherrer and all flat cameras. By positioning the collimating system at a takeoff angle of some 6° below the target plane, the irradiated sample "sees" a foreshortened intense x-ray spot roughly 1 mm² or with the line focus ports a thin 0.1 mm × 12 mm line.

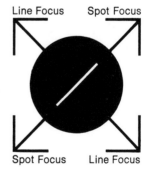

Line Focus   Spot Focus

Spot Focus   Line Focus

Fig. 13 Four window diagram. Opposite the spot focus windows: large or small powder cameras, flat plate cameras, Weissenberg goniometers and precession cameras. Opposite the line focus windows: wide range goniometers and precision symmetrical back-reflection focusing cameras, Guinier cameras. Photo by the courtesy of Phillips Electronic Instruments.

Copper K_α radiation is much used in diffraction studies for two major reasons. First, the relatively long wavelength spreads the diffraction pattern over a wider angular range than does W, Ag, or Mo radiation for example, and yet Cu rays are not appreciably absorbed in air. The greater 2θ interval means that, with a given camera diameter or goniometer, Cu radiation will permit greater accuracy in measurement of d values. Of course, this advantage can be offset to a degree in film work by using a larger diameter camera with shorter λ x rays, but at the cost of much longer exposures. The second reason favoring Cu is that, owing to its outstanding thermal conductivity, such a tube can be safely operated at a higher power loading than most other tubes.

Another factor to be considered in the choice of radiation in applications where film serves as a detector is the possibility of exciting fluorescent radiation in the specimen. K radiation from element A is capable of exciting the K radiation in element B, provided B has an atomic number two or more units lower than A (L radiation may also be produced but this is not usually troublesome).

This fluorescent radiation results in a general background (blackening of the film) and may completely blot out the diffraction pattern sought when B is three or four atomic numbers below A. A specimen containing iron exhibits this phenomenon when examined with Cu K_α radiation. Consequently, one who intends to study ferrous metals (a very important area) would certainly select a tube with target other than Cu. Of course the ideal would be to own tubes with all eight targets, but this is not always economically feasible. It so happens that the region of the periodic table just below Mo contains few elements of great practical interest or importance. On this basis, a Mo target tube is often chosen to supplement the popular Cu tube. For the study of organic and biological compounds, where long spacings are common, Cr radiation is gaining in favor. Diffractometer

TABLE III

Radiations Applicable to Specimens for Back-Reflection Experiments

Sample	Reflection	Relative intensity	$2d$ (Å)	Suitable characteristic radiation	Mean λ of K_α-doublet (Å)
Fe	211	0.38	2.33	Cr K	2.29
	310	0.08	1.81	Co K	1.79
	222	0.03	1.65	Cu K	1.54
	321	0.10	1.53	Zn K	1.44
Cu	222	0.09	2.09	Fe K	1.93
	400	0.03	1.81	Co K	1.79
Al	400	0.02	2.02	Fe K	1.93
	420	0.04	1.81	Co K	1.79
Mg	12.3	0.03	1.80	Co K	1.79
Austenite	220	—	2.53	Cr K	2.29
Martensite	211	—	2.36	Cr K	2.29

users, with pulse height selector equipment can pretty well ignore the preceding admonitions. By setting the PHS window to record only x-ray quanta possessing energy close to that produced by the tube, all extraneous rays will be rejected and fluorescent background is no problem.

It should be obvious that long wavelength radiation is mandatory if back-reflection experiments are contemplated. In this region, $2\theta \approx 180°$ $\theta \approx 90°$, and $\sin\theta \approx 1$. Therefore the Bragg equation becomes approximately $\lambda \approx 2d$. This means that one should select a λ slightly less than twice the value of a fairly strong specimen reflection. Table III records radiation suitable for a number of materials often studied by the back-reflection technique.

D. Health and Safety Precautions

Potentially there are two types of personnel hazards which should be recognized in working around x-ray diffraction equipment, namely, high voltage shock and radiation exposure. Even if one were prone to ignore these hazards there are Federal regulations (and many states have their own rules) covering industrial applications of x rays. One can assume that the manufacturer has met the appropriate regulations regarding equipment he offers for sale. However, this does not guarantee that the user will necessarily employ such equipment in a safe manner. Hence the following comments are in order.

Practically speaking, shock hazard from commercial difffraction equipment is negligible since all high voltages are contained within metal compartments and any access door is supplied with an interlock to cut off power when the door is opened. Radiation exposure is more of a problem. First, anyone working around x rays should have read and understood the recommendations in the NBS Handbook 111 (obtainable from the U.S. GPO, Washington, D.C. 20402). Fortunately, commercial diffraction tubes are rayproof, meaning the only radiation emitted exits from the Be windows. Safety closures are provided to block off any port not in use and fail-safe safety shutters should be attached to each working port.

Commercial cameras and goniometers are designed with radiation safety in mind and equipped with safety devices so that when operating instructions are followed strictly, there is little risk of excessive exposure. A further mitigating factor is that x rays utilized in most diffraction work are not nearly as penetrating as those employed in radiography. Thus, whole body exposure is not generally a likely problem. Even so it is highly recommended that a survey meter be available at all times and frequently used to guard against unforeseen stray radiation. Likewise, workers should wear film badges and/or pocket dosage meters. A ring film badge is particularly valuable since the primary beam near the Be window of a diffraction tube can produce an exposure as high as 10^5 rem/min. Placing a hand or finger in this beam inadvertantly for even a few seconds can lead to a nasty burn. Finally, warning lights should be placed at each entrance to the radiation area and lighted whenever the x-ray equipment is energized.

VI.   X-Ray Diffraction Camera Technique

Prior to commercial introduction of the diffractometer in 1948, photographic film was the exclusive medium for recording x-ray diffraction patterns. Even today the use of camera equipment has its supporters and boasts several advantages, among them simplicity, reliability, moderate capital cost, and minimum sample requirements. A film pattern can often reveal information not readily apparent from a one-dimensional diffractometer trace, such as grain size and preferred orientation, and as noted earlier, by working in the $2\theta \approx 180°$ region (diffractometers are usually limited to $2\theta \approx 165°$) one can determine sharp-line interplanar spacings to a high degree of accuracy. But film techniques also have distinct shortcomings. For peak intensity, peak profile, and quantitative analysis, the diffractometer is far superior, and for routine work where only one or two peaks need be examined to obtain a desired result the diffractometer can run circles (no pun intended) around the x-ray camera. Finally, while practice of the camera technique demands almost constant attention from the operator, e.g., loading and unloading cassettes, processing film, measuring lines and converting same to d values, running densitometer traces, and scanning search manuals and PDF* cards for identifications, by the addition of appropriate software accessories a diffractometer, as already noted, can be programmed to operate around the clock unattended, and through the marvel of modern-day electronic black boxes, can record and store data from many samples, perform complicated mathematical calculations, search memory banks to identify sample components, establish the percentage of each present, and print out the final results ready for the x-ray scientist when he arrives at his desk next morning. But such magnificient robots do not come cheap. For the laboratory with a moderate sample load, film equipment could prove entirely adequate, at least as a starter. Fortunately, basic diffraction units, as the two such depicted in Fig. 14a,b, can be upgraded to a diffractometer if conditions warrant through the

---

* PDF stands for powder diffraction file.

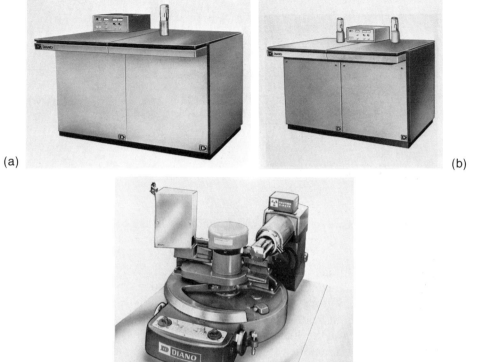

(a)   (b)

(c)

Fig. 14 Two basic x-ray diffraction units designed for use with film cameras. (a) This unit offers a single vertical mount diffraction tube, capable of serving three cameras simultaneously. (b) This unit contains two diffraction tubes. It unit has dual controls and both tubes can be operated simultaneously. In the latter case a total of six cameras can be in use at one time. (c) By means of a horizontal tube mount and spectrogoniometer plus accessory equipment one of the tubes can be converted to a diffractometer, retaining the remaining tube for film work. Photos by the courtesy of Diano Corp.

addition of a horizontal tube mount, a goniometer, Fig. 14c, and necessary additional accessories.

A. Conventional X-Ray Cameras

Many types of cameras have been used in the study of crystals by x-ray diffraction methods. Here we shall consider only those cameras which can be used with powder specimens.

A powder contains myriads of small crystallites oriented in every conceivable direction. When a beam of x rays traverses such a sample some

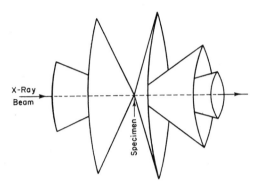

X-Ray
Beam

Specimen

Fig. 15 Diffracted rays from a powder specimen form a series of concentric cones having a common apex at the sample position.

crystallites can always be found which fulfill the Bragg condition for every possible interplanar spacing. The diffracted rays for one particular set of planes will be directed along the surface of a cone having its apex at the specimen and altitude along the incident beam direction. In general there will be many such cones of diffracted rays emanating from the sample, as pictured in Fig. 15. Four schemes are used to record these rays, each having its special field of utilization. They are illustrated in Figs. 16, 17, 18, and 19.

Since it records all the diffraction lines, the conventional "powder camera," the type used for a majority of powder samples, is pictured in Fig. 16. Here a strip of film (35 mm wide) is curved to form a cylinder with the specimen placed in the center and the whole enclosed in a lighttight container. With some powder cameras the film must be covered with black paper or aluminum foil. This proves a disadvantage only for soft radiations such as those from iron and chromium. An important feature of any camera involves the manner in which the beam is conducted to and away from the specimen under study. Long paths in air are to be avoided, since air scattering will produce an unwanted background on the film. The sketch accompanying Fig. 16 depicts one scheme that has been used to minimize air scattering. Also apparent in this sketch is the *guarded* slit or pinhole, a feature common to all final collimating apertures. This arrangement allows for the trapping of rays which might be diffracted by material forming the edge of the defining aperture. The camera diameter is often chosen to be (1) 57.3 mm or (2) 114.6 mm (sometimes a bit larger to allow for film shrinkage). This choice means that (1) 1 mm or (2) 2 mm measured between corresponding lines represents 1° in 2θ. The camera of smaller diameter requires a shorter exposure time, but spacings are more accurately determined with the larger.

The pattern reproduced in Fig. 16b illustrates the Straumanis and Ierins (1935) method of loading film in a powder camera. Following this technique,

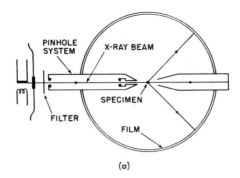

(a)

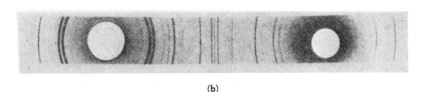

(b)

Fig. 16a,b (a) Schematic diagram showing essential features of powder camera. The exit tube is shown here extending almost to the specimen. This reduces air scattering. (b) Typical film obtained from cylindrical camera. Holes are cut in the film to permit entry and exit of x-ray beam. The Straumanis method of film placement was utilized here. See text. Photo by the courtesy of Picker X-Ray Corporation.

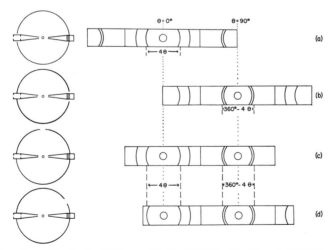

Fig. 16c From *Handbook of X-Rays*, edited by E. F. Kaelble. Copyright 1967 by McGraw-Hill, Inc. Used with permission of McGraw-Hill Book Company. Film mounting in the Debye–Scherrer camera. After (a) Bradley-Jay, (b) van Arkel, (c) Straumanis, and (d) Wilson.

one punches two holes in the film a distance apart equal to one-half the camera circumference and places the film so that the free ends meet roughly half-way between entrance and exit portals. In this way one is certain of recording the complete pattern; there are no missing regions near $\theta = 0$ or $\theta = 90°$. In addition, a doubled back-reflection record is available for accurate measurements. An added advantage accrues from the fact that lines around the exit and entrance ports center about points differing by exactly half the camera circumference. This enables one to deduce the effective camera diameter and correct for possible film shrinkage.

There are three other ways of mounting the film strip in this camera. If the free ends of the film meet at the pinhole system collimator, it is known as the Bradley–Jay method. The exit tube hole is then the center of the film strip and all diffraction lines being doubled, the $\theta = 0°$ position can be readily established. Conversely, the free film ends may be made to meet at the exit tube. This is termed the van Arkel mounting scheme and permits precise location of $\theta = 90°$ from doubled back-reflection lines. When a specimen displays no lines in the back-reflection region, one can arrange to have the free ends of the film strip meet reasonably close to the exit tube hole, a modification of the Straumanis mounting introduced by Wilson. Here some of the front-reflection lines will be paired around the pinhole collimator tube as well as around the exit hole and allow one to deduce the values $\theta = 0°$ and $\theta = 90°$ and the effective camera diameter precisely. All four of these mounting methods are sketched in Fig. 16c.

Figure 17 shows a flat cassette Laue-type camera. The specimen is mounted in front of the second pinhole, and the diffraction pattern is recorded on a flat film. The black disk directly in the x-ray path is a small blob of lead which intercepts the beam and prevents film halation during extended exposure. This camera is particularly suited to samples showing preferred orientation and also to those having large interplanar spacings.

The layout pictured in Fig. 18 is termed a back-reflection camera, since the recorded rays make an angle greater than 90° with the incident beam direction. This technique is ideal for large, thick samples. Provision is made for rotating both sample and film to ensure smooth lines. The outstanding advantage of the back reflection camera is its extreme sensitivity to small lattice changes. This can be seen as follows:

The Bragg equation states

$$n\lambda = 2d_{hkl}\sin\theta \tag{11}$$

To ascertain how θ varies with d_{hkl}, we differentiate Eq. (11), obtaining

$$0 = 2d(d_{hkl})\sin\theta + 2d_{hkl}\cos\theta \, d\theta \tag{12}$$

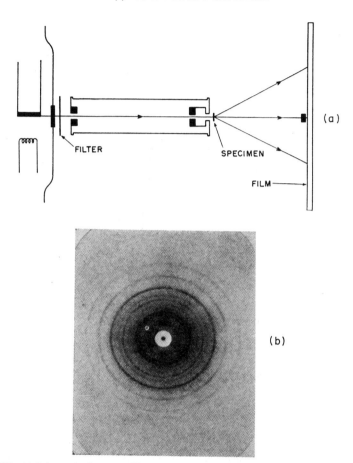

Fig. 17 (a) Schematic diagram of Laue camera arrangement. (b) Typical pattern obtained with this arrangement. This camera is especially suitable for samples showing preferred orientation or having large interplanar spacings. Photo by the courtesy of Picker X-Ray Corporation.

which can be written

$$\frac{d\theta}{d(d_{hkl})} = -\frac{\tan \theta}{d_{hkl}} \tag{13}$$

For values of θ near $90°$ (the back-reflection film covers this region since $2\theta = 180°$), $\tan \theta$ has a very large value. This means that a small change Δd_{hkl} will cause a large θ change in this region.

For this reason lattice parameters can be evaluated with high accuracy from back-reflection patterns, and the method is much used in this connection.

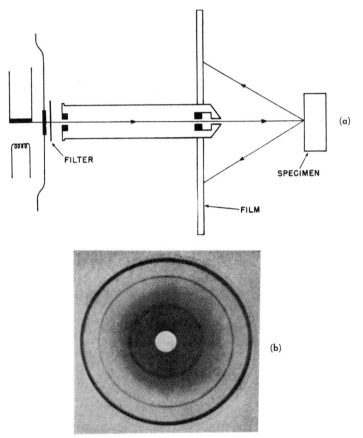

Fig. 18 (a) Sketch of the back-reflection arrangement. In some cameras of this type the collimating pinhole nearest the specimen may be moved independently, parallel to the beam direction. This adjustment makes it possible to satisfy the focusing conditions of Fig. 19 for any desired reflection, thus permitting very sharp rings. (b) Typical pattern taken with a back-reflection camera. Photo by the courtesy of Picker X-Ray Corporation.

B. Special Cameras

Figure 19 illustrates the principle of the Seemann–Bohlin focusing camera which can be used for rapid recording of patterns. Here one bathes a large area of sample with x rays and depends on a geometric principle to achieve sharp lines. If the slit (A), specimen (B), and film (C) lie on the circumference of a circle, all rays diffracted by the same family of planes in the specimen will be brought to a focus at the same point on the film. To see that this is so, let the rays shown represent reflections from a common set of planes. Then

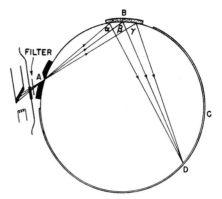

Fig. 19 Sketch to illustrate the focusing principle involved in focusing type of camera.

$\alpha = \beta = \gamma$, since all are supplements of the same scattering angle, 2θ. But on any circle *equal angles inscribe equal arcs*. Since all three angles have point A in common, they must meet the circle at another common point such as D. The same argument holds for any other family of planes. Hilger and Watts Ltd. offer a miniature Seemann–Bohlin camera recommended for use with their microfocus (0.04-mm) x-ray unit. Exposure times can be as little as 3 min with this combination.

Others who have contributed worthwhile modifications and improvements to the Seemann–Bohlin concept include Westgren (1931), Cauchois (1932), and Jette and Foote (1935).

Guinier (1937, 1939) pioneered the idea of adding a curved crystal monochromator to the Seemann–Bohlin arrangement, with the pure K_α beam from the x-ray source brought to a focus by the crystal at the camera slit (A in Fig. 19). This combination has given rise to a whole family of cameras which can be designed to yield sharper diffraction lines and exhibit double the dispersion (in millimeters of film per degree θ) as compared with Debye–Scheerer cameras of the same diameter. Furthermore, by using a sufficiently deep (perpendicular to the beam plane) curved crystal, as many as four Guinier cameras may be stacked on top of one another, all utilizing the same x-ray source. In addition, both a transmission (thin sample) and back-reflection camera can be mounted in tandem as depicted in Fig. 20. For all these reasons, so-called Guinier cameras have proven highly popular, and a sizeable body of literature has grown up around their use. For example, see deWolff (1948), Hoffmann and Jagodzinski (1955), Hellner (1954), Sas and deWolff (1966), Fisher (1957), Corbridge and Tromans (1958), and Hägg and Ersson (1969).

Often one wishes to study the diffraction patterns from samples available only in micrograms or from selected small areas of larger samples. Here microcameras are mandatory, employing collimated beams down to a few

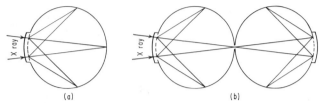

(a)   (b)

Fig. 20   The sketch in Fig. 19 illustrates the Seemann–Bohlin focusing principle applied asymmetrically to a thick (reflection) specimen. The sketch in (a) demonstrates how this same scheme can be applied symmetrically to a thin (transmission) specimen. The sketch in (b) shows how one can combine two cameras in tandem (a thin transmission sample) followed by a thick reflection sample, both using the same monochromatic beam of x rays. (From *Handbook of X-Rays*, edited by E. F. Kaelble. Copyright 1967 by McGraw-Hill, Inc. Used with permission of McGraw-Hill Book Company.)

microns in diameter. Specimen film distances are reduced correspondingly (to a few millimeters) so that exposure times are not prohibitive. Fankuchen and Mark (1944), Chesley (1947), Fried and Davidson (1948), and Cahn (1951), among others, have applied the microbeam technique to a variety of interesting problems.

Since the effect of temperature and pressure on matter is of major significance, studied attention has been paid over the years to designing equipment permitting studies of specimens at high and low temperatures as well as at high pressures. In fact, Kaelble's very fine "*Handbook of X-rays*" (Kaelble, 1967) devotes three chapters (13, 14, and 15) to these matters.

As might be surmised, a major difficulty present here involves finding a design that will ensure a suitably uncluttered path for the diffracted beam over an adequately wide range of 2θ. A salient feature of the energy-dispersive technique is that only two narrow pathways are needed; one for the incoming white radiation and a second for the diffracted (and fluorescent) rays. Another important benefit is the speed of the energy-dispersive diffraction technique (EDD) in producing usable data, thus allowing one to follow phase changes and other short-time phenomena *in situ*, an accomplishment literally unthinkable by any conventional diffraction procedure. Therefore it seems plausible that in the future EDD procedures may well preempt the x-ray diffraction investigation of simple structures (mainly cubic) at other than STP conditions.

C.   Sample Preparation

In preparing a powder specimen the material should be ground in an agate mortar until it will pass through a 325-mesh screen. If the specimen is metallic, fine filings will suffice. There is always danger of lattice distortion here, so light pressure should be used in the filing operation. Of course the back-reflection method may be applied to the undisturbed bulk metal.

There are several ways of fabricating powder into a finished specimen. For a Laue camera it may simply be pressed between metal blocks to form a thin cake which is placed in front of the second pinhole. If the powder cake is too fragile, a small amount of amorphous binder (starch or gum tragacanth) can be added.

For the cylindrical camera the powder may be introduced into a thin walled glass or cellophane capillary (about 0.4-mm i.d.), the tube serving as a sample holder. If long wavelength ($\lambda > 1$ Å) radiation is used, a paste of the powder and collodion can be pressed into a specimen tube and then partially extruded by means of a small wire ramrod. The tube supports the specimen, and the bare section is irradiated. This avoids absorption of x rays in the walls of a tube. If the material is in block or sheet form it can be mounted in the powder camera so that an edge intercepts the beam. Scattering will then result from a thin wedge-shaped section. This camera also has means for rotating the sample during exposure to smooth out spotty diffraction lines. Sometimes, however, it proves advantageous not to rotate the sample, inasmuch as variation in texture of the lines may assist in separating those due to different phases present in the sample. Various aids such as centering screws and fluorescent windows in the exit portal are usually incoporated for accurately centering the sample in the camera.

It will be noted that schemes shown in Fig. 16 and Fig. 17 are in a large measure transmission methods. Therefore, in the preparation of samples care must be exercised that the absorption be not too great. It is easy to show that for maximum diffracted intensity (forward direction) the optimum sample thickness is $1/\mu$, where μ is the linear absorption coefficient of the sample. This thickness would reduce the incident beam intensity to e^{-1} or about 37% of its original value. One can always dilute a highly absorbing sample with an amorphous material like flour in order to approach the above condition, using only a single capillary tube size.

D. Exposure Time and Film Processing

The time required to secure a satisfactory pattern depends on many factors but in general will run from 30 min to 3 hr. The focusing camera is considerably faster (10 min perhaps); the back-reflection camera may require times running to 10 hr or more, since rays are diffracted at large angles with lower intensity. The commercial x-ray units are equipped with an automatic clock which shuts off the apparatus after a predetermined time. Once the exposure is complete, the film is removed from the camera and developed.

X-ray film is orthochromatic, so the processing can be carried out under a ruby lamp. Special x-ray developers are available commercially. The usual time in developer is 4 min at 68°F. A 10- to 15-min fix is suggested (twice the time to clear), followed by a wash of 30 min to 1 hr. The film is then allowed to dry and is at last ready for study.

In recent years fast speed Polaroid film has become very popular with camera diffractionists as a recording medium. The major factors accounting for this fact are very short exposure times and immediate development of the film even without darkroom facilities. Fifteen seconds after the exposure is complete the analyst has the finished negative in his hand ready for viewing. The first users of this technique were single crystal structure workers who used it as a preliminary means of checking on crystal orientation and possible strains and twinning defects present. But the results approached so closely in quality those attained with wet film that many have swung over to the Polaroid Land diffraction cassette and corresponding film as standard procedure. This flat film cassette attachment usable on regular Laue transmission and back-reflection cameras is offered by several manufacturers.

The most popular procedure appears to be the use of Polaroid 107 (formerly type 57 film—ASA 3000) placed in close contact with a suitable backing phosphor screen (duPont's CB-2 is one of several recommended), the latter reducing the exposure appreciably with no deterioration in quality of image. Polaroid also produces a superfast type 410 film (ASA 10 000) which permits short exposure times without an enhancing fluorescent screen. However this is naturally a coarse grain film and a majority of workers seem to prefer the first approach.

Even powder photographs obtained using the Polaroid film have been found quite satisfactory for many purposes. It is difficult to generalize on the anticipated savings in time realizable here since setups vary from case to case, but when one considers both exposure and developing time saved, it has been the experience of many workers that an experiment carried out utilizing the Polaroid technique can be completed in 5% to 10% of the time required using wet film.

E. Measurement of Pattern

Various comparators and viewing screens are marketed that permit accurate measurements on the diameters of the diffraction lines. The writer has found that a ground-glass viewing screen, together with a pair of sharp pointed dividers and a calibrated metal scale, provides satisfactory equipment for measuring routine patterns. The dividers are set on corresponding lines of the film and then transferred to the scale for reading. The line diameter plus a knowledge of the camera geometry gives θ. As stated earlier, a camera diameter of 57.3 mm or 114.6 mm allows θ to be deduced without detailed calculation. Sin θ is then combined with the wavelength λ in the Bragg equation to yield an interplanar spacing d. A tabulation is made of the d values from all lines of the pattern. A relative intensity value is also assigned to each line, commensurate with its strength on the film. This is usually done semiquantitatively by selecting a suitable number between 0.1 and 1.0, 0.1 corresponding to a very weak line and 1.0 to the strongest. The reader may wonder why one does not attempt to place relative intensities on a firm quantitative footing by photometering the film. The answer is that the best microphotometer will miss lines easily visible to the naked eye and thus cannot provide a complete story.

F.   *Errors and Their Elimination*

The following major factors serve to limit the accuracy attainable in deducing interplanar spacings from powder patterns: (1) inadequate knowledge of specimen–film distance, (2) shrinkage of film during processing, (3) absorption of x rays in sample, and (4) eccentricity of specimen with respect to axis in cylindrical cameras.

These first two factors can be corrected for quite simply. This entails two knife edges built into the camera so that they form sharp limits to the exposed film. The camera is calibrated by securing a pattern from a standard material. This pattern allows one to deduce the effective Bragg angle θs for a hypothetical diffraction line occurring at these limits. With other patterns from this camera the Bragg angle θ for any diffraction line is given by

$$\theta/\theta s = D/Ds \tag{14}$$

where D is the distance on the film from one arc of the line to its mate, and Ds is the distance between the knife-edge shadows.

When it is necessary to know lattice parameters with high accuracy all four factors should be taken into account. This can be accomplished in the following manner: a small amount of standard material such as sodium chloride or quartz is mixed with the unknown sample, and a pattern is recorded from the mixture. The well-known diffraction angles from the standard substance are plotted against measurements on the corresponding lines to give a calibration curve. The curve is then used to find accurate values of θ for lines due to the unknown. This procedure is more tedious than the first method but is considerably more accurate. The current procedure is used in conjunction with a back-reflection camera when the ultimate in accuracy is required. Such a combination will give d values that are reproducible to ± 0.0001 Å. For some types of powder work it may not prove feasible to incorporate a standard material with the experimental specimen. In this event the sample tube can be divided into two compartments, the unknown occupying one section, and the standard in the other. Collimating slits serve to define the beam here, replacing the usual pinholes. A septum can be installed in the camera if desired to prevent superposition of the spectra. The two spectra will then appear side by side on the same film. The flat back-reflection camera contains a metal sector disk which exposes only a portion of the film at a time. This likewise permits side-by-side comparison spectra on the same film.

VII.   X-Ray Diffractometer Technique

A.   The Basic Diffractometer

An x-ray diffractometer is far more complex than a camera system. In its simplest configuration a diffractometer is constructed from the four following building blocks:

(a)   An x-ray source. This item is essentially the same as the film unit counterpart except as earlier noted line voltage stabilizing equipment must be introduced ahead of the diffractometer to assure constant x-ray output over long periods of time.

(b)   A calibrated scale single circle goniometer such as that shown in Fig. 14c containing a sample holder, a radiation detector mounted on and movable around the calibrated circle, the detector mount actuated by a worm drive mechanism, which via an appropriate differential linkage also rotates the sample holder through $\angle\theta$, while the detector is traversing $\angle 2\theta$.

(c)   A system of collimating aids situated between x-ray source and detector to retain adequate resolution, while realizing maximum diffracted intensity at the detector.

(d)   A preamp and linear amplifier following the detector, a pulse shaper, a pulse height selector (PHS), also known as a pulse height analyzer (PHA), the latter feeding a rate meter or scaler plus recorder.

B.   Comparing Vertical and Horizontal Goniometer Configurations

One of the major features to be examined in comparing commercial diffractometers is whether the detector moves in a vertical or horizontal plane. For example, the original Norelco unit, pioneered by North American Philips (now Philips Electronic Instruments), employed a vertical plane goniometer whereas the Diano Corp. (who took over the General Electric line in 1972) favors a horizontal plane mount. This difference is by no means trivial. A vertical unit permits the x-ray tube to be mounted with the long axis vertical to assure a *horizontal* line focus source for the diffractometer. With this geometry the remaining tube ports may be used for other experiments while the diffractometer is in operation. The major disadvantage of the vertical mount is the uneven stresses created by the detector head weight as it moves around the circle. Even a slight shift of the mount load will affect the accuracy of 2θ readings.

A horizontal circle mount has excellent structural stability since the head weight load is uniform throughout the traverse (one can even accomodate the weight of a solid-state detector and accompanying Dewar without great difficulty), but this arrangement requires that the x-ray tube be positioned with long axis horizontal to provide the needed *vertical* line focus source. This accounts for the addition of a tube adaptor to convert a conventional film unit to diffractometer use. With x-ray tube long axis horizontal, the two spot focus ports are essentially unavailable inasmuch as one points downward into the equipment housing and the other points toward the ceiling.

Figure 21 illustrates the path taken by the incident and diffracted rays in a vertical type diffractometer, showing the array of collimating devices employed. Horizontal beam spread is limited by the divergence and receiving parallel "Soller" slit assemblies. Vertical spread is limited by the divergence, receiving, and scatter slits. The first of these determines the area of sample bathed by radiation. The receiving and scatter slits define the diffracted beam reaching the detector. A series of interchangeable slits is provided so that the most desirable compromise between resolution and intensity can be achieved, depending on the particular demands.

Not shown here is the filter normally placed over the line focus port to eliminate K_β radiation. Alternatively, one can omit the β filter and receiving parallel slit assembly, placing instead a monochromator (a cylindrically bent,

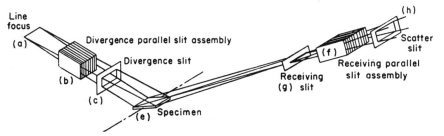

Fig. 21   Schematic diagram depicting x-ray optical arrangement. Courtesy of North American Philips Co. (Philips Electronic Instruments). (Note: They have taken over x-ray equipment of North American Philips Co.)

highly oriented pyrolytic graphite mosaic crystal sheet is popular) after the final scatter slit, focusing the monochromator output onto the detector window, offset from its original position. This scheme can increase diffracted intensity by a factor of two with no loss in resolution.

Though not apparent from the sketch, a variation of the Seemann–Bohlin focusing principle, known as Bragg–Brentano parafocusing is at work here with point A in Fig. 19 representing the tube target line focus. In this setup, as the radiation detector head shifts to a new position, the diameter of the focusing circle changes. To keep matters in proper relationship, the sample must also turn so that its surface remains tangent at all times to the focusing circle. This accounts for the $\theta : 2\theta$ ratio between sample and detector angular motions.

C. Two Prime Detectors Used in Diffractometer Work

The first diffractometers employed the familiar Geiger–Müller (GM) tube as detector, borrowed from the nuclear field. This counter had two valuable attributes; high quantum counting efficiency and a huge output voltage. In fact the latter was so great that the simplest kind of amplifier would actuate a mechanical recorder or rate meter. But the GM counter suffered two serious disadvantages. First, all output pulses, irrespective of the triggering quantum, were the same size since in a GM counter any ionizing event creates a momentary gaseous discharge. Second, GM counters take forever (by electronic microsecond standards) to recover from the discharge state. Thus it had an intolerable "dead time." Actually a GM counter will start missing counts at rates as low as 100/sec. Therefore the GM counter has been abandoned and present-day diffractometers almost invariably depend on a gas-filled, side window proportional counter or a scintillation–photomultiplier assembly as detector. The essentials of these two devices are pictured in Figs. 22a and b.

In Fig. 22a, a potential difference of several hundred volts is set up between the central wire anode and metal cylinder cathode, creating a strong, uniform electric field. Whenever an x-ray quantum enters the Be window it gives up its energy to a photoelectron, which latter produces ionization in the gas. The negative ions (or electrons) are accelerated toward the wire and in turn produce additional ion pairs. The counter voltage is set so the ion pair gas amplification factor is independent of the number of primary ion pairs. Thus the resulting voltage pulse picked up by the wire is proportional to the energy of the quantum responsible.

In the scintillation-photomultiplier detector Fig. 22b the incoming x-ray quantum is absorbed by a fluorescing single crystal which emits the energy

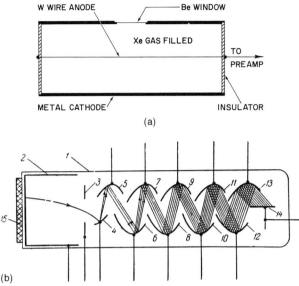

Fig. 22 (a) Sketch of a simple sealed-off proportional counter. A potential difference of several hundred volts is established between the metal cathode and tungsten wire anode (0.1 mm diam.). The x rays enter through the thin (≈ 0.25 mm) Be window. By using a metal window the electric field is undisturbed. The most popular filling gas is xenon since its absorptive efficiency peaks in the Cu radiation region. However other gases can be used and for the softer radiations flow counters boasting extremely thin Al-coated Mylar windows and argon–methane mixtures are standard. (b) A scintillation crystal-photomultiplier counter. X rays penetrate a light-tight cover and cause scintillations in the Tl (Thallium) doped sodium iodide (NaI) single crystal 15. These flashes are picked up by the photocathode 2, usually a semitransparent cesium–antimony (Cs–Sb) layer. The photoelectrons released by 2 are focused by diaphram 3 onto the first dynode 4. As stated in the text, successive dynodes 5–13 is each at a higher positive potential than the preceding and thus electron multiplication occurs. The final pulse is received by anode 14. Reproduced by permission of the Pergamon Press, New York.

as light photons. These strike the photomultiplier cathode 2 which in turn releases a few photoelectrons. These are accelerated to successive dynode surfaces 4–13, each maintained at a higher positive potential than the preceding. This accelerating voltage between dynodes is such that additional electrons are released at each dynode with the final burst collected by anode 14 and passed along to an amplifier. This pulse, like before, is proportional to the energy of the incident x-ray quantum. For a more detailed discussion of these two types of detector the reader should consult Chapter 3 of Kaelble (1967) or pp. 313–330 of Klug and Alexander (1974). Not only do these detectors produce pulses proportional to the initiating x-ray energy but both exhibit very short dead times of the order of 0.2 μsec. Thus high counting rates are feasible (up to 100 000 per sec) with little concern about counting losses. Unfortunately the output pulses from both are at best a few millivolts, so high gain amplifiers are needed here. The scintillation detector with its large effective sensitive mass has high quantum detection efficiency across the x-ray spectrum used in diffraction work. The gas filled counter, already at a disadvantage owing to its mass wise short path length through gas, is even less efficient for detecting hard x rays (< 1 Å) since these may pass right through the counter unscathed. By selecting the proper gas this deficiency can be ameliorated to a degree. On the other hand, if one compares the *resolution* of the two, as measured by the recorded full width at half maximum (FWHM) of a particular diffraction peak, the proportional counter wins hands down. Naturally this means the possible overlapping of neighboring lines will be much more of a problem using the scintillation detector.

D. Pulse Amplifier, Shaper, and Height Selector

As reported, the weak pulses must be greatly amplified. So first a preamp is followed by a stable linear amplifier with gain such that the output pulses are now measured in volts. These amplified pulses go to a pulse "shaper" which converts them to rectangular shape, but unchanged in amplitude. The rectangular pulses then enter a three-component electronic black box known as a pulse height selector (PHS). The first component (lower-level discriminator) passes all pulses having amplitude greater than a preset value V_L. The second component (upper-level discriminator) passes all pulses greater than some higher value V_H. The third component, an anticoincidence circuit, cancels all pulses passed by both discriminators. Thus the only pulses passed on to the recorder are those falling within the "window" ($V_H - V_L$). To appreciate the utility of a PHS let us cite an example. Assume the pulse height amplifier level is set so that Cu K_α quanta fall near 6 V. By setting the lower-level discriminator at 5.8 V and the upper one at 6.2 V we have created a 0.4-V window which should pass all Cu K_α x rays entering the detector. Since the more energetic Cu K_β quanta would produce pulses of 6.6 V at this amplifier setting, none would get through to the recorder. Thus, even without a Ni filter in the beam or a monochromator crystal focused on the detector, we have effectively achieved the same end result without any sacrifice in intensity. Recall a Ni filter would reduce the Cu K_α beam by 50%.

Figure 23 depicts a diffraction pattern of α quartz secured from a diffractometer hooked up to a chart recorder. The excellent resolution is pointed up by the distinct α_1, α_2 doublet for all lines at $2\theta > 35°$ and separation of the five lines in the 1° interval around $2\theta = 68°$. The best Debye–Scherrer powder cameras do not resolve α_1, α_2 doublets below $2\theta = 90°$.

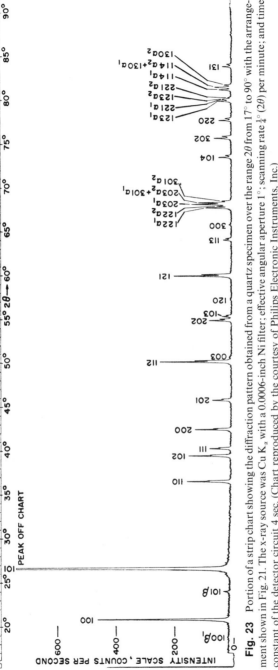

Fig. 23 Portion of a strip chart showing the diffraction pattern obtained from a quartz specimen over the range 2θ from 17° to 90° with the arrangement shown in Fig. 21. The x-ray source was Cu K_α with a 0.0006-inch Ni filter; effective angular aperture 1°; scanning rate $\frac{1}{4}$° (2θ) per minute; and time constant of the detector circuit 4 sec. (Chart reproduced by the courtesy of Philips Electronic Instruments, Inc.)

E. Converting a Ford into a Robot Rolls

What has been described here is the diffractometer in its basic form. While many of these so-called Model T diffractometers are performing yeoman's service, recall this unit is essentially a vintage 1950 machine. To trace all the modifications, additions, and improvements leading to a 1980 model would require a tome all its own. One relatively simple modification which permits automating the manual unit to a considerable degree is replacing the synchronous motor which drives the detector head with a stepping motor. This enables one to step scan; that is, stop the detector at a desired 2θ position, take a reading (fixed time or fixed count), move incrementally, take another run, continue to a third position, etc. This modification also allows "slewing," the ability to shift the detector rapidly from one spot to another many degrees away. This is particularly valuable when the approximate locations of the desired peaks are known and are relatively far apart.

In addition to digesting the literature readily available from diffractometer makers, the reader may wish to consult the following references to learn how individual laboratories have modified and how manufacturers have revolutionized the design of this most unique instrument: Cole *et al.* (1963), Kelly

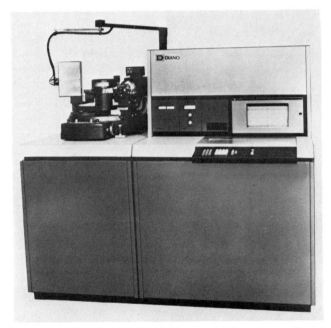

Fig. 24 Diano Corporation XRD 8100 AD Automatic Diffractometer. Photo by the courtesy of Diano Corporation.

and Short (1972), Segmüller (1972), Slaughter and Carpenter (1972), Jenkins *et al.* (1972), Jenkins and Westburg (1973), and Data Sheet XRD 8100 AD, Diano Corp. (1975).

Figure 24 depicts the Diano machine of the prior reference while Fig. 25 pictures the comparable Philips APD3501 instrument successor to the APD3500 described by Jenkins in the above references.

With the purchase of a diffractometer each maker supplies a standard sample whose diffraction pattern is well established. This sample can be inserted anytime to check on the performance of the instrument. The Philips standard is a slice of Arkansas stone (α quartz). They state: "We have found this standard to be free from orientation problems and very stable—it can be washed with soap and water if necessary." Diano's comparable standard sample is known as "permaquartz."

Recall we warned you originally that the diffractometer, while a remarkable example of man's technological ingenuity, is not a simple toy. For example there are a multitude of factors which influence the positions,

Fig. 25 Philips Electronic Instrument APD3501 Automatic Diffractometer. Photo by the courtesy of Philips Electronic Instruments, Inc.

profiles, and peak maxima of diffraction patterns recorded on diffractometers. These can be broken down under six major headings. This listing, with minor additions, is taken from Chapter 9 of Kaelble (1967), with an excellent discussion on pp. 9-16–9-37. We thank McGraw-Hill Book Co. for permission to reproduce this list:

1. Physical effects
   a. Wavelength
   b. Dispersion, Lorentz, and polarization
   c. Filters and monochromators
   d. Refraction
2. Geometrical effects
   a. Flat specimen and horizontal divergence
   b. Axial (or vertical) divergence
   c. Absorption
   d. Source profile and width
   e. Detector slit width
3. Instrumental effects
   a. Calibration of goniometer scale
   b. Accuracy in $2\theta : \theta$ for all values of θ
   c. Backlash
4. Possible alignment errors
   a. Focal-spot displacement
   b. Specimen surface not coincident with diffractometer axis
   c. X-Ray beam not in diffractometer plane
   d. $\theta \neq 0$ when $2\theta = 0$
   e. $2\theta = 0$ not on beam path
5. Measurement effects
   a. RC (time constant) of electronic circuits
   b. Peak position vs centroid
   c. Scanning speed
   d. Scanning direction
   e. Counting statistics
6. Specimen effects
   a. Particle size
   b. Sample thickness
   c. Preferred orientation
   d. Strain or cold work
   e. Surface flatness

While this check list may seem formidable, most of these factors, once resolved, do not represent recurring or insurmountable problems. The diffractometer has proven itself in normal day to day operation to be a tireless, dependable, reliable, and versatile scientific instrument in hundreds of laboratories here and around the world. Without doubt as the tool of choice, it ranks above the others among diffractionists everywhere.

VIII.   Energy-Dispersive X-Ray Diffraction Technique

The digital timepiece has been widely heralded as the watch with no moving parts. The same stationary quality can be assigned the revolutionary energy-dispersive diffraction technique (EDD). And interestingly enough both devices above owe their existence to developments in semiconductor physics and modern electronics.

A.   An Interesting Experiment

As an introduction to EDD consider the following experiment. Let us expose a polycrystalline sheet of platinum (Pt) to a collimated beam of white radiation from an Fe target x-ray tube operating at 45 000 V. Looking at Fig. 26 we ask ourselves what rays might strike a detector placed at point B ($2\theta = 22°$).

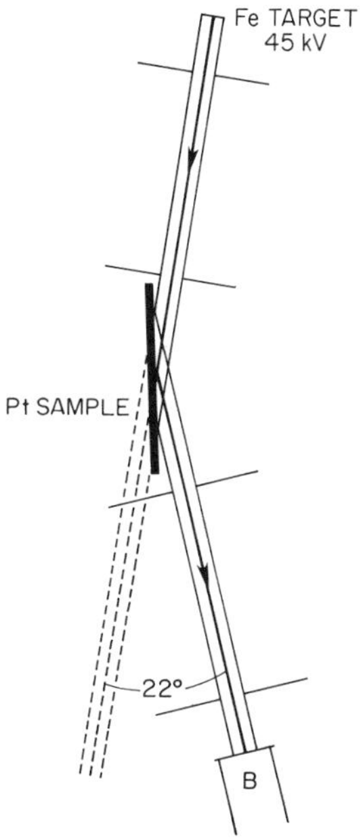

Fig. 26 Polycrystalline sheet of platinum exposed to a collimated beam of white radiation from an Fe target x-ray tube.

According to Eq. (1) the incident beam, in addition to Fe characteristic radiation, will contain a continuum of x radiation with:

$$\lambda_{min} = 12\ 396/45\ 000 = 0.27\ \text{Å} \tag{15}$$

and extending up to around 2.0 Å. Any rays with $\lambda > 2$ Å will be so soft as to be absorbed in air before reaching the Pt specimen. Now Pt is a face-centered cubic crystal with $a = 3.916$ Å. According to Fig. 33 in Section XI.A, the first (hkl) diffraction line expected from Pt would be (111) with $d = 3.916/\sqrt{3} = 2.26$ Å. And from the Bragg equation, the slice of radiation responsible for this reflection ($\theta = 11°$) would be

$$\lambda = 2(2.26)(0.186) = 0.84\ \text{Å} \tag{16}$$

or in terms of the voltage required to produce this wavelength [solving Eq. (1) for voltage]:

$$V_{(111)} = 12\ 396/0.84 = 14.76\ \text{keV} \tag{17}$$

Similarly for other Pt diffraction lines (considering the aforementioned Fig. 33) producible by a continuum of x rays down to a cutoff at $\lambda = 0.27$ Å we would expect to find the values listed in the accompanying tabulation.

hkl	$(h^2 + k^2 + l^2)^{1/2}$	d_{hkl} (Å)	λ_{hkl} (Å)	V_{hkl}(keV)
(200)	4	1.96	0.73	16.98
(220)	8	1.39	0.52	23.84
(311)	11	1.18	0.44	28.18
(222)	12	1.13	0.42	29.52
(400)	16	0.98	0.36	34.44
(331)	19	0.90	0.33	37.57
(420)	20	0.87	0.32	38.74
(422)	24	0.80	0.30	41.33
(333)	27	0.75	0.28	44.28
(511)	27	0.75	0.28	44.28

From the tabulation we see 11 possible diffraction lines. Since there are myriads of small, randomly oriented crystallites in the Pt sample, statistically speaking, a sufficient number should be oriented to satisfy the Bragg relation for every possible reflection. But there will also be other peaks show up at B. If one checks a table of excitation potentials for the elements he will find that while 45 kV x rays will not excite Pt_K characteristic radiation (it takes 78 kV to do this), 45 kV x rays are sufficient to excite $Pt_{L\alpha}$ (1.31 Å $\approx$ 9.40 keV), $Pt_{L\beta}$ (1.09 Å $\approx$ 11.4 keV), and $Pt_{L\gamma}$ (0.94 Å $\approx$ 13.2 keV). The Pt_M series will also be excited, but since $\lambda_{Pt_M} \approx 6$ Å, these rays will be absorbed in air shortly after leaving the sample.

Consequently, if we only had some magic detector at B capable of sorting out and displaying closely spaced x-ray lines according to wavelength (or equally well, according to energy), we would anticipate finding the following sequence of lines starting at the low energy end and working up to λ_{cutoff} at 0.27 Å or 45 kV:

1. $Pt_{L\alpha} = 1.31$ Å or 9.4 keV
2. $Pt_{L\beta} = 1.09$ Å or 11.4 keV
3. $Pt_{L\gamma} = 0.94$ Å or 13.2 keV

followed in sequence by the 11 Pt diffraction lines listed earlier.

Figure 27 shows the results of an actual experiment employing the setup just described. It so happens this is the classical experiment carried out by Giessen and Gordon (1968) which caused tremendous excitement by showing the EDD concept was truly a workable reality. Independently, four Polish scientists (Buras *et al.*, 1968) also published similar findings that same year. *You will note that our predictions are realized to the letter.* Before we inquire into the magic detector and analyzer which made all this possible, consider what will happen it we moved the detector B to a new position, say $2\theta = 60°$. The first three lines would appear at the same positions on the plot since they are fluorescent rays characteristic of that element only and do not depend on any Bragg relationship. However, the 11 *diffraction* lines must satisfy Bragg and would shift. Calculations show all would now fall in a narrow

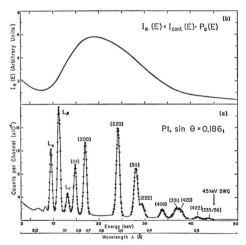

Fig. 27 (a) Diffraction pattern and L-fluorescence spectrum of platinum sheet, obtained by x-ray spectroscopy: SWC, shortwave cutoff of x-ray beam. (b) Response of detector to undiffracted polychromatic x-ray beam. Taken from B. C. Giessen and G. E. Gordon, *Science* **159**, 974 (1968). Copyright 1968 by AAAS. Reproduced by permission of authors and copyright owner.

energy band between 5.5 and 16.0 keV, which would throw them right on top of the fluorescent peaks. Thus utter confusion would result. So it should be clear that in practicing the EDD technique, it is important to choose fixed ∠ 2θ judiciously, so as to keep the fluorescent and diffraction lines separated as far as possible.

Mention should be made of the upper curve in Fig. 27 which is an intensity versus energy plot of the white radiation emitted by the Fe tube, the energy scale being the same as that of the curve below. Note that the slice of spectrum available to produce the low (*hkl*) reflections is much larger than that available for the high (*hkl*) values. Hence the diffracted intensities from an EDD experiment will differ greatly from those obtained on film or with a diffractometer using the same sample.

But now to that magic detector analyzer display system. What we require are three things: (a) a high-resolution detector–amplifier whose output is proportional to the energy of the quantum causing the event; (b) a "pair of trousers" with several hundred marked pockets where one can store these output pulses, each pocket holding pulses whose energy is known and but slightly different from those stored in the adjacent pocket; and (c) a ready means of reading how many pulses have been stashed in each pocket anytime we desire this information.

B. Solid-State High Resolution Detector and FET Preamplifier

The answer to (a) is a Si(Li) or Ge(Li) drifted *PIN** detector. To the trade these semiconductor wafers are known as silly and jelly. They are prepared by taking a single crystal of Si or Ge (*p*-type, i.e., doped to be "hole rich"), coating the top surface with lithium, and then diffusing the latter into the body of the wafer at high temperature. The loosely bound outer Li electrons pair up with atoms of the doping agent (gallium or indium) to form a large *intrinsic* volume. The top surface becomes *n*-type (electron rich); the bottom remains *p*-type so what we have is a *PIN* sandwich, most of which is *I*. By definition, an intrinsic region is one with balanced charge; i.e., it has no excess of either free electrons or positive holes. If a thin gold layer is plated or evaporated on the *p* and *n* surfaces to form electrodes and a reverse bias of 600–900 V applied across the sandwich, a strong electric field is set up throughout the *I* region. Now if a quantum enters this region it undergoes photoelectric absorption and the resulting energetic electron expends its energy in forming many electron–hole pairs. The electrons will migrate undiminished and vice versa for the holes, producing a pulse at the electrodes proportional to the number of electron–hole pairs formed. The remarkable

---

* *PIN* stands for positive-intrinsic-negative.

resolving power of the Si(Li) detector can be attributed to the large number of electron–hole pairs formed in contrast to the relatively low number of free charges created when the same quantum is absorbed in a proportional counter or scintillation–photomultiplier assembly. Even the most mono-chromatic x-ray beam will be recorded as a Gaussian peak owing to the unavoidable statistical spread. The actual peak width is proportional to the square root of the number of primary charges forming each pulse. For example, when a Cu K_α quantum is absorbed in a proportional counter ≈ 350 ion–electron pairs are formed. When the same quantum enters a scintillation setup, only 13 electrons on the average reach the first photo-multiplier dynode. But when a Si(Li) detector absorbs a Cu K_α quantum an average of 2100 electron–hole pairs are created and all these charges are collected. The resolution of a quantum detector is normally expressed as the ratio: full width of recorded peak at half maximum (FWHM) divided by the peak height energy. Using this definition, and the preceding data, we find the resolution for a proportional counter in the Cu K_α region to be $\approx 17\%$; the scintillation device $> 50\%$, while the Si(Li) counter is under 5%. Another way of assessing counter resolution capability is the FWHM ex-pressed in electron volts (eV). Since a Cu K_α quantum packs an energy of 8040 V, the Si(Li) device can be said to have a FWHM of $8040 \times 0.05 = 400$ eV. in the 8 keV region. More recent improvements in field effect tran-sistor preamps utilizing optoelectronic and pulsed feedback techniques have brought the FWHM values for Si(Li) amplifier systems as low as 100 eV.

It is of interest that achieving such sharp lines enables one to separate the characteristic radiation (same series) from adjoining elements. This has given great impetus to the popularity of EDS energy-dispersive *spectrometry,* identification of the elements in a sample from their characteristic fluorescent lines excited by a high energy beam of primary x rays.

The relatively inferior resolving power of proportional and scintillation counters is usually not overly serious in diffractometer work since the diffraction peaks are normally spread out over a large 2θ interval and peak overlap is the exception rather than the rule. But if one attempted to use a scintillation counter in an EDD experiment, the poor FWHM (>5 keV) around the 10 keV region would be devastating. Referring to the 10–20 keV region in Fig. 27, one would record only a broad smear instead of five distinct lines.

Now one seldom gets something for nothing in x-ray work, as well as life, so the Si(Li) device is not a panacea for all ills. Because of high noise at ordinary temperatures, the Si(Li) detector must be operated at liquid nitrogen temperature, which is a nuisance to say the least. This also holds true for the preamplifier, invariably a field effect transistor. Frequently the detector and preamp are formed on the same silicon chip. Even aside from the noise factor, Li ions tend to migrate at room temperature, so a Si(Li) drifted counter must be maintained at low temperature from the moment of manufacture to the end of its days. Progress is being made in developing semiconductors of such purity (Ge in particular) as to be intrinsic without Li drifting, and thus such a unit could be left at room temperature when not in use. However, to achieve an adequate signal to noise ratio, even these advanced devices must be operated at cryogenic temperatures. The EDD main amplifier, aside from the linearity requirement, is essentially conventional.

C. The Multichannel Analyzer

Now for the other magic components, those pants with the marked pockets plus ready means for determining the contents of each. Scientifically, this combination is an electronic marvel called a multichannel analyzer, and consists basically of five major components, as follows: (a) an input condenser (rundown capacitor), (b) an address register, (c) constant frequency oscillator, (d) memory unit, in which each address corresponds to a specific channel (pocket), and (e) one or more readout or display devices for data retrieval and presentation. Actually the multichannel analyzer is not new. Wilkinson (1950) built his "Ninety-Nine Channel Pulse Amplitude Analyzer" some 30 years ago. Of course current components make his apparatus appear ancient, but the basic principles remain the same.

In operation the peak potential of the linear output pulse produced by an x-ray quantum is used to "gate" the oscillator. The unit operates as follows: The incoming pulse charges the input condenser to the peak potential of that pulse. The condenser then discharges at a constant rate. During the discharge period the address register receives and counts pulses from the oscillator. At the end of the discharge period, the register, based on the number of oscillator pulses received, decides which memory address (pocket, or in electronic parlance channel) is to be increased by one count. It should be clear that the greater the input pulse, the larger the number of oscillator pulses counted and the higher the channel number which is increased by one. At this point the next amplifier pulse is admitted and the procedure is repeated, pulse by pulse. Since the pulses are handled sequentially, there is a limit to the count rate that can be accomodated with negligible losses (usually ≈ 5000 counts/sec). Most commercial multichannel analyzers include a *dead time* meter which allows one to ascertain the count rate tolerable without distorting the results. For convenience we have been talking about an analyzer with 100 channels. However even the simplest commercial analyzer offers 256 channels and elaborate models boast over 8000. Irrespective of the number of channels, the pockets can be sized to cover the full

Fig. 28 This photograph illustrates the complex nature of a modernday EDS setup. This on-line computer controlled system incorporates magnetic tape, a large area CRT display oscilloscope, a function keyboard and control panel keyed into a DEC-PDP-8 computer. Photo by the courtesy of R. L. Heath (1972), Aerojet Nuclear Company.

output voltage of the amplifier. Normally a magnetic core memory serves to store the data. Practically all multichannel analyzer units provide a cathode ray tube (CRT) to view the data as it appears and in final form. For a permanent record one can choose a teletype printer, x-y plotter, or magnetic or punched paper tape (for later off-line analysis on a large central computer).

The ED system as described was first pioneered by nuclear physicists for measuring the energy spectra of gamma rays and charged particles. They have carried this development to a high degree of sophistication, particularly with the introduction of on-line computer control into the system which permits far greater versatility than in the "hard-wired" system described above. The on-line feature is particularly valuable in EDD and EDS work because of the rapidity with which the data accumulates. Figure 28 shows a typical EDS on-line complex, while Fig. 29 depicts an ED acquired spectrum of 303 stainless steel. This pattern was obtained in a total time of 17 sec.

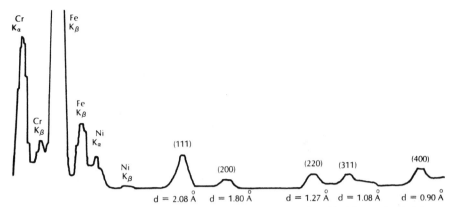

Fig. 29 Spectrum of 303 stainless steel. This sample was analyzed in 17 sec using 1 mA at 46 kV (tungsten tube). This plot is by the courtesy of Nuclear Equipment Corp., San Carlos, CA.

As noted, ED was developed primarily for its *spectrometry* capability, but after Giessen and Gordon demonstrated its value for diffraction work, considerable attention has been paid to this particular aspect by many, including Freud and La Mori (1969), Chwaszezewski *et al.* (1971), five papers by Laine *et al.* (1972a, b, Laine and Lähteenmäki, 1971, 1973; Laine and Tukia, 1973), Martin and Klein (1972), Sparks and Gedcke (1972), Fukamachi *et al.* (1973), Lin (1973), and Wilson (1973). Jenkins and Westburg (1973) coupled an ED unit to a conventional Norelco diffractometer and others have shown that a Si(Li) detector can replace conventional counters on a diffractometer with beneficial results, namely Drever and Fitzgerald (1970) and Carpenter and Thatcher (1973).

Weighing all the work published to date there does appear to be a proper niche for EDD. However it will never displace the diffractometer just as the latter has not made film units obsolete. The small cross-sectional area of the Si(Li) wafer ≈ 30 mm^2 enables the detector to be positioned close to the specimen, subtending a large solid angle and thereby receiving a large fraction of the diffracted rays. Other major advantages offered include: speedy results, literally seconds in many cases; the need for but a single incident and exit beam path, and at no extra effort one obtains elemental information (fluorescent peaks) along with the diffraction pattern. Thus the technique should prove especially useful in rapid, exploratory work, revealing metastable phases in chemical reactions or alloy order–disorder transformations, and also in high temperature, high vacuum, and high pressure experiments (Skelton, 1972) where the auxiliary equipment present makes a broad 2θ exit pathway infeasible if not impossible.

Now for the disadvantages. The cost of a high performance EDD system, the liquid nitrogen demand, the problems of measuring absolute diffracted intensity realizing the variable white source, and the fact that lattice parameters can still be measured more accurately by the diffractometer by an order of magnitude are all drawbacks to be recognized and will likely limit EDD mainly to cubic and other simple structures. However, since most metals and alloys take the cubic form, this latter is not as detrimental as it sounds.

D. Novel Low Angle Diffraction Technique
Utilizing Stationary Position-Sensitive Counter

When we speak of low angle diffraction we are usually referring to 2θ values from $0°$ out to $2°$ or $3°$ maximum. To appreciate the small region involved, let it be noted than with a normal flat plate film sample distance of 5 cm and a round beam stop 2 mm in diameter, the beam stop blanks off an area out to $2\theta = 1°$. It is clear that extraordinary measures and equipment will be needed to study this region. But why, one might ask, would we wish to explore this territory in the first place? Well, it so happens that in certain organic, particularly biochemical specimens, diffraction spacings in the 50–100 Å range are not at all unusual. Second, there is another type of x-ray scattering prominent at low angles which has nothing to do with Bragg diffraction but has a lot to do with the particle sizes of the irradiated sample. Since the latter phenomenon does not involve true diffraction we will not pursue the subject here but an interested reader can refer to the following references: Biscoe and Warren (1942), Jellinek et al. (1945), Shull and Roess (1947), and Guinier and Fournet (1955).

Whether measuring long periodicities or small particle sizes, low angle techniques have changed little over the years with the main features being the need for monochromatic radiation, very fine slits, long specimen–film (or counter) distances, this distance normally comprising an evacuated tube to minimize air scatter and long, long exposure times. Two recent developments have enabled one to obtain low angle scattering or diffraction data in times literally orders of magnitude shorter than was required a few years ago. The first of these we have met above, our friend the multichannel analyzer, although strangely enough its prime value here is a recorder of distances rather than energies as in EDD work. The other development is a revolutionary "position-sensitive" proportional counter, first proposed by Borkowski and Kopp (1968). In their counter design, the central collecting wire is replaced by a high resistance 0.01-mm quartz fiber with a pyrolytic graphite coating. The resistance runs roughly 40 000 Ω/mm. A second wire of low resistance nickel, known as the pickup electrode is spaced 1.5 mm behind and parallel to the quartz fiber and is maintained at a potential so as not to distort the field around the central fiber. When an x-ray quantum enters the counter and produces an ionizing event, the pulse induced in the pickup wire travels rapidly to the end of the counter. The rise time of the pulse traveling along the quartz fiber is much slower. Both pulses are amplified, double differentiated by two RC circuits to form two bipolar pulses. At the fast pulse (Ni wire) crossover point, a fixed frequency oscillator is gated, and at the similar point for the delayed (quartz) pulse the oscillator is stopped. The number of oscillations received and accumulated is converted to a voltage which is proportional to the elapsed time between the crossover points of the two pulses. This voltage is likewise a measure of the counter position at which the pulses were generated. A calibration curve is established linking position to voltage and the time to amplitude voltage is passed along to the appropriate analyzer channel. All this may sound unduly complicated, but the authors assure us the electronics are quite conventional and very dependable. The beauty of it all is that with a counter 400 mm long, one can determine the entry position of the x-ray quantum to a high degree of accuracy. To quote figures, the FWHM position for a beam of 22-keV x rays can be spotted within 0.5 mm of its true location anywhere along the full length of the counter. With such a counter-multichannel analyzer, small angle measurements involving long repeat distances can be determined quickly. For example, DuPont et $al.$ (1972) published a small angle diffraction pattern (a major repeat distance of 97 Å) of a phosphatidic acid–lysozyme–water phase obtained in the incredibly brief time of 1.2 sec. Small angle workers usually expose film for hours or even days. These same researchers (DuPont) were also able to carry out kinetic order–disorder studies among lipids, lipoproteins, and certain biological membranes never before possible.

Andre Guinier,* in his 1974 Fankuchen Memorial Lecture at Berkeley before the American Crystallographic Association alluded to this advance in x-ray instrumentation and displayed two equivalent diffraction patterns of crystalline cholesterol myristate showing spacings of 49.7, 25, and 16.5 Å, respectively. The first pattern was recorded on film at an exposure time of 3 hr. The second pattern, equally informative, was obtained with a position-sensitive counter-multichannel analyzer setup in 100 sec. This striking advance should open up new opportunities for small angle studies of all kinds in the years ahead.

IX. Qualitative Analysis by X-Ray Diffraction

Just as an individual possesses a unique set of fingerprints, crystalline structures yield unique x-ray diffraction patterns if the material under study is composed of small crystallites as is true for most powders and metals.

---

* See Guinier (1975).

The diffraction "fingerprint" for a particular component irradiated by a monochromatic beam of x rays will consist of a sequence of lines or more correctly a series of cones. Using apparatus already described, these lines can be measured and reduced to a list of d values (spacings expressed in Å) and corresponding relative intensities. For convenience, the intensity of the strongest line is assigned the value 100 and others are scaled accordingly. We do not include the EDD procedure here, inasmuch as the line intensities recorded in the standard reference file (more on this later) were obtained exclusively with monochromatic radiation. Hence there would be no relation among standard file and EDD line intensities. Now, how do we take these tabulated data and using them deduce the elements, compounds, or alloys comprising the sample?

A. JCPDS and PDF

Returning to the fingerprint analogy, the FBI compares an "unknown" set of prints with its voluminous master file of known prints until it discovers a match. Fortunately, the x-ray analyst also has a master file of powder diffraction patterns at his command. It is called the Powder Diffraction File (PDF), issued and updated annually by the Joint Committee on Powder Diffraction Standards (JCPDS). Space in this chapter does not permit recounting the tremendous amount of work which went into producing the present PDF, but a diffractionist should reverently utter the names Hanawalt, Frevel, Rinn, Davey, and Fink every time he makes use of this invaluable tool. The complete file at last count contained over 33 000 patterns of crystalline materials and for chronological purposes is composed of 29 sets, each set divided into an inorganic section consisting of inorganic compounds, metals, alloys, and minerals and an organic section consisting of organic and organometallic compounds.

The PDF is available in various formats. For a complete description and information on PDF availability and prices write to the Joint Committee on Powder Diffraction Standards, 1601 Park Lane, Swarthmore, PA 19081. First let us describe the basic $3'' \times 5''$ data cards which are the heart of the system. Figure 30 is a reproduction of the card for sodium chloride (NaCl). As noted at the top of the card NaCl happens to be the 628th pattern in set No. 5. In the first row are the three strongest lines in the pattern listed in order of decreasing intensity (as shown beneath in row I/I_1). The fourth figure in row d lists the longest spacing found and below its relative intensity. The next lower compartment describes the radiation used in securing the pattern, the filter, the instrument, and the source of the information. The following block includes crystallographic data, relatively simple for NaCl; the succeeding section contains optical and physical data; and the final

5-0628

d	2.82	1.99	1.63	3.258	NaCl	✶
I/I₁	100	55	15	13	Sodium Chloride	Halite

		d Å	I/I₁	hkl	d Å	I/I₁	hkl
Rad.Cu λ 1.5405 Filter		3.258	13	111			
Dia. Cut off Coll.		2.821	100	200			
I/I₁ d corr. abs.?		1.994	55	220			
Ref. Swanson and Fuyat, NBS Circular 539, Vol. II, 41 (1953)		1.701	2	311			
		1.628	15	222			
Sys. Cubic S.G. O_H^5 - FM3M		1.410	6	400			
a₀ 5.6402 b₀ c₀ A C		1.294	1	331			
α β γ Z 4		1.261	11	420			
Ref. Ibid.		1.1515	7	422			
		1.0855	1	511			
δα nωβ 1.5428 γ Sign		0.9969	2	440			
2V Dx 2.164 mp Color		.9533	1	531			
Ref. Ibid.		.9401	3	600			
		.8917	4	620			
An ACS reagent grade sample recrystallized		.8601	1	533			
twice from hydrochloric acid.		.8503	3	622			
X-ray pattern at 26°C.		.8141	2	444			
Replaces 1-0993, 1-0994, 2-0818							

Fig. 30   PDF card for NaCl. Card reproduced by the courtesy of JCPDS.

section indicates sample preparation, temperature at which the pattern was recorded and miscellaneous information. The right-hand vertical columns record the complete diffraction pattern in order of decreasing d values, the corresponding relative intensities I/I_1, and the Miller index (hkl) for each line (when known). The presence of a star on a card signifies data of the highest quality and reliability.

The PDF is also available on microfiche, again one section of 236 fiche for inorganic materials, the other consisting of 91 fiche for the organic section. If one purchases the PDF on microfiche, he also receives a metal filing container, a three-ring binder for filing fiche, the respective search manuals (to be described shortly), and a microfiche reader.

Card sets 1–22 can also be had in book form (six inorganic and six organic volumes) with three patterns to the page.

B. Hanawalt and Fink Search Methods

When one purchases the PDF Inorganic (cards or microfiche), one also receives an alphabetical index listing compounds by their chemical names. This listing is rotated and the fragments of the chemical name entered alphabetically providing added convenience when prior knowledge is known. Where applicable, the chemical name is followed by the mineral name. The three most intense d spacings, intensities, chemical formula, PDF card number, and microfiche grid position are given for each entry. A Hanawalt numerical search section is also available listing some 2500 frequently encountered phases. In addition one receives a Hanawalt Method inorganic search manual. Eight d spacings are given for each chemical compound and the three most intense d spacings permuted so that each compound is entered at least three times at three different locations. Intensities are shown as subscripts and the chemical formula, PDF number, and microfiche grid position are given for each entry.

An alternate though less popular search method using PDF data has been developed by W. L. Fink and bears his name. JCPDS also offers for sale a Fink search manual for the inorganic file.

The Fink inorganic search manual is based on searching for one of the four most intense d spacings and matching the remaining d spacings in descending numerical order. Eight d spacings are given for each chemical compound with one of the four most intense d spacings selected for each entry. Intensities, although not imperative for searching with this manual, are shown as subscripts. The chemical formula, PDF number, and microfiche grid position are given for each entry.

When one acquires the organic PDF set, he receives an organic search manual. This is divided into a numerical section, an alphabetical section, and an empirical chemical formula section. The numerical section is arranged and searched in the same manner as in the Hanawalt inorganic method mentioned above. The alphabetical section provides one entry for each compound in alphabetical order of its chemical name. The three most intense d spacings, intensities, chemical formula, PDF number, and microfiche grid position are also given. The empirical formula section lists compounds by carbon number order. Hydrogen is the second element and all others are listed alphabetically.

Because of the specialized needs of mineralogists and geologists, JCPDS also offers a two volume publication, "Selected Powder Diffraction Data for Minerals." One volume contains reproductions of 1900 PDF cards classified as minerals while the other is a search manual similar to the Hanawalt inorganic search manual described above. For an extra $25 one can secure a Fink Method search manual applicable to the mineral data.

Detailed instructions for routine use of both search methods have been prepared by JCPDS.

If the sample is a single crystalline entity and this entity appears in the PDF file, the search is rapid and success almost guaranteed. Even a bicomponent crystalline mixture is usually readily resolved when one fraction predominates, say represents 75% of the sample, since the stronger lines will normally be associated with the major component. Once this pattern has been subtracted from the complete set of data covering the unknown, the minor constituent should be identified in short order. However, nature is seldom this cooperative and it is not uncommon to be faced with identifying a sample comprising three, four, or even a half dozen components.

C. Identifying Complicated Multicomponent Mixtures

In complicated situations manual searches are difficult at best, with odds favoring several superposed lines which make straightforward solutions practically impossible. There *are* a few tricks to this trade. If the diffraction data are recorded on film, look for lines that are more grainy, or conversely, broader or sharper than others. There is a fair chance these kindred lines belong to a specific constituent. Also, if the material has been subjected to elemental analysis, either chemically or spectrometrically (even EDS), such information may offer helpful clues in cracking multi component mixtures. A permanent magnet can be run across the sample and with luck, *one* component will prove magnetic. And on rare occasions *one* component will be water soluble and thus easily removed from the conglomerate. But do not count on such good fortune. In general, for complex samples a more sophisticated approach is mandated. For this reason the PDF is likewise available from JCPDS as a data base on magnetic tape. Two files are present on the tape. The first is comprehensive; the second lists the PDF numbers of: (a) 2500 frequently encountered phases; (b) 3000 minerals; (c) 5700 metals and alloys; (d) 1000 National Bureau of Standards; (e) 200 common materials; and (f) 200 common minerals. Available parameters for the data base tapes are: 7 channel 556/800 bpi or 9 channel 800/1600 bpi with BCD/EBCDIC/ASCII codes. Univac users may also request 6250 bpi. These tapes are leased and a choice of retrieval programs is offered. The most popular is the Johnson–Vand Search/Match Retrieval Program (1968) available in three modifications; one suitable for the IBM 360/370, one for UNIVAC 1108 series, and one applicable to the CDC 6000 series computers. The desired computer program and a copy of the operations manual is supplied as part of the lease agreement.

For laboratories lacking in-house computer capability, JCPDS has a service called 2dTS: Diffraction Data Tele-Search, where, by means of data from the user's laboratory via teletypewriter, complete access to the PDF Data Base using the Johnson–Vand program is achieved.

D. Defects Inherent in the PDF Method

One might assume that, if the diffraction data from the unknown are of adequate quality and are complete, the utilization of computer searches practically ensures an accurate and unequivocal answer in every instance.

Unfortunately this is not quite the case. To see why, just recall that over a half million organic compounds are known, yet the PDF contains only 8400. Actually the odds are not as bad as they seem since the 8400 are compounds one is more apt to come in contact with in organic work. Nevertheless, one cannot expect to identify a component not in the file. Of course, negative information is sometimes quite worthwhile. There are also numerous less obvious reasons which prevent a correct answer. These include such things as spurious lines from an x-ray target contaminated by tungsten evaporated from the filament, a high fluorescent background might well blot out certain weak lines, change in sample composition owing to deliquescence, efflorescence or oxidation, introduction of contaminants during sample grinding or ball milling to reduce particle size, solid solutions with lattice parameters within experimental error equal to other compounds in the standard file listings, and last, polymorphism can sometimes be a problem. For an interesting critique of these and other complicating factors, read Klug and Alexander (1974, pp. 506–531).

One of the most intractable families of compounds to challenge the x-ray diffractionist is the clays. Most of these consist of layer structures akin to the micas and chlorites with pronounced basal cleavages and are lamellar in nature. Frequently there may be imperfections arising from random substitution of one moiety (Al, Mg, Ca) for another, from irregular superposition and from stacking of layers of different composition. Anyone planning to probe this area via x rays would be well advised to first study the following references: MacEwan (1949), Brown (1961), and American Petroleum Institute (1950).

The author would not wish to overemphasize the drawbacks cited in this section. True, they exist and cannot be ignored. But the brilliant record compiled over the years by a myriad of workers using the PDF and its predecessor data to great advantage should not be forgotten. Their successes far overshadow the limited number of failures. In fact, a major triumph for PDF is that it served as the major analytical tool to identify the rock samples brought from the moon. JCPDS might use this to argue PDF is out of this world!

E.   Diffractometers with Built-In Computers to the Rescue

The material in the above section has been written primarily for those workers still operating film cameras or manual diffractometers. If one is fortunate enough to have acess to a modern automated diffractometer with a memory for data storage, much of what has been said in the last few pages is probably superfluous. For instance, one of these remarkable machines allows for the storage of up to 73 separate compound patterns in its memory. Thus, an unknown sample can be analyzed automatically for the presence of any or all of these 73 compounds. And since this unit boasts an automatic 36 sample holder, it can conduct a maximum of 2628 compound analyses unattended and unassisted. While this sounds most impressive, as it is, the law of averages argues that some of those 36 samples will contain compounds not included among the select 73. So for these samples, the keeper of the genial genie is in the same boat as the rest of us less fortunate mortals and must resort to some of the procedures recounted earlier. However, another diffractometer manufacturer has anticipated us here. His automated machine, which can handle 34 samples at a time, prints out the collective peak data (intensity and 2θ) for each sample on punched tape in such form that the latter can be fed directly into a Johnson–Vand programmed computer and theoretically at least, the results come flowing out, untouched by human hand. One might thus conclude that the only role played nowadays by the diffractionist is to load the sample holder and push the starter button. Fortunately for those of us who still like a few challenges, this is not always the case. Johnson–Vand, while usually spotting the true components (up to seven no less) will frequently throw in a number of spurious possibilities owing to small errors in the stored or input data. Even computers are human! Usually an elemental analysis will enable the user to eliminate the erroneous structures.

X.   Quantitative Analysis by X-Ray Diffraction

Now that we have identified the constituents in our sample, someone is bound to want to know how much there is of each. Its an easier question to ask than to answer—a lot of hard work is involved. A cynic might suggest that since we keep referring to difficulties at every turn, why go to all the bother. Here we have a good answer. There are many important chemical problems which can only be attacked via x-ray diffraction. To name just a few:

(a)   Identifying and measuring allotropic and polymorphic forms of the same substance, e.g. [TiO_2 (anatase) and TiO_2 (rutile) or quartz α-SiO_2 and cristobolite α-SiO_2].
(b)   Amount of quartz in silica dust.
(c)   Amount of asbestos in various clays.
(d)   True structure of many solid solutions.
(e)   Amount of retained austenite in steel.

So even our cynic should admit x-ray diffraction *does* have a useful role to play in the chemical laboratory. But back to the original question of quantitative analysis. Since these results depend on accurate measurements of diffraction peak areas associated with the component in question, a diffractometer is the *sine qua non* here. But even if one had his unit operating perfectly and took great pains to prepare uniform, flat surface, nonoriented samples with particle size ≈ 5 μm, he would still have problems. The *big* problem is matrix absorption. The matrix is defined as collectively all the components which are present in the sample. If you are interested in measuring the amount of component a, it is clear that the other matrix constituents will absorb some of the x rays that would otherwise reach the detector if a were a pure sample. So, since 1936, much perspiration and inspiration have been exuded seeking ways of overcoming matrix absorption effects in order that one *can* know truly how much a there is in his sample.

We will present several successful solutions to this problem. One of the first serious approaches was that by Alexander and Klug (1948).

A. Alexander and Klug Method

They investigated the problem of x-ray intensities diffracted by a small-particle-size flat powder specimen, sufficiently thick to give maximum diffracted intensities. This type of sample is commonly employed in diffractometer work. For a uniform multicomponent mixture, they deduce the following general expression:

$$I_1 = K_1 x_1/\rho_1 [x_1(\mu_1{}^* - \mu_M{}^*) + \mu_M{}^*] \qquad (18)$$

where I_1 is the diffracted intensity for a chosen line from component 1; x_1, ρ_1, and $\mu_1{}^*$ represent the weight fraction, density, and mass absorption coefficient of the desired component; $\mu_M{}^*$ is the effective mass absorption coefficient of the matrix; and K_1 is a constant, being a function of the apparatus geometry and component 1.

In making practical use of the above equation, the authors consider three special cases, depending on the number of components present and the relative values of $\mu_1{}^*$ and $\mu_M{}^*$.

Case 1.   Mixture of n components: $\mu_1{}^* = \mu_M{}^*$. Here Eq. (18) reduces to

$$I_1 = (K_1/\rho_1\mu_M{}^*)x_1 = Kx_1 = (I_1)_0 x_1 \qquad (19)$$

which states that the diffracted intensity from component 1 is proportional to its concentration in the sample. Thus, all the prior calibration information needed is a line intensity measurement from a pure sample of 1 to fix

$(I_1)_0$. Admittedly, there will be few practical cases where all components in the sample have equivalent absorption coefficients.

Case 2. Mixture of two components: identifications known, but $\mu_1^* \neq \mu_2^*$. In this instance the intensity–concentration curve will not be linear, since the absorbing powers of the unknown and diluent are unequal. A calibration curve can be prepared experimentally from synthetic mixtures of the two materials, covering the full range of composition. Since the respective mass absorption coefficients can be obtained from reference to tables (e.g., "Handbook of Chemistry and Physics"), an equivalent calibration curve can be calculated from the expression

$$\frac{I_1}{(I_1)_0} = \frac{\mu_1^* x_1}{x_1(\mu_1^* - \mu_2^*) + \mu_2^*} \tag{20}$$

here $(I_1)_0$ is the line intensity from a pure sample of material 1. Figure 31 illustrates typical results for cases 1 and 2.

Case 3. Mixture of n components ($n > 2$): $\mu_1^* \neq \mu_M^*$. In this case the absorbing power of the desired component differs from that of the matrix,

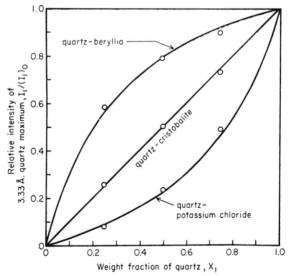

Fig. 31 Calibration curves to illustrate cases 1 and 2 (see text). Since quartz and cristobalite are polymorphs of silica, the criterion that $\mu^* = \mu_M^*$ is fulfilled, and a linear calibration curve results. The quartz–beryllia and quartz–potassium chloride examples correspond to case 2. Here the solid lines were calculated from Eq. (20), with $\mu_{SiO_2}^* = 34.9$, $\mu_{BeO}^* = 8.6$, $\mu_{KCl}^* = 124$. The experimental points show the excellent agreement found for synthetic mixtures containing known weight fractions of quartz. Reprinted with permission from Alexander and Klug, 1948, *Anal. Chem.*, **20**, 886. Copyright by the American Chemical Society.

the latter usually being unknown. Here one must resort to use of an internal standard. It can be shown that when a weight fraction, x_s, of internal standard is added to the sample under analysis, the concentration of the desired component is given by

$$x_1 = \frac{K'}{1 - x_s} \left(\frac{I_1}{I_s}\right) \qquad (21)$$

with I_1 and I_s being the diffracted intensities for a chosen line from component 1 and the internal standard, respectively,

$$K' = \frac{K_s \rho_1 x_s}{K_1 \rho_s}$$

Values of K_1 and K_s can be established from line intensity measurements $(I_1)_0$ and $(I_s)_0$ on the pure materials. Provided a constant weight fraction of internal standard is always added, Eq. (21) simplifies to

$$x_1 = K'' \left(\frac{I_1}{I_s}\right) \qquad (22)$$

and a linear calibration curve of I_1/I_s vs x_1 results. A sample consisting of pure component 1 with the constant proportion of internal standard added suffices to establish K''. Figure 32 depicts the results obtained when 20% by

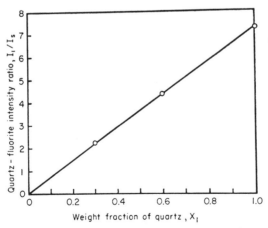

Weight fraction of quartz , X_1

Fig. 32 Plot to illustrate case 3, with 20% by weight CaF_2 as an internal standard. Each experimental point is the average of ten determinations from a sample containing a known amount of quartz. Note the linear relationship thereby confirming Eq. (22). Reprinted with permission of Alexander and Klug, 1948, *Anal. Chem.*, **20**, 886. Copyright by the American Chemical Society.

weight of CaF_2 (the internal standard) was added to known mixtures of quartz and calcium carbonate. The intensities of the 3.34-Å quartz line and the 3.16-Å fluorite line were used to determine the ratio (I_1/I_s).

Naturally our explanation of the Alexander–Klug studies have been abbreviated. For a more detailed account, read Klug and Alexander (1974, pp. 531–562).

B. Chung Matrix Flushing Method

More recently Chung (1974, 1975) has taken a refreshingly different approach to the matrix absorption headache. In essence he says, let us flush away the matrix problem, and this he proceeds to do. Space in this chapter does not permit a full description of his theory, but in fact he does add a "flushing agent" [his preference is corundum since it is offered by the National Bureau of Standards as a Standard Reference Material (SRM) and is of the correct crystallite size] to the mixture under study. The only stipulation is that the flushing agent not be present in the matrix (original sample). Chung insists this addition is not to be viewed as the usual internal standard since he does not require a calibration curve, rather only a single comparison point. Chung has developed the following simple equation to get the desired results:

$$x_i = \left(\frac{x_c}{k_i}\right)\left(\frac{I_i}{I_c}\right) \tag{23}$$

where x_i is the percentage (to be determined) of component i in the multi-component sample, x_c the percentage of flushing agent corundum in the total sample, I_i the strong diffraction peak count for i, and I_c the corresponding strong peak count for corundum from the sample. k_i is Chung's "reference intensity value," a constant obtained by measuring the ratio I_i'/I_c' from the strong diffraction lines obtained from a 50%–50% weight mixture of component i and corundum, respectively. This ratio should be obtained using the same sample and instrumental conditions as in the main experiment. Chung uses diffraction peaks, not areas under the peaks, and yet his cited results are amazingly accurate. It is my opinion that this happens because his synthetic mixture samples which he reports and the reference intensity values were obtained from the same source, so the particle sizes, etc. are identical in both determinations. In a general case I feel area measurements would be desirable.

Let us take a look at one of his experiments: The accompanying tabulation is for a three-component sample of ZnO, KCl, and LiF.

There are numerous samples besides this tabulation shown in his publications, all giving surprisingly excellent results.

Sample components	Weight (g)	Ref. int. value	40 sec. peak count I_i	Composition	
				Found (%)	Actual (%)
ZnO	1.8901	4.35	5968	41.14	41.49
KCl	1.0128	3.87	2845	22.04	22.23
LiF	0.8348	1.32	810	18.40	18.32
Flushing agent Al_2O_3	0.8181	1.00	599(I_c)	17.96 (known)	17.96
				99.54	100.00

Two final points should be made. When one is working with a binary mixture, Chung claims no flushing agent is needed since one component can serve as flushing agent for the other. Second, when all the experimental component percentages are added, if the total is less than 100% it suggests the presence of some amorphous material and the discrepancy indicates the amount.

C. Sahores' Method Using Compton Scattering

Sahores (1973) also proposed to eliminate the matrix absorption bugaboo but in fashion different from that of Chung. What Sahores did is set up a second detector on his diffractometer at $2\theta = 160°$ and measure the Compton scattered x rays. At this angle a Cu K_α x ray will have an increased λ of some 0.04–1.58 Å. Now, according to the author (Sahores) the diffracted peak area I_i from compound i of a multicomponent sample will be

$$I_i = K_1 C_i/\mu \tag{24}$$

where C_i is the weight fraction of the compound desired, μ the mass absorption coefficient of the matrix, and K_1 an undetermined constant. The Compton scattering from the same sample will be given by

$$I_c = K_2/\mu \tag{25}$$

where K_2 is another constant involving many factors and μ is as before, the mass absorption coefficient.

Dividing (24) by (25) we find:

$$I_i/I_c = K_3 C_i \tag{26}$$

So now μ has been eliminated and the I_i/I_c ratio is a linear function of C_i with a slope equal to K_3. This prediction was tested by mixing 20% quartz with a whole series of compounds (individually), whose μ values varied from as low as 8 to 220 cm²/g. When I_i/I_c was plotted versus μ for the various mixtures all points fell on a horizontal line, proving the ratio to be independent of absorption variables. Next, a series of calibration curves was established varying the quartz content from 0 to 100% using the preceding gamut of widely varying absorbers. When I_i/I_c was plotted versus % quartz in sample all points fell on a single straight-line calibration curve.

So Sahores was in business and has automated the equipment so it can handle 300 samples on its own. The diffraction data are recorded on magnetic tape simultaneously with the Compton scattering data on a separate channel and then the ratio is determined by an off-line computer. Like the Chung case, this technique is useful in measuring the amount of amorphous material present, very important in mineral studies which is Sahores' area of interest. As an added bonus, smaller than usually adequate samples can be accomodated since both the diffracted and Compton intensities are affected equally, and the ratio yields a valid result.

D.   Computers with Brains

The final quantitative technique we will mention involves those memory-laden, computer-controlled diffractometers such as pictured in Figs. 24 and 25. The Diano unit allows one the choice of three weight percent calculation equations and up to four corrections for absorbing compounds to take care of the μ problem. Without doubt the Philips machine is equally versatile in this respect. After sample introduction and button pushing to initiate the program, all necessary measurements will be carried out in proper order without fail; the indicated corrections will be applied, the chosen equation solved in a matter of milliseconds and the final results printed on a teletype for all to see. And the irony of it all is that, while these marvelous machines may on occasion blow a diode or develop a short in the microprocessor, not one has ever been known to exude a single drop of perspiration!

XI.   Crystal System Identification and Unit Cell Determinations

A.   Identifying Crystal System from Powder Pattern

In Fig. 33 are shown those diffraction lines obtained from the various kinds of cubic crystal, with the same unit cell size assumed for each. The expressions under the root sign are values of $h^2 + k^2 + l^2$, which allows one to deduce the reflections as specified. It is seen that in the range covered, a simple cubic crystal shows a diffraction line for every possible value of $h^2 + k^2 + l^2$. Seven, fifteen, and twenty-three are absent because there exists no three integers the sum of whose squares yields these totals. In body-centered and face-centered cubic crystals certain additional lines are missing.

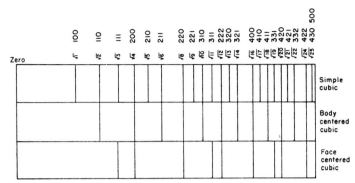

Fig. 33 Powder spectra expected from various cubic crystals, with the same unit cell size assumed. The figures under the root sign correspond to the sum $h^2 + k^2 + l^2$. The reflections *hkl* for each line are also listed.

This means there must be parallel planes containing an equal number of lattice points half-way between the set under consideration. When this condition exists, x rays scattered by the adjacent planes are 180° out of phase and complete destruction results. For example, in Fig. 10 no first-order reflection would be expected for (100) planes ($h^2 + k^2 + l^2 = 1$), since there are (200) planes interspersed between this set. There are no lattice points between adjacent (200) planes and thus no chance of interfering planes. Consequently, a line would be expected from the (200) set ($h^2 + k^2 + l^2 = 4$), and Fig. 33 shows this to be correct. In fact, it is by the absence of certain spectra from the diffraction pattern that one makes crystal class identifications.

The following general statements can be made with regard to body- and face-centered lattices. These statements hold true for all systems.

(1) Body-centered crystals give reflections only when $h + k + l$ is an even integer.

(2) Face-centered crystals yield reflections only for indices which are all odd or all even.

All crystals having the same type of cubic lattice give similar patterns. It must be remembered, however, that although the line *sequence* will be identical the actual positions will differ if the crystals have unit cells of different size. Likewise, the ratio of intensities from corresponding planes will not be the same for all lines of two different cubic crystals. These points can be illustrated by comparing the patterns from sodium chloride and potassium chloride. Both are face-centered cubic crystals. Yet their patterns as sketched in Fig. 34 differ on two counts. In the first place, corresponding lines from sodium chloride are spaced farther apart than those from potassium chloride. This tells us that potassium chloride has a larger unit cell.

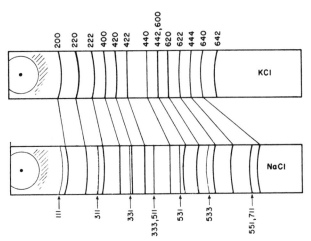

Fig. 34 Comparison of powder patterns from potassium and sodium chloride (a_{KCl} = 6.29 Å; a_{NaCl} = 5.64 Å).

Then there are a few weak lines on the sodium chloride film that do not show up at all for potassium chloride. This happens because K^+ and Cl^- have practically the same scattering power. Complete interference results, and thus certain legitimate reflections show zero intensity.

Suppose one wished to analyze a powder pattern having the following line sequence, moving outward from the central beam spot: A pair of lines in close proximity, a single line, another pair, a single line, a third pair, ... This would be strong evidence favoring a face-centered cubic structure. It would then require but a few minutes to index each line according to Fig. 33. Next, Eq. (8) could be applied to give a value for a, the lattice constant. If one obtained the same a using d and hkl values for every line, he could feel confident that his interpretation was correct.

Except for isometric (cubic) crystals this simple approach is not available to us. In general $a \neq b \neq c$ and $\alpha \neq \beta \neq \gamma$. Hence, there will be no unique line sequence, and the indexing poses a real problem. The situation is not too difficult for tetragonal, hexagonal, and rhombohedral crystals (rhombohedral crystals can be referred to hexagonal axes) since the angles are known, $a = b$, and one needs only a and the ratio c/a to index the pattern properly.* The procedure here has been greatly simplified by the publication

---

* Lipson (1949) has proposed an analytical procedure for indexing orthorhombic crystals which has been used by numerous workers with good success. Ito (1950) has developed a method which can in theory be applied to any lattice. Though extremely ingenious, these techniques are rather involved, and a detailed explanation is beyond the scope of this review. One interested in becoming familiar with these methods should consult the original papers.

of so-called Hull–Davey charts for the above systems. The ordinate represents c/a, and the abscissa corresponds to relative d values (based on $a = 1$) plotted on a logarithmic scale. Various curves indicate how all hkl spacings within the range covered vary with c/a. All planes parallel to the Z-axis (hko) have a constant spacing, as would be expected. To use these charts one places a strip of paper along the calibrated base line and a mark opposite each d value calculated from a powder pattern of the unknown. In most of the plots the base line covers a spacing decade, i.e., 0.25 to 2.5 Å, 2.5 to 2.5 Å, etc. For any spacings that cannot be accommodated at a single setting, the strip is shifted along the scale one full decade, effectively extending the range another order of magnitude. Once all spacings are recorded, the strip is moved over the chart to various horizontal positions until the marks coincide with curves on the chart. This amounts to trying different c/a and a values until the predicted pattern matches the observed one. When this is accomplished the lines are identified as to hkl, and the lattice constants can then be calculated. For example, if we tested this technique using the d values obtained from a powder pattern of zinc metal we would find an excellent match on the Hull–Davey plot for the hexagonal system at a c/a value of 1.85. Furthermore we would see at a glance that the 2.47 Å spacing was a $(00\cdot2)$ reflection, the 2.29 Å spacing a $(10\cdot0)$ reflection, the 2.08 Å spacing a $(10\cdot1)$ reflection, and so on. We would now possess sufficient data to determine the unit cell dimensions for zinc.

B. Unit Cell Determination

Noting that the 2.29 Å spacing arises from the $(10\cdot0)$ set of planes, we use Eq. (10) to find a. This gives

$$a = \sqrt{4/3}\, d = 2.66 \text{ Å}$$

Since $c/a = 1.85$, we know that

$$c = (1.85)(2.66) = 4.93 \text{ Å}$$

Thus the unit cell of zinc is completely determined.

C. Number of Atoms per Unit Cell

It is always prudent to check unit cell values by calculating the number of atoms or molecules per unit cell. A face-centered cubic crystal contains four equivalent points, a body-centered cubic crystal contains two points, a hexagonal crystal contains one point per unit cell, etc. The number of atoms (for elements) or molecules (for compounds) will be equal to or be some multiple of these numbers. If our calculations show this to be true, we may feel confident that our unit cell constants are probably correct. The number of atoms (molecules) per unit cell is determined as follows: The density of a body is

$$\text{density} = \rho = \frac{\text{mass of body}}{\text{volume of body}} \tag{27}$$

Consider a zinc unit cell. The mass it contains is

$$\text{mass} = (n)(65.38)(1.66 \times 10^{-24}) \, \text{g}$$

where n is the number of atoms (molecules) in the cell, 65.38 the atomic weight of zinc (for a compound, the molecular weight is used), 1.66×10^{-24} the mass in grams of unit atomic weight. The volume of a unit hexagonal prism is

$$\text{volume} = (\sqrt{3}/2) \, a^2 c = (0.866)(7.12)(4.93) = 30.5 \, \text{Å}^3 = 30.5 \times 10^{-24} \, \text{cm}^3$$

For zinc ρ is $7.14 \, \text{g/cm}^3$.

Putting the above figures into Eq. (27) we find that $n = 2.02$. Thus we conclude that zinc contains two atoms per unit cell. This is consistent with the previous statement and is in fact what one expects for an element whose structure is close-packed hexagonal.

In general, merely finding the number of atoms per unit cell does not uniquely determine the structure since there will be numerous ways to arrange this number of entities in the given volume. A knowledge of spectral intensities and the absence of certain reflections are usually required to make this final step. These data normally involve single crystal studies and will not be discussed here.

XII. Crystallite and Particle Size from Powder Patterns

An important property of powders and metals in their *particle* or *grain size*. Since an x-ray powder pattern is markedly affected by the size of the crystallites* producing the pattern, diffraction methods can be applied to problems of crystallite size.

From an x-ray diffraction point of view, the crystallite size (diameter) spectrum is divided into three ranges: (1) above 10^{-3} cm, (2) 10^{-5}–10^{-3} cm, (3) 0–1000 Å.

---

* A clear distinction should be made between particle and crystallite. A crystallite is taken to mean a small single crystal. A particle *may* consist of many crystallites in close contact. In metals terminology *grain* and crystallite are used interchangeably.

A. Crystallite Size from Spottiness of Pattern

In region 1 each crystallite in diffracting position contributes a spot to the x-ray pattern. (Such a pattern is depicted in Fig. 53b.) If one has calibrated a camera by measuring the spots produced by samples of known crystallite size, he can use this procedure on unknowns to gain some idea as to the dimensions of their crystallites. As an alternative he can calculate the volume of sample exposed to the x rays (transmission method) and count the number of spots on any diffraction ring. Then, allowing for the multiplicity of equivalent planes in the particular family giving rise to the ring, he can deduce an average crystallite size for the specimen. Microscope methods are to be preferred in this region because of their greater accuracy. For subsurface particles, however, or for measurements on the crystallite size of one component in a heterogeneous mixture, the x-ray method has a definite utility.

In range 2 $(10^{-5}–10^{-3}$ cm) the x-ray pattern shows only smooth lines (see Fig. 53a), since the individual spots are no longer resolved on the film. Furthermore, the pattern is insensitive to size changes within this region. Hence region 2 represents a blind spot for the x-ray method.

B. Crystallite Size by Line Broadening

Below 1000 Å (region 3) the diffraction lines begin to broaden because the minute crystals lack resolving powder. This phenomenon is quite analogous to the well-known optical case of diffraction maxima produced by line gratings. If the grating contains many lines to the inch the diffraction maxima are quite sharp, whereas a grating composed of but a few lines yields broad, ill-defined maxima. A crystal behaves as a three-dimensional grating toward a beam of x rays, and the breadth of a resulting diffraction line will depend critically on the number of diffraction centers in each crystallite, i.e., the crystallite size. The broadening effect is very pronounced once the average crystallite diameter falls below 200 Å. Since this is just where the electron microscope begins to lose sensitivity, there is real need for a method applicable to this region.

Many people have examined theoretically the problem of line broadening as a function of crystallite size. As long as one confines his attention to a size 200 Å and less, the average crystallite dimension is related to the line broadening by the formula

$$l = K\lambda/B\cos\theta \tag{28}$$

where l is the crystallite thickness normal to the (hkl) plane responsible for the diffraction line, λ the x-ray wavelength, θ the Bragg angle, B the angular broadening due to size of scattering units, and K a numerical constant. For normal three-dimensional crystals K is taken as 0.89. The value of B is normally computed from a photometer trace of the line in question, the width at half-maximum intensity divided by the camera radius being the desired quantity. Since all diffraction lines possess a finite width due to camera geometry, etc., B is more correctly given by

$$B = (B_x^2 - B_s^2)^{1/2} \qquad (29)$$

where B_s is the angular width at half-maximum for a line from a standard material known to fall in region 2. The standard may be mixed with the unknown so long as important lines do not overlap. B_x is the angular width for the unknown.

Figure 35 shows powder patterns for magnesium oxide of different crystallite size. These samples were produced by heating magnesium carbonate in air at (a) 450°C, (b) 550°C, (c) 650°C, and (d) 800°C. The effect of crystallite size on line breadth is clearly shown. Figure 36 presents a plot relating particle size in magnesium oxide samples with the heating period of magnesium carbonate at various temperatures. Here the sizes were determined by both x-ray broadening and the electron microscope. The two methods show good agreement in the region above 100 Å.

At this point a word of caution should be injected. Line broadening can also arise from other factors, chief among these being lattice distortions within crystallites. These distortions are caused by internal stresses which are not uniform throughout the grain. Therefore, it is sometimes a problem to assign correctly the cause of the observed broadening. If the sample has been annealed and is known to be free of internal stresses, then one can assert with some assurance that the effect is due to particle size. This is also true when the broadening is quite large, as is the case for the 450°C and 550°C magnesium oxide samples of Fig. 35. Microstrains cannot induce such extreme broadening as this except under very special conditions. Thus the work by Biscoe and Warren (1942) and Clark and Rhodes (1940) on carbon blacks and that by Clark *et al.* (1925) on nickel catalysts, to cite only a few cases, seem to be on a firm basis. In borderline cases where the broadening is not very marked, one must be conservative in interpreting the observed results. Dehlinger and Kochenörfer (1939) and Smith and Stickley (1943) have endeavored to distinguish between particle size broadening and distortion broadening in metals, but neither group claimed very positive results.

Oftentimes crystallite sizes obtained by line broadening differ considerably from particle sizes for the same sample as observed in electron micrographs. There is a simple explanation for this apparent discrepancy. It hinges on the difference between a crystallite and a particle. As noted on p. 242, a

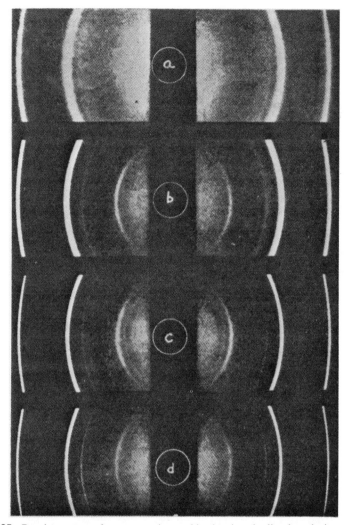

Fig. 35 Powder patterns from magnesium oxide showing the line broadening resulting from decreased particle size. Magnesium oxide specimens produced by heating magnesium carbonate at (a) 450°C, (b) 550°C, (c) 650°C, and (d) 800°C. Average crystallite diameter: (a) 42 Å, (b) 90 Å, (c) 142 Å, and (d) 225 Å. (From Warren, 1941.)

discrete particle viewed in an electron micrograph may be composed of many crystallites in close contact. Thus there is no *a priori* reason why one should demand agreement between the the two measurements. When agreement is obtained, as is the case in Fig. 36 it must mean that each particle is likewise a single crystallite.

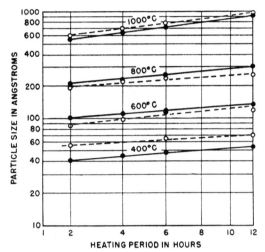

Fig. 36   Comparison of x-ray line broadening and electron microscope determinations on the particle size of magnesium oxide samples.—X-ray line broadening. - - - Electron microscope. (From Birks and Friedman, 1946.)

XIII.   Metals

Metals constitute the sinews of modern civilization and, as such, richly deserve the great amount of scientific attention that has been devoted to them. It is the atomic structure and crystallite arrangement in a metal or alloy and not merely the chemical composition which determine in large measure the unique properties it may possess. Since x-ray diffraction methods reveal structures, it is not surprising that these techniques have contributed greatly to our knowledge of metals and their technological use.

A.   Crystal Structure of Metals

A metallic crystal may be pictured as a lattice of positive ions submerged in a free electron gas. Such a model admirably explains many of the usual properties associated with metals and alloys, including their ductility, high thermal and electrical conductivity, ready combination in other than stoichiometric proportions, and closeness of packing. Judged by their large densities, it is not too surprising that many metals crystallize so as to make the most efficient use of the space available. This leads to one of two lattice structures, the face-centered cubic or the hexagonally close-packed arrangement. The third common metal structure, body-centered cubic, is not quite so efficient in its use of space. The ideas underlying the efficient packing of

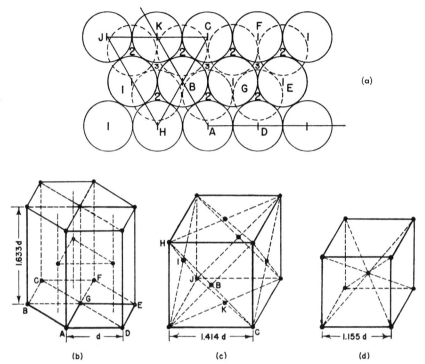

Fig. 37 (a) Diagram which illustrates the close packing of equal spheres. Those marked 1 form the first layer. The dotted circles (2) represent the second layer. The third layer can be formed in two ways either by placing spheres on position 1 or on 3. The first arrangement leads to (b), a hexagonally close-packed structure. The second plan yields (c), a face-centered cubic lattice; (d) represent a third common structure found in metals, the body-centered cubic lattice. (b) Two atoms per unit prism. Volume of unit prism equals $1.41d^3$. (c) Four atoms per unit cell. Volume of unit cell equals $2.82d^3$. (d) Two atoms per unit cell. Volume of unit cell equals $1.54d^3$.

equal spheres can be understood by reference to Fig. 37a. We start with a layer of spheres as indicated by the full circles. The second layer is formed by placing a sphere over each point marked 2. To complete a third layer we have two alternatives, to fill either those positions marked 1 or those labeled 3. It we choose the first course it is clear that spheres in the third layer are directly above those in the first, and the repeating sequence will be 1, 2, 1, 2, 1, 2, etc. If we adopt the second plan, however, the fourth layer is the first having positions which coincide with the initial layer and the sequence becomes 1, 2, 3, 1, 2, 3, 1, 2, 3, etc. It can be seen that the first scheme fits the close-packed hexagonal arrangement depicted in Fig. 37b, whereas the second corresponds to the face-centered cubic (f.c.c.) cell shown in Fig. 37c. The labeled points in each cell correspond to similarly marked spheres in the bottom layer of Fig. 37a. This shows that, although the bottom layer forms the base of the unit hexagonal, it is really the (111) plane in the f.c.c. cell. The body-centered cubic (b.c.c.) cell is shown in Fig. 37d. If it is assumed that each atomic sphere has a diameter d, the unit cell dimensions will be as indicated for the three modes of packing. Noting that the hexagonally close-packed (h.c.p.) unit prism contains two spheres, the f.c.c. cell four spheres, and the b.c.c. cell two spheres, we easily see that 74% of the available space will be filled in the first two arrangements, and 68% in the third. It is important to us that most metals crystallize in one of these ways, because all three possess high symmetry. This means that we can usually deduce the unit cell of a metal from its powder pattern. Bulk metals and alloys are normally composites of extremely minute crystallites and can be induced to form large single crystals only with difficulty.

B. Alloys

If two metals are thoroughly mixed in the molten state and then allowed to cool, several things can happen.

(1) They may separate as crystals of each metal. In this case, the x-ray diffraction pattern will exhibit the lines due to each component.

(2) They may dissolve one in the other to form a continuous series of solid solutions (alloys). Here a single pattern will be obtained, but the positions of the lines will be shifted to indicate a different size of unit cell.

(3) Under the right conditions a special kind of solid solution, termed an intermetallic compound, may be formed. In the x-ray pattern the shifting of atoms to form an intermetallic compound will be evidenced by changes in the line intensities and the possible appearance of new lines.

(4) Because of limited miscibility two distinct phases may exist side by side. Each phase will exhibit its own characteristic pattern on the diffraction film.

Which of these possibilities actually occurs depends on several factors, including identity of the two metals, amount of each in mixture, and heat treatment given mixture. For example, two metals will form a continuous series of solid solution only if they have the same lattice structure, have similar electronic configuration, and consist of atoms differing in size by no more than 12%. Au–Cu, Au–Ag, and Cu–Ni are examples of binary systems fulfilling these conditions. In such alloys atoms of one kind replace at random atoms of the other at lattice points. Gold and copper are f.c.c. crystals having cube edges a equal to 4.07 and 3.61 Å, respectively. Consequently, an alloy of gold and copper would be expected to show a f.c.c.

pattern over the whole range. According to Vegard's law of addivity, the lattice parameter will be linearly related to the atomic percentage of one of the components. Though generally true, numerous exceptions to this law are found (Mehl, 1934; Westgren and Almin, 1929). In any event, from x-ray patterns of known mixtures it is possible to establish a curve relating lattice constant to composition. It then becomes a simple matter rapidly to analyze unknown compositions of the alloy system by powder diffraction techniques. Trzebiatowski *et al.* (1947) describe how this is done for chromium–molybdenum alloys.

This general behavior, which corresponds to event 2 above is premised on an alloy quenched from the melt. If we should take a 50–50 (atomic percentages) gold–copper molten mixture and allow it to cool very slowly, a new phenomenon would result. The change is depicted in Fig. 38. Fig. 38a represents the quenched condition, with an equal chance of finding gold and copper atoms at any lattice point. Slow cooling brings about the condition displayed in Fig. 38b. There are now gold atoms at the corners of each cell plus top and bottom face centers, with copper atoms at the center of the side faces (or vice versa). This structure is called a *superlattice* and is a natural consequence of the crystal's endeavoring to assume the state of lowest potential energy.

The change from (a) to (b) will be readily noted in the x-ray pattern. Line intensities will change, owing to a difference in scattering power of the layers. Because of the different radii of gold and copper atoms, the new lattice will not be strictly cubic and new lines will appear. The transition between states (a) and (b) is known as an order–disorder transformation. Johansson and Linde (1925) were the first to make an extensive x-ray study of order–disorder transitions in the gold–copper system.

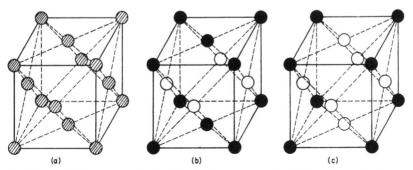

(a)   (b)   (c)

Fig. 38   (a) Gold–copper alloy in disordered state (quenched). There is equal chance of finding either atom type at any lattice point. (b) Gold–copper alloy in ordered state (super lattice). ●—Au positions; ○—Cu positions. (c) Gold–copper alloy in ordered state. ●—Au positions; ○—Cu positions.

The geometry of the ordered 50% gold–50% copper configuration requires a 1:1 ratio for the number of gold and copper atoms. It is termed an *intermetallic compound*, although it depends for existence on geometrical and not on chemical considerations. Another indication that the word compound is not strictly applicable arises from the fact this ordered phases is retained over a rather broad range of composition. In the neighborhood of 25% gold–75% copper, another such compound, $AuCu_3$ is found, with gold atoms at cube corners and copper atoms at the six face centers (Fig. 38c).

The system copper (f.c.c.)–zinc (h.e.p.) represents a combination of events 3 and 4 mentioned above. One would not expect a continuous series of solid solution here, and this is correct. Starting with pure copper one continues to get only a copperlike pattern (α-brass) until 38% zinc by weight has been added. Similarly, if one starts with zinc, 5% copper may be added before the hexagonal zinc lattice is seriously affected. Between these two limits one finds new patterns arising. For instance a 60% copper–40% zinc alloy shows, in addition to the brass pattern, a second phase (β-brass) which exhibits a b.c.c. lattice. Normally the two atoms occupy any lattice point in random fashion. When equal numbers of each are present, however, an ordered stated state is possible consisting of copper atoms at cube centers with zinc atoms at the cube corners (or vice versa). Since each corner contributes one-eighth atom to the unit cell, this corresponds to a "compound," $CuZn$. Other phases in this system are γ-brass (cubic) and ε-brass (hexagonal), represented in their most ordered states by Cu_5Zn_8 and $CuZn_3$, respectively.

Numerous other binary systems are found to yield a similar sequence of β-, γ-, and ε-structures. Hume-Rothery (1926) pointed out that the important condition governing each phase is the ratio

$$\frac{\text{Number of valence electrons}}{\text{Number of atoms}}$$

For the β-phase this is $\frac{3}{2}$, for the γ-phase it is $\frac{21}{13}$, and for the ε phase it is $\frac{7}{4}$. Note that the copper (one valence electron)–zinc (two valence electrons) compositions fit these ratios. It turns out that the atomic structure possessed by an alloy is a more important criterion of its physical properties than the atoms it contains. Therefore by studying the structure of a new alloy with x rays one can predict its properties with considerable accuracy.

All the previous cases refer to "substitutional" solid solution. An alternate arrangement is known as "interstitial" solid solution. This takes place when one of the alloying elements (H, B, C, N) has a small atomic radius, being possible for these to take up interstitial positions among the atoms of the solvent lattice. Above 800°C 1% carbon dissolves in iron in this way to form austenite. Interstitial solution always leads to an enlargement of the host lattice.

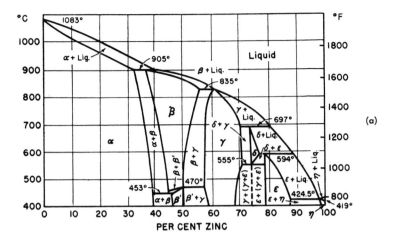

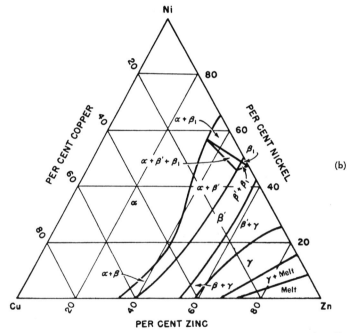

Fig. 39 (a) Constitution diagram for copper–zinc system. (b) Constitution diagram for copper–zinc–nickel system at 775°C. (From *Metals Handbook*, 1939, pp. 1367, 1375.)

XIV.   Constitution Diagrams

A.   *Methods Used to Establish Constitution Diagrams*

Information regarding the phases of an alloy as a function of composition and temperature can be presented most compactly in a "constitution" or "equilibrium" diagram. Figure 39a is such a diagram for the copper–zinc system, and Fig. 39b is a diagram for the ternary system copper–zinc–nickel. The complete ternary diagram is a three-dimensional plot. Shown here is the section corresponding to a temperature of 775°C.

Several methods can be used to establish these diagrams. They include: (1) use of heating and cooling curves, (2) dilatometric measurements, (3) study of physical attributes such as resistivity and magnetic properties, (4) microscopic examination of polished and etched surfaces, and (5) x-ray diffraction patterns.

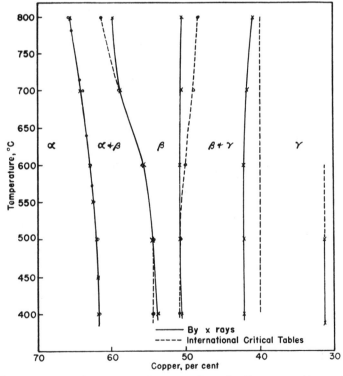

Fig. 40   Phase boundaries in copper–zinc system determined by x rays and by metallurgical methods. In general, the agreement is satisfactory. (From Lipson, 1943.)

The last two methods are used most and supplement each other very well. When care is exercised, the x-ray method yields results in rather good agreement with those obtained by metallurgical techniques Lipson (1943) (see Fig. 40). An advantage of the x-ray scheme is that it determines not only the phases present but also the crystal structures involved. Whereas both microscope and x ray usually examine the specimen at room temperature, depending on quenching to lock in the high temperature structure, special x-ray cameras have been used to study samples at the desired temperature. The high temperature cameras developed by Goetz and Hergemrother (1932) and by Hume-Rothery and Reynolds (1938) may be considered as typical designs. The x-ray method is inferior to the microscope in finding phases present in small amount. Sometimes, because of lattice imperfections, a phase representing as much as 5% of the total sample may be missed in the x-ray pattern. On the other hand, Bradley and Lipson (1938) by careful adjustments have found phases constituting less than 0.2% of the overall sample. Conversely, the microscope method fails for crystallites of very small size.

B.   Determination of Constitution Diagrams in Binary Systems Using X Rays

There are two ways of using diffraction patterns to locate phase boundaries. They are the disappearing phase method and the parametric method.

The first procedure depends on the fact that in a two-phase region the amount of each phase is proportional to the intensity of the diffraction lines from that phase. Hence, one takes x-ray patterns of several compositions in the two-phase region, quenched from a common temperature. The patterns are photometered, and from these data an *intensity versus composition* curve is plotted for one phase. Extrapolation to zero intensity will give the boundary for that phase at the quenching temperature. The boundary is extended by repeating the above, with a different quenching temperature.

The parametric procedure is based on the knowledge that lattice spacings usually vary smoothly with composition in a single-phase region, whereas within a two-phase area of a binary system (at any one temperature) the lattice spacings do not change with composition. Therefore one first establishes a *lattice parameter versus composition* curve by measuring several single (say α) phase samples quenched from a temperature that ensures complete solubility. Now one takes a specimen lying in the adjoining $\alpha + \beta$ region, quenches it from a known temperature, and measures the lattice constant associated with the α-phase. The composition corresponding to this parameter on the above plot represents the boundary for the particular temperature. Again the entire boundary may be established by repeating the second step at different temperatures.

It should be emphasized that these methods are subject to numerous errors, and extreme care is neccessary for accurate results. The reader is referred to an article by Hume-Rothery and Raynor (1941) for an extensive discussion of the precautions that should be taken in such work.

The previous considerations apply equally well to ternary alloy systems, although the possibility of added complications should be quite evident. Nevertheless, much effort has been devoted to ternary systems. Through use of certain properties associated with isothermal diagrams and numerous labor-saving principles that have been established, the work necessary to complete a ternary phase diagram can be greatly decreased. Barrett and Massalski (1966) have published an excellent discussion of method used in determining constitution diagrams.

XV. The Study of Precipitation Hardening with X Rays

Of great technical importance is the hardening of alloys on aging. A good example of this behavior is an aluminum–copper alloy containing about 5% copper, studied by Fink and Smith (1936). When this alloy is given a solution heat treatment followed by a quench, it exists as a supersaturated solution of copper in aluminium and initially shows a yield strength of approximately 25 000 psi. As shown in Fig. 41, after aging at 160°C for two weeks the yield strength has increased to 45 000 psi. This hardening also proceeds at room temperature, though more slowly. With continued aging the yield strength decreases once more. When examined microscopically certain alloys, during aging show a definite growth of new phases along slip lines and grain boundaries. From this picture it was originally believed that the hardening was due to the keying of slip planes by particles precipitated from the supersaturated matrix. A major objection to this theory was the lack of evidence for any precipitate in aluminum–copper specimens aged at low temperatures and studied either with the microscope or with x rays. It might be said that x rays actually hindered the early development of aging theory, because improved metallographic technique finally enabled workers to observe microstructural alterations in certain alloys as soon as the hardness began to increase. The negative x-ray results are explained by the inherent insensitivity of the diffraction method to phases present in minute amount. Precipitation can be well advanced in certain restricted areas while the general matrix is still largely unaltered. In this event the powder pattern may show no change. To avoid this difficulty several workers have resorted to the more sensitive Laue method involving single crystals. It is true that

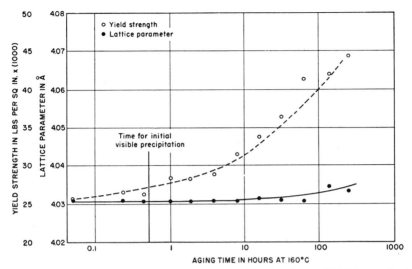

Fig. 41 Yield strength and lattice parameter vary with aging time at 160°C for an aluminum–copper alloy (5.17% copper). It is apparent that the yield strength begins to increase and precipitation of a new phase occurs long before the lattice parameter shows a measurable change. (From Fink and Smith, 1936.)

many alloys exhibit a uniform precipitation of particles throughout the grains, and here the lattice parameter deduced from powder patterns changes continuously during aging.

Once it had been appreciated that the x-ray powder method must be applied with prudence in aging studies, its value increased greatly. The method has now vindicated itself through numerous worthwhile contributions to the theory of precipitation hardening.

As an example, let us review the work by Barrett *et al.* (1941) on the alloy 79.8% aluminum–20.2% silver. They recorded powder patterns in a cylindrical camera at three different stages: (a) As quenched from solution heat treatment; (b) after 5 minutes of aging at 387°C; (c) after 16 hours of aging at 175°C. Figure 42 is a sketch of the patterns they obtained. The first plot shows a single f.c.c. δ-phase, typical of aluminum–silver solid solutions. The second pattern indicates that a new phase has formed, this one a h.c.p. structure which is called γ'. The third pattern still shows the h.c.p. structure but with slightly changed lattice constants. This phase is termed γ. A fact not evident in the plots is that the basal planes of the hexagonal crystallites showed the same orientation as the (111) f.c.c. planes of the bulk matrix. All these facts are nicely explained by the mechanism pictured in Fig. 43. Figure 43a represents f.c.c. δ-phase prior to aging, although the cubic symmetry is not apparent from the drawing. The layers pictured are

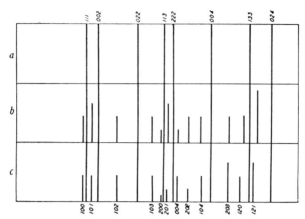

Fig. 42 Powder photograms of alloy of 20.2% silver in aluminum. (a) Spectrum of δ solid solution as quenched from solution heat treatment. (b) Spectra of $\delta + \gamma'$. Quenched and aged 5 min at 387°C. (c) Spectra of $\delta + \gamma$. Solution heat treatment followed by cold work; aged 16 hr at 175°C. (From Barrett *et al.*, 1941.)

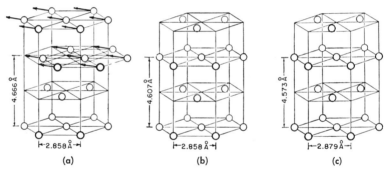

Fig. 43 Crystal lattice of phases present in aged alloys. (a) Structure of solid solution. (b) Structure of transition phase. (c) Structure of equilibrium precipitate phase. (From Barrett *et al.*, 1941.)

(111) planes. At an early stage of aging it is suggested the third and fourth layers shift as indicated. This leads to pattern (b) and a h.c.p. lattice (γ'). As aging continues, more and more planes make this shift, causing the thin precipitate layer to grow. This sets up extreme stresses in the bulk matrix, since at each interface the h.c.p. atoms must come into registry with the (111) planes of the f.c.c. lattice. These stresses undoubtedly contribute greatly to the hardening and are likely more important than any keying action produced by the precipitate. Eventually the layer grows to such a thickness that it is able to overcome the matrix stress. At this point the precipitate breaks

Fig. 44   Polycrystalline wire aged $25\frac{1}{2}$ hours at 387°C. ($\times 2000$). Precipitate is $\gamma' + \gamma$. Etched in 0.5% hydrofluoric acid. (From Barrett *et al.*, 1941.)

away and assumes the equilibrium γ-structure depicted in Fig. 43c. The interfacial stress is relieved, and softening of the alloy begins. The importance of this explanation lies in the far reaching effects that are predicted from relatively trivial shifts in the atom layers. Figure 44 pictures $\gamma' + \gamma$ plates precipitated on the (111) planes of the aluminum–silver matrix.

XVI.   Use of X Rays in Explaining Magnetic Properties

The idea of a strained lattice is also used in explaining the unusual magnetic properties of certain ternary alloys. For example, Fe_2NiAl gives a product having high coercivity and is thus useful as permanent magnet material. The optimum properties are secured not by quenching the melt or by slow cooling, but through a special controlled cooling. Bradley and Taylor (1937)

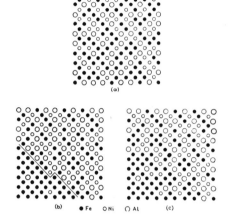

Fig. 45 Fe$_2$NiAl alloy after different heat treatments: (a) quenched, (b) slowly cooled, (c) special controlled cooling treatment. (From *Applied X-Rays* by G. L. Clark. Copyright 1940 by McGraw-Hill, Inc. Used with permission of McGraw-Hill Book Company.)

utilized x-ray powder patterns to explain the mechanism accountable for this result.

At high temperatures, and hence in quenched samples, the alloy shows a single b.c.c. lattice. As pictured in Fig. 45a the nickel and aluminum atoms occupy definite lattice positions, with the iron atoms appearing at random, some replacing nickel units and others aluminum. If the melt is cooled slowly, a pattern showing two phases is obtained, both b.c.c. but differing in lattice constant by a small amount. What has happened is that a phase consisting almost entirely of iron atoms has split off to form grains of essentially α-iron, as indicated in Fig. 45b. At an intermediate rate of cooling the iron atoms start to segregate but do not have time to form definite grains and are forced to assume a lattice which fits that of the bulk material (Fig. 45c). This condition gives rise to a single phase, but the x-ray lines are somewhat broadened. Requiring these islands of iron atoms to conform with the dimensions of the larger bulk lattice introduces large stresses within the alloy. These are held to be responsible for the extraordinary magnetic properties of the finished material.

XVII.   Stress Measurement by X-Ray Diffraction

A.   *Stress Measurement by Film Methods*

The measurement of residual stress is of considerable importance in metals technology. Many schemes have been used including boring-out techniques, sectioning methods utilizing strain guages and photoelastic studies on plastic models. To these can be added x-ray diffraction.

This method is based on the fact that applied stresses produce corresponding strains in a material, the *elastic* portion of the strain revealing itself through a change in the lattice constants. When one is primarily interested in surface stresses such as may be produced by grinding, machining, or flame hardening, the x-ray method is especially applicable. For these studies it is nondestructive. Thus changes resulting from stress relieving treatments or from service conditions may be followed on a single specimen. The extremely small area of specimen that is sampled is an additional advantage of the x-ray method. This is important when high stress gradients are present, which is often the case. The time required for an x-ray stress analysis makes its use rather expensive, and for this reason it is still primarily a laboratory tool.

To illustrate the method, suppose we desire to know the residual stress on the surface of an iron bar near a welded joint. Regardless of the actual stress configuration in an isotropic body, three mutually perpendicular axes can always be chosen so that the stress may be represented by components directed along the axes. These are called the principle stresses σ_1, σ_2, and σ_3. In our case we assume that σ_1 and σ_2 lie in the plane of the surface. Then σ_3 will be zero since a free surface cannot support a stress perpendicular to that surface. It can be shown that under these conditions the resultant strain perpendicular to the surface is

$$\varepsilon_\perp = -(\sigma_1 + \sigma_2)v/E \tag{30}$$

where v is Poisson's ratio and E is Young's modulus for iron. But by definition

$$\varepsilon_\perp = \Delta l/l = (d_1 - d_0)/d_0 \tag{31}$$

where d_0 is the spacing of a set of atomic planes (in the unstressed state) parallel to the surface, and d_1 is the same spacing under stress.

Inasmuch as the spacing change will be very small at best, the back-reflection method is a necessity in this work. The setup is made according to Fig. 18. Here annealed gold or silver powder sprinkled on the iron surface serves as a calibration standard. Should we use Co K_α radiation (see Table III), the (310) iron planes will reflect at a Bragg angle of $80\frac{1}{2}°$. If we choose this line, it means that we are measuring the spacing of planes almost but not quite parallel to the surface. For this reason our value of $\sigma_1 + \sigma_2$ will be too small by about 7% and can be corrected accordingly.

The problem of finding d_0 remains. We may anneal the specimen, or else cut out the small section examined above with a hollow drill. Either method will relieve the stress, and a new pattern of the same area will suffice to give d_0. Figure 46 illustrates results obtained by this method on a welded triangle of steel tubing. Note especially the large differences in stress for

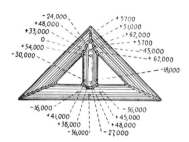

Fig. 46 Stresses in a welded triangle of steel tubing. $(\sigma_1 + \sigma_2)$ in psi. (From Möller and Roth, 1937.)

points in close proximity. These gradients would be difficult to detect by conventional methods.

The procedure described is weak in that it gives only the sum $\sigma_1 + \sigma_2$. A more elegant procedure has been devised which permits the determination of σ in any direction and furthermore does not require a measurement of d_0. It involves two x-ray exposures, one with the surface normal to the incident beam as above, the other with oblique incidence. The reader is referred to Barrett and Massalski (1966, p. 480) for a discussion of this technique.

B.   Stress Measurement in Hardened Steel Using Diffractometer

It should be pointed out that the film method described here is applicable only to samples yielding sharp diffraction lines. The probable error in careful measurement of a sharp back-reflection line is roughly ± 0.0001 Å or $\pm 0.01\%$. Since the residual stress is proportional to Young's modulus E (30×10^6 psi for steel), our stress measurements here could be off as much as ± 3000 psi. Therefore when considering stresses in hardened steels which yield broad, diffuse diffraction peaks (for example, high carbon martensite lines may spread over a 2θ interval of $8°$–$12°$ FWHM), the photographic film method is helpless. Enter here the diffractometer. Numerous workers have had remarkable success in obtaining reproducible results in measuring surface stresses in hardened steels in spite of the broad diffraction peaks. The experimental approach follows basically the two-exposure procedure mentioned above and discussed in detail in Barrett and Massalski (1966). Practicing this procedure, one run is made involving a symmetrical setup where the normal to the sample surface subtends an equal angle between incident and diffracted beams. This is known as the $\psi = 0$ condition. Next the sample rotated so the normal is displaced $40°$–$60°$. This destroys the original parafocusing condition, so the mounting must be such as to allow moving the detector toward the sample until it satisfies the new focus position.

Now a second trace is obtained. Data treatment is rather involved here. Because of the large 2θ peak spread, the recorded intensities must be corrected for absorption and Lorentz polarization factors (only the latter comes into play when $\psi = 0$). The corrected intensities close to the peak maximum are then further manipulated to fit a three or five point parabola. The vertex of the parabola is then deemed to be the true diffraction maximum. The care required in this work can be appreciated when, while working with a martensite (211) reflection where $2\theta = 10°$ FWHM in one case, the $\psi = 0$ run produced a final 2θ value of 155.124° whereas the $\psi = 60°$ result turned out to be $2\theta = 155.498°$. However this small 2θ difference of 0.374° denoted the presence of a stress amounting to 22 000 psi.

As usual, the tedium and operator time involved in this work has encouraged attempts at automation. Weinman *et al.* (1969) and Kovea and Ho (1964) describe two such developments, but in each instance the data are recorded off-line and must be processed further. Kelly and Eichen (1973) have extended the concept to include on-line computer control. Once the sample is positioned and the program initiated, the entire procedure of peak location, calculation of the three data angles (parabola vertices) and stress, and printout of the data are controlled by the computer. As a demonstration of the reproducibility of this setup, one hardened steel sample was measured 13 times. The average stress value was 16 900 psi and 12 of the 13 measurements fell within a 2σ of ± 2000 psi.

XVIII.   X Rays as an Aid in the Processing of Metal

A mass of metal undergoes many physical operations between its extraction from the ore and appearance in a finished product. All these operations induce profound changes in its physical characteristics, and each change must be understood and controlled if efficient processing is to be achieved. X-ray diffraction has played a notable role both in explaining mechanisms involved and in supplying information leading to improved procedures.

A.   Deformation

The first operation on a metal ingot is usually one of rolling, drawing, or extruding. The available evidence supports the contention that no matter how a single crystal may be deformed the general mechanism is the same. Since an ingot is a conglomerate of small single crystals, the picture here will be altered only in so far as neighboring crystals interfere with the motion of one another. The deformation mechanism within a crystal consists of slip or glide of thin lamellae parallel to certain atomic planes, the motion

TABLE IV

Glide Components of Metal Structures

Structure	Glide plane[a]	Glide direction[a]
Face-centered cubic	(111)	$[10\bar{1}]$
Body-centered cubic	(112)	$[111]$
Close-packed hexagonal	(00·1)	$[11·0]$

[a] These really refer to the whole family of equivalent planes and directions. For example, the (111) form includes (111) ($\bar{1}11$) ($1\bar{1}1$) ($\bar{1}\bar{1}1$) and four other planes parallel to these. Similarly each plane contains three different directions of the form $[10\bar{1}]$. Thus there are 12 *slip systems* associated with (111) $[10\bar{1}]$.

being in one particular crystallographic direction. A special method of deformation is distinguished when a thin layer within a crystal takes up such a position that its lattice is a mirror image of the rest. This process is known as *twinning*.

Normally the *slip plane* is the atomic plane most heavily populated with atom centers, and the slip direction is likewise that of closest packing.

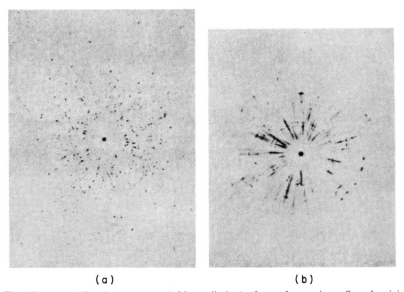

(a)   (b)

Fig. 47 Laue diffraction patterns (white radiation) of transformer iron. Sample giving pattern (a) is composed of perfect crystallites and will show good magnetic properties. Specimen giving pattern (b) contains distorted grains as shown by the asterism (streaks) present. This sample will have inferior magnetic properties. Photo by the courtesy of Picker X-Ray Corporation.

Hence both quantities will depend on the type of space-lattice present. Table IV gives the glide components for the common metal structures.

As a result of the movement of lamellae relative to one another, plus constraints induced by neigboring grains, many of the atoms lying near the slip planes no longer fit a perfect lattice arrangement. In other words, the lattice is distorted. Slip planes are no longer so well defined and the grain shows increased resistance to further deformation. This behavior is known as *work hardening.*

X-ray diffraction is a valuable tool for revealing distortion within crystal grains (see Seeger, 1956). A specimen composed of perfect crystallites gives sharp diffraction spots on Laue photographs (white radiation), whereas distorted grains yield long radial streaks, a phenomenon termed *asterism.* This effect is clearly illustrated in Fig. 47, which shows patterns from two samples of transformer iron.

B. Preferred Orientation

If further stress is applied, the deformed grains begin to break up into smaller entities. When the cold work is considerable, these fragmented grains assume preferred orientations. The appearance of sharp arcs in the x-ray pattern rather than smooth rings is evidence of preferred orientation of the crystallites.

The orientation in metals depends on the metal being worked and the method of working.

Table V lists the preferred orientations in several common metal structures for three types of cold work.

For drawn and compression (uniaxial strains) it is necessary to specify only the crystallographic *direction* lying parallel to the direction of work, since the crystallites assume random orientations around this axis. For

TABLE V

Ideal Preferred Orientations in Metal Structures

			Rolling	
Structure	Drawing	Compression	Direction	Plane
Face-centered cubic	$\begin{cases}[111]\\ [100]\end{cases}$	$[110]$	$\begin{cases}[\bar{1}12]\\ [111]\end{cases}$	$\begin{cases}(110)\\ (112)\end{cases}$
Body-centered cubic	$[110]$	$\begin{cases}[111]\\ [100]\end{cases}$	$[011]$	(100)
Close-packed hexagonal	Variable	$[00\cdot1]$	$[10\cdot0]$	$(00\cdot1)$

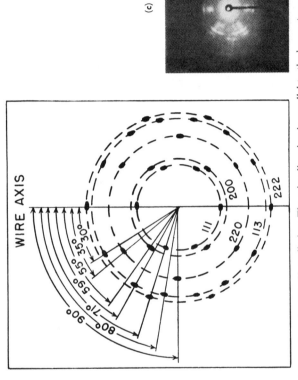

Fig. 48 (a) Powder pattern from aluminium wire using Mo K_α radiation. The wire direction is parallel to the beam stop support. (b) Predicted pattern from aluminium wire, with all the crystallites assumed to be oriented with their [111] axis parallel to the wire direction. This is to be compared with the actual pattern shown on the left. Fairly good agreement is apparent. There is some evidence for a slight amount of [100] orientation, even though Table VII claims that aluminium normally shows only [111] orientation. (c) Powder pattern from drawn copper wire (Mo K_α radiation). Both [111] and [100] orientation are evident here, in agreement with the expected behavior listed in Table VII.

example, Fig. 48a is a powder pattern secured from specimen of drawn aluminum wire, the sample being normal to the x-ray beam and wire direction as indicated. According to Table VII all the aluminimum crystallites should oriented with their [111] direction along the fiber axis. Let us now calculate the kind of pattern to be expected from a wire made up of f.c.c. crystallites with this orientation and see how well it agrees with that actually found.

In Fig. 49, AOX represents the x-ray beam, WW the aluminum wire, and F the photographic film. If a random arrangement of crystallites obtained, the family of planes represented by H could assume all directions in space and would thus produce a uniform diffraction ring R on the film. Since we are requiring that the [111] axis for all crystallites be parallel to WW, the planes H and plane normal NO can, at most, assume a position about WW wherein α remains constant. Only for specific positions will the Bragg conditions be fulfilled for H, and hence ring R will be replaced by a series of diffraction spots. Let us assume that with H as shown rays will be diffracted to P, which makes an angle β with the vertical CXS. Symmetry considerations require that a spot corresponding to P be found in each of the other quadrants on the film. Therefore, we need consider only the pattern formed in one quadrant such as CXT, utilizing the fact that CS and TV form lines of symmetry to fill in the remainder. Under the conditions cited the following relation can be shown to hold:

$$\cos \beta = \cos \alpha / \cos \theta \qquad (32)$$

If relatively hard radiation (such as Mo K_α) is used, the Bragg angle for the most important reflections is small, and so $\cos \theta \approx 1$. To this approximation we see that $\beta = \alpha$. From the lattice geometry of cubic crystals one can deduce the following expression for the angle α between the normal NO

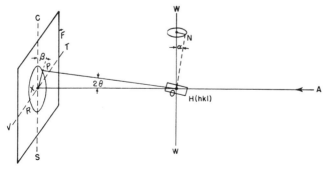

Fig. 49 Preferred orientations may be evaluated from the resulting diffraction pattern.

TABLE VI

Angular Positions of Diffraction Maxima from Cold
Drawn Face-Centered Cubic Materials

	α (deg)	
Reflection *hkl*	$[uvw] = [111]$ parallel to wire axis	$[uvw] = [100]$ parallel to wire axis
111	0, 71	55
200	55	0, 90
220	35, 90	45, 90
113	30, 59, 80	25, 72
222	0, 71	55

to any set of planes (*hkl*) and any zone axis $[uvw]$:

$$\cos \alpha = \frac{uh + vk + wl}{(u^2 + v^2 + w^2)^{1/2}(h^2 + k^2 + l^2)^{1/2}} \tag{33}$$

Since a knowledge of α gives us β directly, we are able to use Eq. (33) to locate the angular positions of the spots on any diffraction line assuming a particular crystallographic axis to be aligned with the wire axis.

Proceeding with the synthesis of our x-ray pattern from aluminum wire, reference to Fig. 33 for the face-centered cubic line sequence permits us to identify the planes responsible for the rings of the aluminum pattern. These are depicted in Fig. 48b. Table VI lists the angles α (also = β) calculated from Eq. (33) for these planes, with the fiber axis assumed to be [111]. Inasmuch as some f.c.c. crystals orient with [100] parallel to the wire axis we also include the angles expected in this instance. In Fig. 48b we have inked in spots at the calculated positions. This pattern shows good agree-

TABLE VII

Orientation Habit of Face-Centered Cubic Materials
Subjected to Unidirectional Stress

Metal	Percent crystallites showing [111] axis parallel to wire direction	Percent crystallites showing [100] axis parallel to wire direction
Al	100	0
Cu	60	40
Au	50	50
Ag	25	75

ment with that displayed in Fig. 48a and thereby confirms the fibering habit of aluminum wire as being that originally assumed. Figure 48c is a diffraction pattern obtained from drawn copper wire, another f.c.c. metal. Note that in addition to the [111] *texture* there are spots indicative of a second type of orientation. The latter corresponds to the [100] texture mentioned previously. The amount of each type of orientation present varies from one metal to another as shown in Table VII.

Samples that have been subjected to cold rolling present a more complicated type of preferred orientation. Figure 50 shows the diffraction pattern obtained from a sheet of cold rolled aluminum. Comparison with Fig. 48a illustrates the fact that the preferred orientation is somewhat different in the two cases. For rolled samples, two indices are necessary to describe the situation: those specifying orientation in the rolling direction and those giving the crystal planes which lie parallel to the rolling plane. Drawing and compression yield a *true fiber* structure, whereas rolling leads to a so-called *limited fiber* structure. It should be emphasized that the data in Table V represent an idealized picture. Actually cold work never develops perfect orientation in a specimen but at best only approaches this condition. In practice, all gradations between complete randomness and perfect alignment occur. So, although it is true that the orientation occuring with greatest frequency can be described by one or two ideal orientations, the details of the scatter about these ideal alignments can be specified only by proposing a considerable number of less prominent, somewhat arbitrary orientations. It is of considerable technical importance, therefore, to possess a means

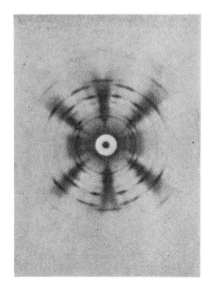

Fig. 50 Pinhole diffraction pattern of rolled aluminium sheet, showing referred orientation. Photo by the courtesy of Picker X-Ray Corporation.

for quantitatively describing the true texture of any sample. This is achieved through the medium of *pole figures.*

C. Pole Figures

The construction and significance of a pole figure may be understood by reference of Fig. 51. Here the x-ray beam AOX intersects normally a thin flat specimen, O. A reference sphere of radius OX is drawn about O as center. Let us imagine the plane of the specimen to be extended cutting the sphere in a great circle, C. This plane is to serve later as our projection plane. Assume now that plane H represents a concentration of, say (111) planes, which are in proper position to diffract. Thus a strong spot will be registered on film F at point P. We now desire to construct a normal to plane H from O to the point of intersection with the reference sphere. This can be done since it is clear that:

(1) The incident beam, diffracted beam, and plane normal all lie in a plane tipped at an angle α to the vertical.

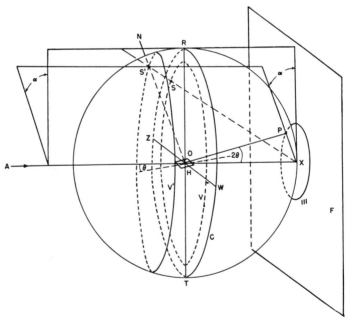

Fig. 51 Diagram to help elucidate the plotting of a pole figure. The plane of sheet specimen is mounted perpendicular to the incident beam of x rays. H represents a concentration of (111) planes which satisfy Bragg conditions for reflection.

(2)   The plane normal must make an angle $90° - \theta$ with the incident beam.

From the diagram it is seen that the plane normal strikes the reference sphere at S′. S′ is known as a *pole* for the (111) family of crystal planes. In general, there will be several other spots on the (111) diffraction ring, and the pole for each is constructed in a similar manner. These additional poles will fall somewhere on circle V′, since a line drawn to the circle from O makes a constant angle of $90° - \theta$ with the incident beam. The next step involves the stereographic projection of S′ onto a sheet of paper coincident with the extended specimen plane. This is accomplished by viewing S′ from point X, giving point S as the projection. The other (111) poles will fall somewhere on V, which is the projection of V′ on the projection plane. V is termed the (111) reflection circle. V is concentric with C only when the incident beam is normal to the specimen plane.

The specimen is now rotated some 20° around RT (the projection plane and great circle C also rotate) and a second diffraction pattern secured. The poles for the (111) planes are again constructed and projected onto the same sheet of paper. This procedure is continued until the (111) reflection circle has moved across the projection plane. Even now there will be areas near R and T not covered. To fill in these regions one restores the sample to its original position and then rotates about ZW (which is mutually perpendicular to RT and AOX) in small increments, plotting the projections of the (111) poles as above. The finished plot is termed a pole figure for the (111) family of planes. Similar plots are made for two or three other planes having low indices. Taken together they provide detailed quantitative information on the texture of the specimen (Decker 1945).

Figure 52 illustrates such a plot for the (111) plane of a rolled copper sheet. Had all the crystallites aligned themselves ideally, as recorded in Table V, i.e. [$\bar{1}12$] (110), the plot of Fig. 52a would have resulted. The experimental results give Fig. 52b, the areas of greatest density representing the most favored positions of the projected poles. Although rough agreement is found between the two plots, it is clear that there is by no means a unique orientation of the (111) planes in the copper sheet.

Figure 52c depicts the pole figure for the (110) poles of mild steel reduced 85% in thickness by cold rolling. The rolling plane is also the projection plane, with the rolling direction as shown. Superposed on the pole figure are two (110) reflection circles. The circle near the periphery is that for a beam of x rays normal to the rolling plane, and the other is the reflection circle for a beam of radiation parallel to the rolling plane. Note that the former circle nowhere passes into regions of greatly differing pole concentration. This implies that the (110) reflection will show little evidence of preferred orientation when a pattern is made with the beam normal to the

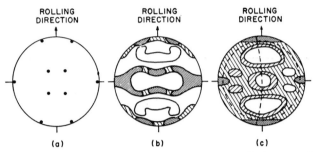

Fig. 52 (a) Pole diagram expected for the (111) plane of a rolled copper sheet showing perfect [$\bar{1}12$] (110) alignment. (b) Actual pole diagram for a rolled copper specimen. From Iweronowa and Schdanow (1934). (c) Pole figure for (110) poles of mild steel reduced 85% by cold rolling. Two reflection circles are indicated to show how the direction of the x-ray beam affects the orientation noted in the resulting pattern. (From Gensamer and Mehl, 1936.)

rolling plane. Conversely, a pattern taken with the beam parallel to the rolling plane should show strong evidence of preferred orientation.

Intensity measurements for pole figures can also be obtained using a diffractometer. The detector is positioned at $\angle 2\theta$ corresponding to the chosen set of (*hkl*) planes and the specimen is rotated stepwise about two perpendicular axes so as to permit the measurement of the intensities diffracted by these particular planes oriented at various angles within the specimen. The recorded data are plotted as before.

While the pole figure is indispensible in revealing the precise, detailed texture existing in a metal specimen, the complete plot (requiring both transmission, and reflection data replete with numerous corrections (absorption, defocusing, varying exposed sample volume, etc.) demands so much time and effort that for the past half century workers have been busy devising schemes to reduce the labor involved. Several manufacturers (Diano, Philips, Siemens, etc.) have developed semiautomated pole figure goniometers. Others have married these pole figure goniometers to printers or plotters, some even computer controlled, making the operation fully automatic and therefore painless. For the reader interested in exploring this fascinating, albeit complicated, subject the following references are recommended, listed in chronological order: Dawson (1927), Kratky (1930), Barrett (1931), Decker *et al.* (1948), Schulz (1949), Geisler (1953), Holden (1953), Holland *et al.* (1960), Eichhorn (1965), Alty (1968), Chao (1969), Desper (1969), Segmüller (1969), Segmüller and Angelello (1969), Baro and Ruer (1970), Klappholz *et al.* (1972), Viana and Ferran (1974), and Morris (1975).

Fully automated pole figure devices can be quite expensive. Therefore an alternative approach to reducing the total effort involved has been the use of composite samples prepared from the gross metal piece under study. These techniques are particularly useful for many metals of such size or shape as to be completely opaque to x rays, wherein only the reflection method is applicable. The following references describe some of the more interesting "composite" approaches: Norton (1948), Mueller and Knott (1954), Meiran (1962), Lopala and Kula (1962), and Leber (1965).

D. *Annealing*

In the oriented state a metal is quite anisotropic. For a few applications anisotropic qualities are desirable. For uses involving further forming operations, however, such qualities are highly detrimental. Therefore the next step is normally an annealing operation designed to restore the original properties. This relieves residual stresses and brings about a growth of the grains without affecting the external form of the material. Figure 53 shows the effect of heating on worked metal. Figure 53a is a pattern taken from a specimen of rolled brass strip. Figure 53b is from the same strip after it was subjected to an anealing treatment. The recrystallization is very evident.

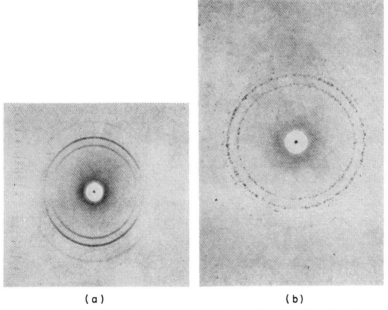

(a)   (b)

Fig. 53 Diffraction patterns from brass strip: (a) after rolling operation; (b) rolled sample has been given an annealing heat treatment. Recrystallization is graphically demonstrated. Photo by the courtesy of Picker X-Ray Corporation.

Often the grains of a recrystallized metal do not possess random orientation but assume new orientations which may or may not be the same as those of the cold worked state. These are known as *recrystallization textures*. The actual behavior again depends on the metal or alloy, the temperature of annealing, and the type of cold work originally employed. Here also diffraction patterns and pole figures constitute an important method of study.

General References

"Handbook of X-Rays"—For Diffraction, Emission, Absorption, and Microscopy. E. F. Kaelble (ed.). McGraw-Hill, New York, 1967.

An excellent reference guide not only for description but containing much valuable tabular data found only in crystallographic tables on the reference shelf. Twenty-three of the forty-eight chapters are devoted to x-ray fundamentals and x-ray diffraction. A monumental work.

"X-Ray Diffraction Procedures for Polycrystalline and Amorphous Materials," 2nd ed. H. P. Klug and L. E. Alexander. Wiley, New York, 1974.

This is an updated version of the classic volume familiar to all x-ray diffractionists. Much new subject matter has been added and the descriptions of various procedures and techniques are extremely detailed and readily followed.

"Interpretation of X-Ray Powder Diffraction Patterns." H. Lipson and H. Steeple. Macmillan London and St. Martin's, New York, 1970.

A succinct but very readable account of x-ray powder technology. A diffractionist concentrating on powders should have this volume on his shelf.

Advances in X-Ray Analysis (Vols. 1–22) Plenum, New York.

For many years the Denver Research Institute and the University of Denver have jointly sponsored a three day summer conference on the above subject. The 1958 and all subsequent conferences have been published in book form by Plenum Press. The bound volume normally appears within a year of the conference. Usually a particular subject is emphasized at each meeting but papers of general interest are also welcomed. Access to and study of these volumes will keep a researcher abreast of the x-ray state of the art.

"Principles and Practices of X-Ray Spectrometric Analysis" 2nd ed. E. P. Bertin. Plenum, New York, 1975.

While this large book is primarily concerned with x-ray fluorescence spectrometry (elemental analysis), it contains an excellent treatment of the new energy-dispersive technique which is also assuming greater importance in x-ray diffraction analysis. The approach is mainly descriptive rather than mathematical but the explanations are lucid and very well done.

"Structure of Metals," 3rd. ed. C. S. Barrett and T. B. Massalski. McGraw-Hill, New York, 1966.

Another classic volume which has stood the test of time and is found on the bookshelves of both x-ray enthusiasts and metallurgists. The only possible criticism is that the volume is now more than a decade old and does not report developments occuring within the past few years. But the basics remain valid and nowhere are they better set forth than in this book.

"Polycrystalline Book Service." P. O. Box 11567, Pittsburgh, Pennsylvania 15238.

This firm specializes in books on crystallography as the name suggests, and can supply practically any book in print devoted to this subject. It also offers bound volumes of such periodicals as *Acta Crystallographica, Journal of Crystallography*, ACA monographs and transactions, even crystal model kits. If you can't find a particular book on crystallography, chances are they can supply it.

Analytical Chemistry's Laboratory Guide and *Science* magazine's annual Guide to Scientific Instruments.

For information on these guides see text.

References

Alexander, L., and Klug, H. P. (1948). *Anal. Chem.* **20**, 886.

Alty, J. L. (1968). *J. Appl. Phys.* **39**, 4189.

American Petroleum Institute. (1950). "Project 49: Reference Clay Minerals." Am. Pet. Inst. New York.

Baro, R., and Ruer, D. (1970). *J. Phys. E* **3**, 541.

Barrett, C. S. (1931). *Trans. AIME* **93**, 75.

Barrett, C. S., and Massalski, T. B. (1966). "Structure of Metals," 3rd ed. McGraw-Hill, New York.

Barrett, C. S., Geisler, A. H., and Levenson, L. H. (1940). *Trans AIME* **137**, 112.

Barrett, C. S., Geisler, A. H., and Mehl, R. F. (1941). *Trans. AIME* **143**, 134

Birks, L. S., and Friedman, H. (1946). *J. Appl. Phys.* **17**, 687.

Biscoe, J., and Warren, B. E. (1942). *J. Appl. Phys.* **13**, 364.

Borkowski, C. J., and Kopp, M. K. (1968). *Rev. Sci. Instrum.* **39**, 1515.

Bradley, A. J., and Lipson, H. (1938). *Proc. R. Soc. London Ser. A* **167**, 421.

Bradley, A. J., and Taylor, G. I. (1937). *Nature (London)* **140**, 1012.

Bragg, W. H., and Bragg, W. L. (1934). "The Crystalline State," p. 74. Bell, London.

Brown, G., ed. (1961). "X-Ray Identification and Crystal Structures of Clay Minerals." *Mineral. Soc.*, London.

Buras, A., Chwaszezewski, J., Szarras, S., and Szmid, Z. (1968). Inst. Nucl. Res., Rep. 894/11/PS. Warsaw.

Cahn, R. W. (1951). *Acta Cryst.* **4**, 470.

Carpenter, D., and Thatcher, F. R. (1973). *Adv. X-Ray Anal* **16**, 322.

Cauchois, Y. (1932). *C. R. Acad. Sci.* **195**, 298.

Chao, H. C. (1969). *Adv. X-Ray Anal.* **12**, 391.

Chesley, F. G. (1947). *Rev. Sci. Instrum.* **18**, 422.

Chirer, E. G. (1967). *J. Sci. Instrum.* **44**, 225.

Chung, F. H. (1974). *J. Appl. Crystallogr.* **7**, 519; **7**, 526.

Chung, F. H. (1975). *J. Appl. Crystallogr.* **8**, 17.

Chwaszezewski, J., Szarros, J., Szmid, Z., and Szymezak, M. (1971). *Phys. Status Solidi A* **4**, 619.

Clark, G. L. (1940). "Applied X-Rays," p. 413. McGraw-Hill, New York.

Clark, G. L., and Rhodes, H. D. (1940). *Ind. Eng. Chem. Anal. Ed.* **12**, 243.

Clark, G. L., Asbury, W. C., and Weeks, R. M. (1925). *J. Am. Chem. Soc.* **47**, 2661.

Cole, H., Okaya, Y., and Chambers, F. W. (1963). *Rev. Sci. Instrum.* **34**, 872.

Corbridge, D. E. C., and Tromans, F. R. (1958). *Anal. Chem.* **30**, 1101.

Davey, W. P. (1934). "Study of Crystal Structures and Its Applications." McGraw-Hill, New York.

Davidson, W. L. (1950). *In* "Physical Methods in Chemical Analysis," Vol. I, 1st ed. (W. G. Berl, ed.) Academic Press, New York.

Davidson, W. L. (1960). *In* "Physical Methods in Chemical Analysis,"Vol. I, 2nd ed. (W. G. Berl, ed.) Academic Press, New York.

Dawson, W. E. (1927). *Physica* 7, 302.

Decker, B. F. (1945). *J. Appl. Phys.* 16, 309.

Decker, B. F., Asp, E. T., and Harker, D. (1948). *J. Appl. Phys.* 19, 388.

Dehlinger, U., and Kochendörfer, A. (1939). *Z. Kristallogr.* 101, 149.

Desper, C. R. (1969). *Adv. X-Ray Anal.* 12, 404.

deWolff, P. M. (1948). *Acta Crystallogr.* 1, 207.

Drever, J. I., and Fitzgerald, R. N. (1970). *Mater. Res. Bull.* 5, 101.

DuPont, Y., Gabriel, A., Chabre, M., Gulik-Krzywicki, T., and Schechter, E. (1972). *Nature (London)* 238, 331.

Eichhorn, R. M. (1965). *Rev. Sci. Instrum.* 36, 997.

Fankuchen, I., and Mark, H. (1944). *J. Appl. Phys.* 15, 364.

Fink, W. L. (1974). *J. Test Eval.*, 2, 26.

Fink, W. L., and Smith, D. W. (1936). *Trans. AIME* 122, 284.

Fisher, D. J. (1957). *Z. Kristallogr.* 109, 73.

Freud, P.J., and La Mori, P. N. (1969). *Trans. Am. Crystallogr. Assoc.* 5, 155.

Fried, S., and Davidson, N. (1948). *J. Am. Chem. Soc.* 70, 3539.

Fukamachi, T., Hosoya, S., and Terasaki, O. (1973). *J. Appl. Crystallogr.* 6, 117.

Geisler, A. H. (1953). *Trans. Am. Soc. Met.* 45A, 131

Gensamer, M., and Mehl, R. F. (1936). *Trans. AIME* 120, 277.

Gensamer, M., and Mehl, R. F. (1943). *In* "Structure of Metals" (C. S. Barrett, ed.), p. 162. McGraw-Hill, New York.

Giessen, B. C., and Gordon, C. E. (1968). *Science* 159, 973.

Goetz, A., and Hergenrother, R. C. (1932). *Phys. Rev.* 40, 643.

Guinier, A. (1937). *C. R. Acad. Sci.* 204, 1115.

Guinier, A. (1939). *Ann. Phys. (Leipzig)* 12, 161.

Guinier, A. (1975). *Phys. Today Feb.*, p. 23.

Guinier, A., and Fournet, G. (1955). "Small Angle Scattering of X-Rays." Wiley, New York.

Hägg, G., and Ersson, N. O. (1969). *Acta Crystallogr. Sect. A* 25, S64.

Hanawalt, J. D., Rinn, H. W., and Frevel, L. K. (1938). *Ind. Eng. Chem. Anal. Ed.* 10, 457

Heath, R. L. (1972). *Adv. X-Ray Anal.* 15, 1.

Hellner, E. (1954). *Z. Kristallogr.* 106, 122.

Hilley, M., Larson, J. A., Jalczak, C. F., and Ricklefs, R. E. ed, (1971). *SAE Automotive Inf. Rep.* No. J784A.

Hoffmann, E. G., and Jagodzinski, H. (1955). *Z. Metallkd.* 46, 601.

Holden, A. N. (1953). *Rev. Sci. Instrum.* 44, 225.

Holland, J. R., Engler, N., and Powers, W. (1960). *Adv. X-Ray Anal.* 4, 74.

Hull, A. W. (1917). *Phys. Rev.* 9, 564.

Hume-Rothery, W. (1926). *J. Inst. Met.* 35, 295.

Hume-Rothery, W., and Raynor, G. V., (1941). *J. Sci. Instr.* 18, 74.

Hume-Rothery, W., and Reynolds, P. W. (1938). *Proc. R. Soc. London Ser. A* 167, 25.

Ito, T. (1950). "X-Ray Studies on Polymorphism." Maruzen, Tokyo.

Iweronowa, W., and Schdanow, G. (1934). *Tech. Phys. USSR* 1, 64.

Jellinek, M. H., Solomon, E., and Fankuchen, I. (1945). *Ind. Eng. Chem. Anal. Ed.* 18, 172.

Jenkins, R., and Westburg, R. G. (1973). *Adv. X-Ray Anal.* 16, 310.

Jenkins, R., Haas, D., and Paolini, F. R. (1972). *Norelco Rep.* 18, 12.

Jette, E. R., and Foote, F. (1935). *J. Chem. Phys.* 3, 605.

Johansson, C. H., and Linde, J. O. (1925). *Ann. Phys. (Leibzig)* 78, 539.

Kaelble, E. F., ed. (1967). "Handbook of X-Rays." McGraw-Hill, New York.

Kelly, C. J., and Eichen, E. (1973). *Adv. X-Ray Anal.* 16, 344.

Kelly, C. J., and Short, M. A. (1972). *Adv. X-Ray Anal.* **15**, 102.
Klappholz, J. J., Waxman, S., and Feng, C. (1972). *Adv. X-Ray Anal.* **15**, 365.
Klug, H.P., and Alexander, L. E. (1974). "X-Ray Diffraction Procedures for Polycrystalline and Amorphous Materials," 2nd ed. Wiley, New York.
Kovea, G., and Ho, C. Y. (1964). *Norelco Rep.* **11**, 99.
Kratky, O. (1930). *Z. Kristallogr.* **72**, 529.
Laine, E., and Lähteenmäki, I. (1971). *J. Mater. Sci.* **6**, 1418.
Laine, E., and Lähteenmäki, I. (1973). *Kem. Teollisuus* **30**, 381.
Laine, E., and Tukia, L. (1973). *X-Ray Spectrom.* **2**, 115.
Laine, E., Lähteenmäki, I., and Kantola, M. (1972a). *X-Ray Spectrom.* **1**, 93.
Laine, E., Lähteenmäki, I., and Kantola, M. (1972b). *Acta Crystallogr., Sect. A* **28**, S245.
Leber, S. (1965). *Rev. Sci. Instrum.* **36**, 1747.
Lin, W. (1973). *Adv. X-Ray Anal.* **16**, 298.
Lipson, H. (1943). *J. Inst. Met.* **69**, 3.
Lipson, H. (1949). *Acta Crystallogr.* **2**, 43.
Lopala, S. L., and Kula, E. B. (1962). *Trans. AIME* **224**, 865.
MacEwan, D. M. C. (1949). *Research (London)* **2**, 459.
Martin, G. W., and Klein, A. S. (1972). *Adv. X-Ray Anal.* **15**, 264.
Mehl, R. F. (1934). *Trans. AIME* **111**, 91.
Meiran, E. S. (1962). *Rev. Sci. Instrum.* **33**, 319.
"Metals Handbook" (1939). Am. Soc. Met., Cleveland, Ohio.
Miller, M. H. (1967). *In* "Handbook of X-Rays" (E. F. Kaelble, ed.), Chap. 19. McGraw-Hill, New York.
Möller, H., and Roth, A. (1937). *Mitt. Kaiser-Wilhelm-Inst. Eisenforsch. Duesseldorf* **19**, 127.
Morris, P. R. (1975). *Adv. X-Ray Anal.* **18**, 514.
Mueller, M. H., and Knott, H. W. (1954). *Rev. Sci. Instrum.* **25**, 1115.
Norton, J. T. (1948). *J. Appl. Phys.* **19**, 1176.
Sahores, J. J. (1973). *Adv. X-Ray Anal.* **16**, 186.
Sas, W. H., and deWolff, P. M. (1966). *Acta Crystallogr.* **21**, 826.
Schulz, L. G. (1949). *J. Appl. Phys.* **20**, 1030.
Segmüller, A. (1969). *J. Appl. Crystallogr.* **2**, 259.
Segmüller, A. (1972). *Adv. X-Ray Anal.* **15**, 114.
Segmüller, A., and Angelello, J. (1969). *J. Appl. Crystallogr.* **2**, 76.
Shull, C. G., and Roess, L. C. (1947). *J. Appl. Phys.* **18**, 295.
Skelton, E. F. (1972). Rep. NRL Prog. NRL Problem No. P03-06.
Slaughter, M., and Carpenter, D. (1972). *Adv. X-Ray Anal.* **15**, 145.
Smith, C. S., and Stickley, E. E. (1943). *Phys. Rev.* **64**, 191.
Sparks, C. J., and Gedcke, D. A. (1972). *Adv. X-Ray Anal.* **15**, 240.
Straumanis, M., and Ierins, A. (1935). *Naturwissenschaften* **23**, 833.
Trzebiatowski, W., Ploszek, H., and Labzowski, J. (1947). *Anal. Chem.* **19**, 93.
Viana, C. S., and Ferran, G. (1974). *Adv. X-Ray Anal.* **18**, 514.
Warren, B. E. (1941). *J. Appl. Phys.* **12**, 374.
Weinman, E. W., Hunter, J. E., and McCormack, D. D. (1969). *Met. Prog.* July, p. 88.
Westgren, A. F. (1931). *Trans. AIME* **93**, 13.
Westgren, A. F., and Almin, A. (1929). *Z. Phys. Chem. (Leipzig)* **85**, 14.
Wilkinson, D. H. (1950). *Proc. Cambridge Philos. Soc.* **46**, Part 3, 508.
Wilson, A. J. C. (1973). *J. Appl. Crystallogr.* **6**, 230.

PHYSICAL METHODS IN MODERN CHEMICAL ANALYSIS, VOL. 2

Analytical Aspects of Ion Cyclotron Resonance

Robert C. Dunbar

Department of Chemistry
Case Western Reserve University
Cleveland, Ohio

I. Introduction

Ion cyclotron resonance spectroscopy (icr) is a vigorous new area of spectroscopy, less than a dozen years old, which has been central in the emergence of gas phase ion chemistry as a significant new discipline, and has given rise to an extraordinary expansion of our knowledge of ion properties. The aim of this chapter is to describe icr spectroscopy from a point of view which will illuminate its achievements and its potential strengths as an analytical method. Accordingly, the chapter is not a comprehensive survey; introductory, survey, and review literature is available (Baldeschwieler, 1968;

277

Baldeschwieler and Woodgate, 1971; Beauchamp, 1971, 1974, 1975, 1977; Brauman and Blair, 1973; Drewery *et al.*, 1972; Dunbar, 1974, 1975b; Futrell, 1971; Gray, 1971; Henis, 1972). The book of Lehman and Bursey (1976), in particular, is recommended as an excellent introduction and overall survey. Owing to the important recent advances in instrumental techniques, and the substantial accumulation of data about ions and ion–molecule interactions, this new spectroscopy will move into an important position in analytical chemistry. This chapter represents the author's perception of those aspects of the present state of the technique which are likely to be relevant in this context. Two aspects of the domain of icr spectroscopy can be considered analytical:

(a) Determination of the structures and other important properties of gas phase ions.

(b) Determination of the identity and structure of an unknown neutral molecule, using ionic analytical reagents, ion optical spectroscopic methods, and other icr techniques.

II. Techniques and Theory

The utility of an emerging instrumental analytical technique is closely tied to the capabilities, versatility, and reliability of the instrumentation. A great deal of effort has been invested in the evolution of icr instrumentation, and a fairly comprehensive survey of currently used devices and techniques will be an appropriate starting point for this chapter. In addition to the reviews and book already mentioned, many aspects of icr theory and practice are discussed by Anders (1966) and Beauchamp (1967).

A. Prehistory

In the chemical community, the history of the icr technique can be considered to have begun with the introduction by Baldeschwieler and his colleagues (Anders *et al.*, 1966) of the drift-cell icr spectrometer in 1966. However, at least two instruments predating that of Baldeschwieler were technically successful (their relative lack of popularity was due to a failure to find timely application to important problems). The first was the Omegatron of Sommer *et al.* (1949, 1951), which was used in a variety of configurations and applications taking advantage of its simplicity, low cost, and small size. The second was the resonance detection instrument using a solenoidal magnet of Wobschall *et al.* (1963; Wobschall, 1965). This instrument was used for collision-rate studies of atmospheric gases. Neither instrument has been of significance in the emergence of icr spectroscopy as a method of chemical importance, and they will not be discussed further.

B. *Principles of Operation*

The bedrock phenomenon underlying icr techniques is the nature of the motion of a charged particle in a magnetic field. Motion is free and unconstrained along the field lines, but in the plane perpendicular to the magnetic field motion is constrained to a circular (more generally cycloidal) path. The ion circles with a cyclotron frequency f_c which depends only on the field strength H and on the mass/charge ratio of the ion m/e;

$$f_c = \frac{\omega_c}{2\pi} = \frac{1}{2\pi} \frac{eH}{mc} \quad (Hz) \tag{1}$$

where c is the speed of light (Gaussian cgs units are used throughout). A radio-frequency oscillating electric field in the cyclotron plane at f_c can drive the cyclotron motion of the ion, with a twofold observable result: the ion velocity increases, as does the radius of its cyclotron orbit, while at the same time energy is absorbed from the rf field. Detection methods have exploited both of these phenomena. In the first case, which we can call Omegatron detection or ion current detection, the ion cyclotron orbit at resonance is allowed to increase to the point that the ion strikes a collecting electrode to give a measurable current to an electrometer detector. In the second case, which we can call resonance detection or marginal oscillator detection, the extraction of rf energy from the applied field is detected as a resistive loading of an oscillator circuit (usually, although not always accurately, termed a marginal oscillator), giving a resonance peak when the oscillator frequency matches f_c. A sweep of frequency (or more conventionally magnetic field) yields a spectrum having peaks corresponding to the mass and concentration of the different ions in the icr cell. The spectrum provides a display of the ionic contents of the cell. It is also possible to detect ions by observing the rf radiation from the circling charged particles. Various methods have been based on this idea, the most promising of which is Fourier transform icr spectroscopy.

The spectrum becomes interesting under conditions where chemical events are occurring which affect the ion abundances. Study of the chemistry associated with ion–molecule reactions and more recently with ion photochemistry has turned out to be well suited to the characteristics of the icr spectrometer. A chapter on icr must be concerned with a description of such chemistry and the ingenious tricks which can be played with the spectrometer to elucidate it.

C. Instrumentation: Detection
by Radio-Frequency Energy Absorption

In absorption icr work, the icr cell is constructed with a pair of plates above and below the ions (the magnetic field conventionally being considered horizontal, so that the cyclotron plane is vertical), and the rf field is applied across these plates, creating a vertical rf electric field driving the ion cyclotron motion. Electrically this pair of plates forms a capacitor which is included in the LC elements of an oscillator tank circuit. The ions present in the cell constitute part of the dielectric of the capacitor. Near resonance, the ions contribute both real and imaginary terms to the impedance of the tank circuit. The imaginary impedance gives rise to small frequency shifts in the oscillator, an effect which has apparently never been exploited. The real component of the ion-induced impedance corresponds to the absorption of energy by the ions; it manifests itself as a resistive loading of the oscillator leading to a decrease in the rf amplitude in the oscillator. An oscillator circuit is used (McIver, 1973), either a marginal oscillator (linear feedback) (Pound and Knight, 1950) or a Robinson oscillator (voltage-clipped feedback) (Robinson, 1959) which is designed for sensitive display of small changes in the rf level in the tank circuit, and which in practice provides an output voltage proportional to the rate of power absorption by the ions.

These ideas have been applied in a number of different cell arrangements, the most common of which will be described.

1. Drift Cell

If a dc electric field is applied in a direction perpendicular to the magnetic field, the result is a drift of the ions (the Hall effect) in a direction perpendicular to *both* the magnetic and electric fields, with a drift velocity V_d given by

$$V_d = E_d c / H \qquad (2)$$

where E_d is the electric "drift field" intensity. The traditional three-region drift cell to exploit this effect is shown in Fig. 1a. Ions are produced by electron impact in the source region of the cell. A dc voltage across the source drift plates creates a drift field which causes the ions to drift the length of the source region and into the analyzer region. A similar drift field is applied across the analyzer drift plates. The ions traverse the length of the analyzer region, enter the collector region, and are collected (and if desired observed by electrometer) on the collector plates. Since ion motion along the magnetic field lines is free, the ions are prevented from leaving the cell (in the source and analyzer regions) by applying a repelling potential (the trapping voltage) to the trapping plates. This voltage (positive for cation experiments, negative

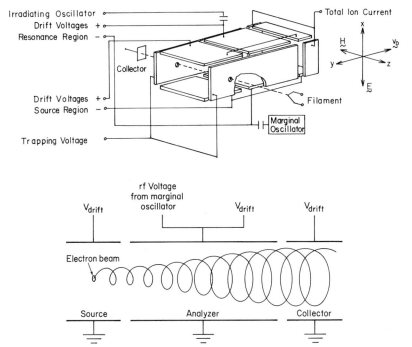

Fig. 1 (a) View of the standard three-region icr drift cell. The magnetic field **H** is along the Z axis, the observing rf field **E** along the X axis, and the ions drift (V_0) in the y direction. [Reprinted, with permission, J. L. Beauchamp *Ann. Rev. Phys. Chem.* **22**, 527 (1971), Fig. 1. Copyright © 1971 by Annual Reviews Inc. All rights reserved.] (b) Motion of a typical ion through the drift icr cell. The magnetic field direction is into the plane of the paper. The ion is in resonance with the marginal oscillator frequency, so that its cyclotron orbits increase in radius during the time it spends in the analyzer region. The drift velocity suggested by the diagram is unrealistically large, and under normal conditions an ion would make $\sim 10^3$ cyclotron revolutions during its transit from formation at the electron beam to destruction in the collector, instead of the dozen or so revolutions drawn.

for anion experiments) creates a potential well along the magnetic field (Z) direction, and prevents the escape of the ions.

The rf detector voltage is applied across the analyzer drift plates. During the time an ion traverses the analyzer region, it is continuously in the rf field and is able to absorb energy. The detecting oscillator "sees" and measures the power absorption of all the ions in the analyzer region at a given instant. A representation of the path followed by a typical resonant ion traversing the icr cell is shown in Fig. 1b. A sweep of oscillator frequency, or of magnetic field, across the region of the cyclotron resonances [Eq. (1)] will thus yield a spectrum with a series of peaks corresponding to the different ionic species. If the field is swept at a constant rf frequency, a peak separation of one atomic mass unit (amu) always corresponds to the same number of gauss; a common choice is an oscillator frequency of 153.3 kHz corresponding to 100 G/amu, so that a 20-kG magnet covers the 0–200 amu mass range. Qualitatively, a strong icr peak corresponds to a large ion abundance at that mass, and a small peak to a small ion abundance. Quantitatively, the situation is a bit more complex, expecially when ion–molecule reactions are involved.

Part of the complexity arises because the rate at which an ion absorbs power is a function of the time that has passed since the ion entered the analyzer region. Two limiting cases can be solved easily. At low pressure, such that the ion undergoes no ion–neutral collisions in the course of its traversal of the cell, the rate of power absorption is (at resonance)

$$A(t) = e^2 E_{rf}^2 t / 4m \qquad (3)$$

where $A(t)$ is in units of power, E_{rf} is the rf electric field intensity, e the charge of the ion, m its mass, and t the time since the ion entered the rf field region. In a drift cell with continuous ion production, this leads [by averaging (3) over all ions in the analyzer] to peak heights (i.e., power absorption at resonance for a given ion) of

$$I_i = \frac{N_i e^2 E_{rf}^2 t_d^2}{8m_i} = \frac{N_{oi} m_i \omega^2 l^2 E_{rf}^2}{8 E_d^2} \qquad (4)$$

where N_{oi} is the rate of production of species i at the electron beam, t_d the ion drift time in the analyzer region, and the last expression in (4) is obtained by inserting Eqs. (1) and (2) into the middle expression in (4) at oscillator frequency ω (where l is the length of the analyzer region). It is seen that, at low pressure, the peak height at constant rf frequency and swept magnetic field increases linearly with ion mass, so that to obtain a quantitative spectrum the peak heights must be divided by mass. The peak heights and the spectrometer sensitivity increase with the square of the operating frequency ω.

In the high pressure limit, the ions undergo frequent collisions with neutrals. The energy gained by the ion from the rf field is rapidly dissipated to the neutral gas molecules and a steady state is rapidly reached in which energy gain from the rf field is balanced by energy loss due to collisions. Equation (3) still applies, but t must now be taken as the collisional energy–relaxation time. The time an ion spends in the analyzer is irrelevant to its rate of power absorption, since its average power absorption does not depend on how long it has been in the analyzer. Quantitation of icr results thus depends on the nature of the ion–neutral collision. If, as a reasonable approximation, the ions are assumed to have a constant mean free time between collisions [see below; also Beauchamp (1967b)] an analysis similar to that for the low pressure case gives

$$I_i = N_{oi}eE_{rf}^2\omega l/4E_d v \qquad (5)$$

where v is the collisional damping rate constant. Thus, the high pressure case in the drift cell gives peaks which have no mass bias but do increase with the operating frequency. It is this rapid increase in signal strength with frequency that makes working at the highest possible frequencies and magnetic fields so advantageous.

We have thus seen that the quantitative relationships of icr peak heights to primary ion current require a mass-dependent correction whose nature depends on the pressure regime. The low pressure limit is appropriate in cases where the icr instrument is used to obtain a simple low-pressure mass spectrum, or where the reactions being studied are so fast that nearly every collision results in a reaction. On the other hand, if a reaction is slow, it may be necessary to work at higher pressure in order to observe products, and the high pressure expression may be applicable. Actually, many drift-icr studies are probably done in an intermediate pressure regime, where accurate quantitation is difficult.

When ion–molecule reactions occur in the cell, the preceding analyses are greatly complicated by the fact that primary ions are being destroyed, and secondary and higher order reaction products are being produced, along the length of the cell. Much effort has been expended to generalize the power absorption expression in order to describe reactive systems with the aim of obtaining accurate reaction kinetic data from the spectra. This is reviewed by Lehman and Bursey (1976, Chapter 4); for more details on icr lineshapes, see Marshall (1971), Comisarow (1971), Dunbar (1971a), Huizer and van der Hart (1977), van der Hart (1973), van der Hart and Van Sprang (1975), Woods et al. (1973), Bloom and Riggin (1974), Riggin and Woods (1974). Although successful, this approach has been supplanted largely by pulsed icr methods for accurate reaction rate determination.

All drift-cell spectrometers use some form of modulation of the ion signal combined with phase sensitive detection for convenient baseline stabilization and signal averaging. While of practical importance, these methods have no fundamental effect on the nature of the information obtained. Common modulation methods include modulation of the magnetic field, the trapping potential, the electron beam voltage or current, and the double resonance oscillator amplitude.

2. Trapping Cell

The pioneering work of McIver (McIver, 1970; McIver and Dunbar, 1971; McIver and Baranyi, 1974; McIver et al., 1975) followed up by several other groups, has shown beyond reasonable question that a pulsed icr mode of operation using a trapped-ion cell has great advantages over continuous operation in a drift cell. The traditional drift cell is now virtually obsolete for most quantitative work.

When operating with high trapping voltage (~ 1 V) and low or zero voltage on the drift plates, the icr cell operates as a Penning ion trap. The field lines from the trapping plates curve toward the drift plates, so that near the periphery of the cell the field lines are directed radially outward. The Hall effect prevents ions from drifting toward the drift plates. Instead, an ion drifts in closed paths around the cell, and in the absence of collisional damping will never strike the drift plates. This effect operates in the normal drift-cell geometry at high trapping voltage, and was noticed and exploited, although not fully understood, quite early (Henis, 1968; Dunbar, 1971b). However, to give better defined trapping field geometry, most trapping cells are now constructed with "drift" plates on all four sides of the cell instead of leaving the ends of the cell open. (The McMahon–Beauchamp cell, to be described, is an exception.) There is no evidence that one trapping design is very superior to another, and any reasonable geometry probably traps ions for comparably long periods, limited by collisional damping to the time needed for a few hundred collisions. Ion trapping for several minutes at low pressure has been described (Dunbar, 1973).

Trapping cells are usually operated in a pulsed mode (see Fig. 2): the electron beam is pulsed on for a brief period to create a group of ions (formation pulse); a period of time is allowed to pass while the ions react (reaction

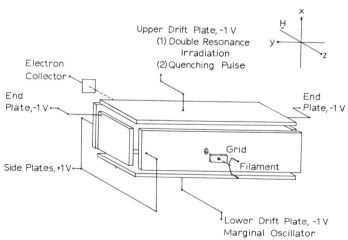

Fig. 2 A standard version of the McIver trapped-ion cell. [Reprinted, by permission, from R. T. McIver, Jr., *Rev. Sci. Instrum.* **41**, 555 (1970), Fig. 1.]

period); the detector is switched on to sample the contents of the cell (detect pulse); and the cell is purged of ions by pulsing the trapping plates (quench pulse). Periods of double resonance and/or optical illumination may also be added to the pulse sequence (see below). Simplicity of analysis is the foremost advantage of pulsed operation: since all the ions are formed at one time, and the analysis of the cell contents takes place at a well-defined later time, a unique and well-defined reaction time can be used in the kinetic equations, eliminating the troublesome integrations over ion formation, detection, and reaction times which make drift-cell kinetics difficult. A second important advantage is that in a trapped-ion situation a quick reaction may go to completion and equilibrium among species may be established; the interesting possibilities presented by this will become clear in Section IV.D.

(a) *McIver Cell*   The McIver cell (McIver, 1970) is the most straight-forward trapped-ion design, being a simple box with normal trapping plates, with side ("drift") plates on all four sides, and with the electron beam traversing the center of the cell (see Fig. 2). The pulse sequence is shown in Fig. 3. Two approaches to controlling the detector pulse have been used: one, the pulsing on and off of the marginal oscillator, is straightforward and attractive in principle, but in practice it is not easy to design a sensitive oscillator circuit which will tolerate the necessary switching on and off of the high impedance tank circuit. The other approach, switched trapping voltage, involves running the marginal oscillator continuously, and moving the ions into resonance during the detect pulse. This is possible because the cyclotron frequency of an ionic species has a weak but significant dependence on the magnitude of the

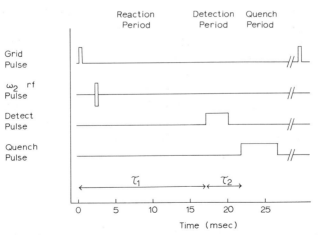

Fig. 3   Typical pulse sequence in pulsed icr experiments. [Reprinted, by permission, from R. T. McIver, Jr., and R. C. Dunbar *Int. J. Mass Spectrom. Ion Phys.* **7**, 471 (1971), Fig. 2.]

trapping voltage, so that in this scheme the detect pulse consists of briefly switching to the value of trapping voltage which brings the ions into resonance. The rest of the time the oscillator is not in resonance with any ion, and therefore does not disturb the events occurring in the cell.

(b) *McMahon–Beauchamp Cell* This cell (McMahon and Beauchamp, 1972) has the geometry of a conventional drift cell. During the detect pulse it functions as a drift cell with the packet of ions drifting through the analyzer and being detected by the (continuously running) marginal oscillator. During the trapping part of the cycle, however, the ions are trapped in the source region by applying ground or negative potentials to all drift plates; during ion trapping the source region thus looks much like a McIver cell, except that there is no plate at the analyzer end of the cell, and the potentials in the source region are defined instead by negative biasing of the analyzer drift plates. While a strong distinction is sometimes made between the McIver and the McMahon–Beauchamp cells, the ion traps used are virtually the same; the distinction lies in ion detection in the trap (McIver) or after extraction into a separate analysis region (McMahon–Beauchamp). Since the two approaches are so similar, the techniques and results of operation are very similar and can be discussed without distinguishing between the instruments used.

(c) *Continuous-Observation Trapping Cell* As already noted, the conventional drift cell operates as an ion trap at high trapping voltage. It is possible, by using a low amplitude of observing rf, to observe the trapped ions continuously without driving them from the cell, providing a continuous record of the analyzer-region ion concentration at a given mass during the many seconds or minutes the ions are trapped (Dunbar, 1973). The capability for continuous observation has considerable advantages in experiments involving very long trapping 'times, in which the large number of repeated cycles needed to plot the entire time behavior of the system using a scheme such as Fig. 3 becomes very time consuming compared with the complete plot obtained in a single continuous observation experiment. This mode of operation has been used in both continuous- and pulsed-ion-production (Riggin *et al.*, 1976) experiments. It has been used extensively in photochemical studies in our laboratory where long ion trapping times are commonly needed.

3. *Fourier-Transform* icr

The recent application of Fourier transform (FT) spectroscopic techniques to icr by Comisarow and Marshall (1974a–c, 1975, 1976) is a development of some promise in terms of bringing to the icr field the much more favorable tradeoff terms between sensitivity and resolution which FT techniques have brought to nmr, ir, and other spectroscopic areas. Analogous to the nmr experiment, ion cyclotron motion is excited by the initial application of a short pulse of high intensity rf voltage. The ions moving in their large cyclotron orbits induce an rf voltage on the cell plates, and this induced transient voltage (analogous to the nmr free induction decay) is amplified, heterodyned to low frequency, and stored. The appropriate Fourier transform of this received transient signal is the icr spectrum, with frequency range and resolution determined by excitation and receiver bandwidths, length of time of observation of the transient signal, memory size, and transform details. The spectrum ultimately obtained should be equivalent to the ordinary frequency-swept icr spectrum. The advantage is one of sensitivity which arises because the entire spectrum is observed simultaneously, rather than one frequency at a time as in swept spectra. This advantage can amount to several orders of magnitude, and in practice it may be used for enhanced operation speed, extensive signal-averaging, enhanced resolution, or a combination of these operational advantages.

Current FT–icr work is done with a cell similar to the McIver cell. The excitation pulse is applied to one drift plate (or a pair of opposite plates), and the observation is made on another plate (or pair of plates). The pulse sequence, double resonance capability, ultimate resolution, trapping times, and other factors are similar to those of the normal pulsed-icr experiment.

4. *Tandem* icr

Futrell and co-workers (D. L. Smith and Futrell 1974a,b) have developed an ingenious spectrometer combining some of the advantages of tandem mass spectrometers with those of icr. The icr component of the spectrometer is a normal drift cell lacking an electron beam. The ion source for the icr cell consists of a small 180° Dempster-type magnetic sector mass spectrometer whose exit slit is at the source end of the icr cell. The sector and the icr cell are both in the same magnet gap in very compact geometry. This arrangement permits formation in the sector of a thermalized (by collisions in the high pressure source), mass-selected (by the sector) set of ions which are introduced into the icr cell in order to study their reaction chemistry by icr techniques.

The practical advantage of this arrangement lies in the use of a high pressure ion source feeding of an icr cell, with the consequent independent control over the internal energy of the reactant ions. Much of the work with this device has been concerned with the effect on the ion–molecule reaction chemistry of varying numbers of thermalizing collisions which the reactant ions undergo in the source. Efficient differential pumping should allow high pressure source production of high-pressure species such as cluster ions, but this possibility has not been widely exploited.

D. Double Resonance

The double resonance technique is the key to the power of icr for elucidation of ion chemistry. The term double resonance encompasses several methods for the selective acceleration and (usually) ejection of specific-mass ions from an icr cell using an rf field, distinct from the observing rf field, at a frequency ω_2 resonant with one of the periodic motions of the ion in question.

1. Drift Cells

The earliest and most usual method of double resonance (Anders et al., 1966) is to apply an ω_2 rf field to the source or analyzer drift plates and to excite the cyclotron motion of the species to be ejected. When the ω_2 frequency matches the cyclotron frequency of an ionic species, the ions are accelerated into progressively larger cyclotron orbits with progressively higher kinetic energy. When a kinetic energy of the order of 10 eV is reached, the cyclotron orbit is large enough for the ions to strike the cell plates and be lost. In a reaction such as

$$A^+ + B \rightarrow C^+ + D$$

the observing oscillator will typically monitor the C^+ signal while the double resonance is swept through the A^+ cyclotron frequency. A change in the C^+ signal when the ω_2 field ejects A^+ proves the existence of the reaction coupling A^+ and C^+. Two effects may occur. If the reaction is exoergic, the double resonance ejection of A^+ will result in a dip in the C^+ signal. If the reaction is endoergic, the acceleration of the A^+ ions by the ω_2 field may actually drive the reaction faster and bring about an increase in C^+ intensity. A sweep of the ω_2 frequency through the frequencies corresponding to various possible reactive precursors of C^+ produces a double resonance spectrum in which the signs of the double resonance peaks can be taken as differentiating between exoergic and endoergic reactions. The capability of the double resonance technique to isolate from a complex reaction system the individual reaction coupling reactant A^+ with product C^+, measure its rate, and check its exoergicity or endoergicity makes the icr technique promising for analytical applications involving the reaction of a specified reagent ion with an unknown neutral to give a specific product ion. When reading Section IV, it should be borne in mind that double resonance observation of the specific reactions described is the key to their analytical utility. The interpretation of double resonance spectra in quantitative terms has been discussed extensively (e.g. Beauchamp, 1967; Dunbar, 1968; Anders, 1969; Goode et al., 1970).

An interesting variation on the conventional drift cell is the four-region cell of Futrell and his colleagues (Clow and Futrell, 1972a,b). Between the analyzer and the source a reaction region is added. This region is free of rf fields and the ions have sufficient time to react before reaching the analyzer. This arrangement greatly enhances the flexibility available in performing reaction studies, since the spectrometer allows a choice of double resonance ejection of primary ions in the source, or of chosen reactively produced ions in the analyzer following traversal of the reaction region. The kinetic analysis of double resonance data is simplified and accuracy is improved.

The other important approach to double resonance is trapping-plate excitation, as described by Beauchamp and Armstrong (1969) and discussed by Freiser *et al.* (1973). In this case, rather than exciting the cyclotron motion of the ions, the double resonance oscillator excites the oscillatory motion the ions undergo in the Z direction (along the magnetic field) under the influence of the trapping potential. The frequency of this trapping oscillation is

$$f_T = 4\pi(V_T e/md^2)^{1/2} \tag{6}$$

where f_T is the trapping oscillation frequency (in hertz), V_T the voltage on the trapping plates, and d the separation of the trapping plates. If the ions are excited by an rf field at this frequency, the amplitude of their trapping-field oscillation is increased until the ions strike the trapping plates, giving a net result similar to that of cyclotron ejection. The trapping ejection rf voltage is applied most efficiently across the trapping plates, but since the electric field from the drift plates curves to give some field component in the Z direction, application of the ejecting rf to the drift plates is also possible.* Trapping ejection frequencies are typically an order of magnitude lower than cyclotron frequencies (tens versus hundreds of kilohertz), and mass resolution for this technique is very poor. Otherwise, this double resonance method differs little in its application from the cyclotron resonance approach. The most characteristic use of this double resonance method, owing to its low mass resolution, is for simultaneous ejection of all the various ions in an extended mass region.

2. Trapping Cells

In a trapped-ion cell, the double resonance techniques are essentially the same, and have the same potential for elucidating ion–molecule chemistry, as in the drift cell. The most important difference in pulsed trapped-cell systems is the greater flexibility afforded by the well-defined pulse sequence, to which

---

* Since the electric field from the drift plates is symmetrical about $Z = 0$, a trapping ejection rf voltage on the drift plates must have twice the frequency [Eq. (6)] of the ejecting rf applied unsymmetrically across the trapping plates.

an ion-ejection period can be added at will. Typically, an ejection pulse consisting of a burst of rf voltage at ω_2 would be applied to the desired plates during or immediately following the ion-formation pulse (as in Fig. 2). In this mode, all reaction sequences which initiate with the chosen ejected primary ion will be suppressed, allowing them to be identified. For instance, in methane where the principal primary ions (low voltage) are CH_3^+ and CH_4^+, ejection at the beginning of the pulse sequence of all CH_3^+ will suppress all reaction product ions (from any order reaction) which originated with CH_3^+.

An alternative mode of operation is to apply double resonance rf voltage throughout the length of the pulse cycle until the detect pulse commences. This will cause the immediate ejection of any chosen ion species formed and suppress all products which involve it as a precursor at any stage. For instance, in methane-containing systems, CH_5^+ is formed reactively by a variety of processes and then undergoes various possible reactions. If CH_5^+ is continuously ejected throughout the cycle, all products resulting from reaction sequences in which CH_5^+ appears will be eliminated.

Other, more elaborate schemes can be evolved for the ejection of various ions at various points in the pulse sequence, but the two operating modes described serve to indicate the power of such techniques. In contrast to the drift-cell double resonance experiments, in which kinetic analysis can become very complex, the interpretation of double resonance ejection data in pulsed icr work is straightforward and unambiguous.

E. Ion Current Detection

The problems associated with the high magnetic field have discouraged combining electron multipliers with icr techniques, but a number of approaches have used detection of ions by an electrometer in various ways.

1. Omegatron

The omegation is in its physical structure and ion-trapping operation very similar to the McIver trapped-ion icr cell, but instead of using resonance detection, an ionic species at resonance is accelerated by the accelerating rf field (as in the double resonance technique) until it strikes a drift plate and is collected and detected by the electrometer. Ledford and McIver (1978) have recently described omegatron-type operation of a trapped-ion icr spectrometer, which may have advantages in mass resolution and sensitivity over the conventional trapped-cell experiment. The extent to which this instrument will be of significance as a high resolution analytical tool is not yet clear.

2. Total Ion Current (TIC) Detection

Many icr drift-cell spectrometers are equipped for electrometer collection of those ions that succeed in traversing the cell length. If an ejecting rf field is applied in the analyzer region such that the ions of a given mass are ejected *before* they reach the collector region, a dip in the electrometer current will be observed; by sweeping the double resonance oscillator across the frequency range of interest while monitoring the total ion current reaching the collector region, a mass spectrum may be traced. This is referred to as the TIC detection mode. An advanatage of this scheme is that the entire experiment may be done at high, constant magnetic field where resolution is best. This advantage is not decisive, and therefore TIC detection is very seldom used.

3. Trapping-Plate Current Detection

It is feasible to monitor the ion current to the trapping plates in conjunction with a trapping-plate double resonance ejection oscillator. The experiment exactly analogous to the omegatron experiment can be carried out, although in itself this has little utility. However, detection at the trapping plates is of considerable interest as a means of detecting photodissociation product ions in the icr cell: since the photoproducts must climb out of the trapping potential well in order to strike the trapping plate, only ions with substantial kinetic energy are collected. This approach can be put on a satisfactory quantitative basis, and provides a means of determining the kinetic energy of photoproduct ions (Orth and Dunbar, 1977; Orth *et al.*, 1977).

III. Properties of Gas Phase Ions

A. Ion Structures

Determination of structures and other properties of gas phase ions is a rapidly growing enterprise and is an important aspect of the attempt to analyze neutral compounds via their ionization products. Spectroscopic and reaction-pattern methods constitute the two principal approaches in which icr has made distinctive contributions.

1. Spectroscopic Methods

(a) *Photodissociation Spectroscopy of Cations*   One has always wanted to apply the powerful methods of optical spectroscopy to gas phase ions, but the extraordinary difficulty of obtaining a sufficient sample concentration has limited absorption spectroscopy to a few heroic studies on simple ions (e.g.

Lew and Heiber, 1973). However, besides measuring the attenuation of a light beam traversing a sample of ions, spectroscopic information can also be obtained by measuring the changes in the sample induced by absorbed light: for gas phase ions, it is convenient to follow the ion dissociation in the sample following the absorption of light. This approach is termed *photo-dissociation spectroscopy* (Dunbar and Fu, 1973; Dymerski *et al.*, 1974): the extent of dissociation in a sample of trapped ions is determined as a function of wavelength, and the resulting spectrum is given absorption-spectroscopic significance through the obvious fact that an ion can dissociate only if it first absorbs a photon. Use of the icr cell in a dual capacity as an ion trap and as a detector of dissociation events has proved to be an exceptionally convenient way of performing photodissociation spectroscopy. Figure 4 shows a typical photodissociation spectrum obtained by icr—that of the toluene radical cation. Several cases will be described in which ion photo-dissociation spectra have yielded useful information about ion structures.

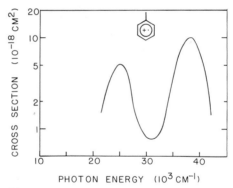

Fig. 4 Photodissociation spectrum of toluene radical ion $C_7H_8^+$

(1) *Olefins* An early photodissociation study was on the radical cations of the butenes (Kramer and Dunbar, 1973; Riggin *et al.*, 1976). The photo-dissociation spectra of all $C_4H_8^+$ isomers examined were similar, with dissociation commencing near 500 nm and rising steadily to shorter wave-lengths with a strong peak near 300 nm. Photodissociation proceeded via two competing channels (7). Of structural interest was the finding that the

$$C_4H_8^{+\cdot} \xrightarrow{\ h\nu\ } \begin{array}{l} \longrightarrow C_4H_7^+ + H\cdot \\ \longrightarrow C_3H_5^+ + CH_3\cdot \end{array} \tag{7}$$

1-butene cation has a distinctly different onset and curve shape from both *cis*- and *trans*-2-butene cations (which gave identical spectra). Previous suggestions of 1-butene ion rearrangement to the more stable 2-butene form were clearly ruled out.

(2) $C_7H_8^+$ *ions* The photodissociation spectra of three isomeric $C_7H_8^+$ species (Fig. 5) provide a more dramatic illustration of the structural capabilities of this spectroscopic technique (Dunbar and Fu, 1973). The mass spectral fragmentation patterns of the parent ions of toluene, cycloheptatriene, and norbornadiene are virtually indistinguishable. It had long been suspected that a single parent ion structure resulted from ionizing any of these compounds. The photodissociation spectra of the three species are entirely dissimilar, however, showing that the nondecomposing ions retained distinct identities. Comparison with theory and with the photoelectron spectra of the neutrals indicated that toluene and cycloheptatriene radical ions retained the structure of the neutral, while the norbornadiene ion most likely rearranged in a manner not yet identified (Fu, 1976).

(3) *Substituted benzenes* The photodissociation spectra of a variety of substituted benzene radical ions have been studied (Dunbar, 1973a,b, 1975d;

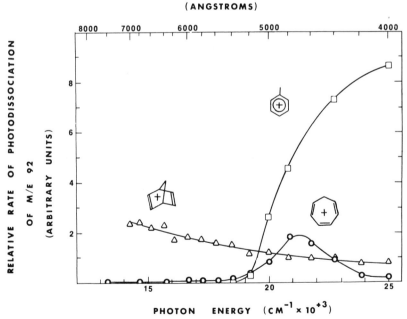

Fig. 5 Photodissociation spectra of three $C_7H_8^+$ isomers, showing their spectroscopic nonequivalence, and presumably structural nonequivalence, on the relevant time scale of several seconds. [Reprinted, by permission, from R. C. Dunbar and E. Fu, *J. Am. Chem. Soc.* **95**, 2716 (1973). Copyright by the American Chemical Society.]

Dymerski *et al.*, 1974b; Freiser and Beauchamp, 1975, 1976a). The spectra were all similar to that shown in Fig. 4. The peak shifts and sometimes the appearance of additional peaks related to side-chain molecular orbitals are common and useful. It is generally easy to distinguish a disubstituted benzene from an isomeric monosubstituted one; for instance, the distinction between xylene and ethylbenzene (Dunbar, 1973b), chlorotoluene and benzyl chloride, bromotoluene and benzyl bromide (Fu *et al.*, 1976), and so forth. These differences are often prominent, and have clear analytical potential. There are also small differences between *o*, *m*, and *p* disubstituted isomers in many cases, as in the xylenes (where para is discernibly different from ortho and meta), the chlorotoluenes, and other cases, but these differences are not sufficiently large to be very promising for analytical purposes unless other methods of isomer distinction cannot be applied. As suggested in the previous paragraph, isomers of benzene compounds which do not retain the aromatic nucleus will normally have qualitatively different spectra, and the appearance of a spectrum resembling Fig. 4 is good presumptive evidence for the presence of a benzene ion chromophore (Dunbar, 1976a).

(4)   *Styrenes*   The styrene ion is qualitatively different from and easily distinguished from its isomer cyclooctatetraene (Fu, 1976), as expected. More interesting are the significant differences found among the spectra of various isomeric substituted styrene ions. Ion **1** is qualitatively distinct from its isomer **2**, but the spectra of **1** and **3** are identical, indicating likely rearrangement of

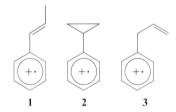

3 to **1**. The spectrum of **4**, however, is very different from those of **5** and **6**, indicating that double bond migration into conjugation with the ring does not take place in this case.

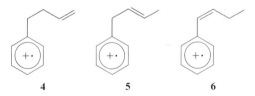

A useful additional capability of the photodissociation analytical approach is illustrated by benzyl chloride (Fu *et al.*, 1976). Benzyl chloride was

ionized in a trapped-ion experiment, and was irradiated for many seconds at 580 nm. Under irradiation the parent ion signal at first decreased rapidly, but then leveled off at a value suggesting that only $\sim 35\%$ of the $C_7H_7Cl^{+\cdot}$ ions initially formed were susceptible to photodissociation at this wavelength. These results were interpreted as indicating that ionization of benzyl chloride yields a mixture of two ion structures: the first in 35% abundance, being readily dissociated at 580 nm and most likely having the benzyl chloride structure **7**; the second, nondissociating at 580 nm, very likely having a chlorotoluene structure **8**.

$$7 \qquad\qquad 8$$

(b) *Photodetachment*   Brauman and Smyth (1969; Smyth and Brauman, 1972) discovered very early that the icr ion trap is a convenient environment for examining the electron photodetachment process

$$A^- \xrightarrow{\ h\nu\ } A + e^- \tag{8}$$

This is comparable to a photoionization process (although the fact that only one of the products is charged has an important effect on the threshold properties), and for many simple anions, theory predicts a clearcut wavelength threshold and a leveling off of the detachment cross section at shorter wavelength. Since in most cases the photodetachment efficiency curve rises with nonzero slope from the baseline, accurate thresholds are readily determined, and from them the electron affinities of neutral molecules are measured. Such experiments have been a source of very high quality electron affinity data for a number of species (Richardson *et al.*, 1973, 1974b, 1975a,c,d; Reed and Brauman, 1974, 1975). Although most of this work has not been of structural importance, two recent publications have had clear relevance to the structures of the anions involved.

(1) NO_2^- Instead of showing a single, well-defined onset in the photodetachment spectrum, the NO_2^- ion (obtained by resonance electron capture in NO_2) shows the expected sharp rise near 500 nm, corresponding to threshold detachment from normal ground state NO_2^-, and a long tail to long wavelength with an apparent onset near 700 nm (Richardson *et al.*, 1974a; Pearson *et al.*, 1974). A strong case is made for the long-wavelength tail corresponding to a second NO_2^- structure, possibly a peroxy form **9** or a triangular species **10**, and in any case to a structure distinct from and not interconverting to the expected ONO^- ion **11**.

$$\begin{array}{ccc} & \text{N} \\ & /\backslash \\ \text{N—O—O}^- & \text{O—O} & \text{O—N—O}^- \\ \textbf{9} & \textbf{10} & \textbf{11} \end{array}$$

(2) *Thresholds* An elegant theoretical and experimental study of the relation of photodetachment threshold behavior (Reed *et al.*, 1976) to molecular orbital symmetry suggested that the shape of the threshold curve is a distinctive property of the symmetry of the highest occupied MO of the anion, and might therefore be a probe of elements of geometrical symmetry (or lack of symmetry) in the anion. It is not yet clear whether such curve shapes are a sufficiently sensitive function of molecular symmetry to be widely useful in a practical sense. The possibility of a spectroscopic structure probe for anions comparable to the photodissociation spectroscopic one for cations is nevertheless an interesting one. Among other anions considered, the photodetachment spectrum of $C_5H_5^-$ was found to show evidence for a high degree of symmetry, and was well matched by the theoretically predicted curve based on D_{5h} symmetry leading to a degenerate highest occupied orbital.

(c) *High Resolution Laser Photodissociation* The recent success in using narrow-band laser light sources in photodissociation promises an analytical capability of unsurpassed specificity, although perhaps of limited applicability. Leading to this optimistic prediction was the observation in the 1,3,5-hexatriene ion at 10 Å optical resolution, using a tunable dye laser, of a photodissociation spectrum with extensive and clearly resolved vibrational structure (Dunbar, 1976b). Two vibrational modes of the cation excited state were identified, a carbon stretch at 1200 cm^{-1} and a carbon bend at 350 cm^{-1}. Vibrational information of this quality permitted a unique identification of the ion structure involved: for instance, the ions from *cis*- and *trans*-1,3,5-hexatriene were clearly shown from their laser photodissociation spectra to be identical, probably with cis configuration (Teng and Dunbar, 1978). Not all ions showed resolved vibrational structure in their photodissociation spectra (Fu *et al.*, 1976), but in favorable cases this approach gives information about the gas phase ion in more detail than any other approach, and the analytical value of unambiguous specific ion identification of an ionic structure seems significant.

2. *Characteristic Reactions with Neutrals*

(a) *Structures of* $C_3H_6^+$ The C_3 hydrocarbon ions have always been of interest to mass spectroscopists, being some of the simplest hydrocarbon systems whose structures and fragmentations are interesting, and icr analysis has contributed importantly to their study. NH_3 is a useful reagent in examining $C_3H_6^{+\cdot}$ (Gross, 1972; Gross and McLafferty, 1971), the observation being that $C_3H_6^{+\cdot}$, from cyclopropane and a variety of other methylene-chain-containing precursors, undergoes the reactions (9) while $C_3H_6^{+\cdot}$ from propylene is unreactive. The former type of $C_3H_6^{+\cdot}$ may derive its reactivity from

$$C_3H_6^{+\cdot} + NH_3 \begin{cases} \longrightarrow CH_5N^{+\cdot} + C_2H_4 \\ \\ \longrightarrow CH_4N^+ + C_2H_5\cdot \end{cases} \qquad (9)$$

internal excitation, but the argument is made that the similarity in behavior for a variety of precursors and ionizing energies would be unlikely in this case, and that a structural difference is more likely. Labeling experiments suggest that the six hydrogens of reactive $C_3H_6^{+\cdot}$ are reactively equivalent, and it is attractive to hypothesize that the reactive species is $\cdot CH_2CH_2CH_2^+$, which may randomize its hydrogens through a cyclic transition state. The ammonia reagent is probably an acceptor for the

$$\overset{|}{\underset{|}{^+C}}-$$

group and might be considered as a selective probe for the carbenium ion feature via strong covalent-bonding interactions (Beauchamp, 1977).

(b) *McLafferty Rearrangement Product Ions* The elegant experiments of Djerassi et al. (Diekman et al., 1969; Eadon et al., 1969, 1970) were an early demonstration of the power of neutral-reagent analysis of ion structures. Rearrangement product ions $C_3H_6O^+$ were generated from several precursors [see (10)–(11)] in order to examine the structures of the product ions **12–14**, specifically the questions of whether any of them reketonize to

(10a)

12

(10b)

13a   **13b**

(11)

14

15, and whether the oxonium ion structure **13a** reverts to **12** or **15**. For comparison, authentic **15** ion was prepared from acetone, and characteristic

15

reactions of the ions were examined with the ketone reagents acetone, 2-hexanone, and 5-nonanone (among others). Ion **15** undergoes a variety of reactions with ketones including the acetylation of acetone, charge transfer to 2-hexanone, and acetylation with loss of 2 C_3H_6 with 5-nonanone. The presumed enol ions, in contrast, undergo a characteristic proton transfer to ketone reagents, obviously facilitated by the ready availability of the oxygen proton: thus **12–14** were clearly shown not to reketonize to **15**. No difference in reactivity was found between **13** and **12** (or **14**), suggesting that **13b** was the species actually formed; careful labeling experiments confirmed this and elucidated and mechanism of **13** formation.

While these cases show retention of the enol form of the product ion, in the case of 2-propylcyclopentanone (Hass *et al.*, 1972) the analysis shows slow reketonization (**16a,b**). If **16** is probed at short times after formation

16a   **16b**

($\gtrsim 10^{-4}$ sec), it shows the characteristic enol behavior of proton transfer to acetone, while if the reaction time scale (adjusted by lowering the neutral pressure and raising the ion residence time) is increased to $\sim 10^{-1}$ sec, this enol reaction disappears and is is replaced by the characteristic ketonic behavior of acetylation of acetone. This clearly indicates that the initially formed **16a** ion rearranges to **16b** on a time scale of 10^{-2} to 10^{-3} sec.

(*c*) $C_7H_7^+$ This ion has long been of extraordinary interest because of the possibility of the 6π aromatic tropylium structure **17** and the attractive looking benzyl structure **18**, as well as less stable structures such as the tolyl ions **19** (*o, m,* or *p*). $C_7H_7^+$ has been examined from a variety of precursors:

17   **18**   **19**

much work has been done with the collisional activation mass spectrometric method, which has apparently distinguished all three structures **17–19** (Winkler and McLafferty, 1973). Icr work has focused on $C_7H_7{}^+$ from toluene. One way of generating the ion is by photodissociation of $C_7H_8^{+\cdot}$ (Dunbar, 1975c), which gives $C_7H_7{}^+ + H^\cdot$ as the only product. The structure of the product ion was probed by its reactivity with toluene neutral (12),

$$C_7H_8^{+\cdot} \xrightarrow[-H\cdot]{hv} C_7H_7{}^+ + C_7H_8 \rightarrow C_8H_9{}^+ + C_6H_6 \tag{12}$$

a reaction which is apparently fast for the benzyl structure (Shen *et al.*, 1974) but not for the tropylium structure. Analysis of the fraction of benzyl by determining the fraction of reactive $C_7H_7{}^+$ indicated that both structures were produced at all wavelengths, with the more stable tropylium structure becoming predominant at the longer wavelengths. This was interpreted as a straightforward manifestation of the effect of internal excitation on the competitive unimolecular decay of $C_7H_8^{+\cdot}$ via two channels to give either benzyl or tropylium (Dunbar, 1975c).

(*d*) $C_7H_8^{+\cdot}$ Complementary to the spectroscopic work on $C_7H_8^{+\cdot}$ was a reactivity study of $C_7H_8^{+\cdot}$ structures (Hoffman and Bursey, 1971): it was shown that at least some of the $C_7H_8^{+\cdot}$ ions resulting from toluene ionization are reactive toward isopropyl nitrite (13), while the parent ions of cyloheptatriene and norbornadiene are wholly unreactive with the same reagent.

$$C_7H_8^{+\cdot} + i\text{-}C_3H_7ONO_2 \rightarrow C_7H_8NO_2{}^+ + C_3H_7O^\cdot \tag{13}$$

Again it is clear that these parent ions do not rearrange to a common structure.

(*e*) $C_8H_8^{+\cdot}$ In a similar study, Wilkins and Gross (1971) established that the parent ions of styrene and cyclooctatetraene are nonequivalent, by showing that the former reacts with neutral styrene (14), while the latter

$$C_8H_8^{+\cdot} + C_8H_8 \rightarrow [C_{16}H_{16}]^{+\cdot} \rightarrow C_{10}H_{10}^{+\cdot} + C_6H_6 \tag{14}$$

does not. Careful labeling and comparison with authentic samples suggested 1-phenyltetralin as the structure of the $C_{16}H_{16}^{+\cdot}$ intermediate.

(*f*) $C_6H_6O^{+\cdot}$ The $C_6H_5DO^{+\cdot}$ ion generated by electron impact from phenyl ethyl ether might have two plausible structures, (15) and (16). Nibbering (1973) showed that the $C_6H_5DO^{+\cdot}$ ion generated in this way transfers exclusively D^+ to the dimethyl pyridine reagent, indicating that there are no equivalent H's and D's in the ion, and ruling out structure **20b** for the fragment ion.

$$+ C_2H_2D_2 \qquad (15)$$

20a

$$+ C_2H_2D_2 \qquad (16)$$

20b

B. Excited States of Ions

While still rather primitive in comparison with knowledge about neutral molecules, our understanding of excited states of polyatomic ions is beginning to increase. As one would expect, the most detailed and specific knowledge of excited electronic states comes from spectroscopic methods, although considerations of chemical reactivity also yield some clues.

1. Spectroscopic Methods

(a) *Energy and Symmetry of Excited States* Photodissociation spectroscopy has made an important contribution in identifying excited ion states. Each peak in a photodissociation spectrum such as that shown in Fig. 4 can be interpreted as a transition to an optically allowed state of the ion, and the photodissociation cross section sets a lower limit on the optical absorption cross section (since each molecule dissociated represents at least one photon absorbed). Careful correlation of the photodissociation spectrum of a radical cation with the photoelectron spectrum of the corresponding neutral and with theory can in many cases give a very good picture of the low-lying excited doublet states of the ion. This has been done with some care for a number of ions, including the parent ions of toluene and butadiene (Dunbar, 1975d), methane (Riggin and Dunbar, 1975), N_2O (Orth and Dunbar, 1977), methylnaphthalene (Dunbar and Klein, 1976), benzene (Freiser and Beauchamp, 1975), and others. Evidence has accumulated suggesting that the photodissociation spectrum of a large ion is (within limitations) a good representation of the optical absorption spectrum in many or most cases.

Photodetachment spectra of anions have been less rewarding in this context, because many appear to show no spectral features beyond a rise from the baseline at threshold, with a leveling off at shorter wavelength. One exception is the identification of spin-orbit split states (Smyth and Brauman, 1972). More significant perhaps is the recent observation of definite peaking in the photodetachment spectrum of the phenoxy anion (Richardson *et al.*, 1975; Richardson and Stephenson, 1975), which is most likely associated with autodetachment from an excited anion state.

(*b*) *Chemical Character of Excited States* Freiser and Beauchamp (1976b) in some elegant experiments have determined the basicity of some gas phase aromatic species in their excited electronic states. It is readily shown that if the basicity (proton affinity) of the ground state molecule is known, and if the excitation peak wavelength for both the neutral molecule and the ion into the excited electronic state of interest is determined, the basicity of the excited state of the molecule is immediately available. The proton affinity is a fairly powerful probe of the electronic distribution within the molecules, and this approach thus provides a useful chemical probe of the molecular excited state properties. These ideas have been applied to benzaldehyde and cyanobenzene, and have been extended to the consideration of Li^+ binding as well as H^+ binding to the excited state (Freiser *et al.*, 1976).

2. Reactivity of Excited Ions

(*a*) N_2^+ *and* CO^+ Bowers and colleagues (Kemper and Bowers, 1974; Bowers *et al.*, 1974, 1975) have found that upon electron impact N_2 and CO can be ionized into excited states which undergo chemical reactions not exhibited by their ground state ions:

$$N_2^{+*} + N_2 \rightarrow N_3^+ + N \tag{17}$$

$$CO^{+*} + CO \rightarrow C_2O^{+*} + O \tag{18}$$

The electron energy at which these reactions begin to be observed gives an idea of the location of the excited states involved. It is not entirely clear whether the excitation in the ions exists as electronic or as vibrational energy, but it does seem clear that at least initially excited electronic states of the ions are formed, which might then undergo internal conversion to give vibrationally excited ions.

(*b*) *Tandem icr Results* Using the tandem icr spectrometer, Futrell and his colleagues have extensively studied the effect of internal vibrational energy on simple ion–molecule reactions, and the rate of relaxation of such energy by collisional deactivation, as well as examining metastable decomposition (R. D. Smith *et al.*, 1975; R. D. Smith and Futrell, 1975a, 1976a–g; D. L. Smith and Futrell, 1974a, 1975, 1976; Fiaux *et al.*, 1974).

IV. Analysis of Neutral Compounds

A. Electron Impact Ionization Fragmentation

This is of course the traditional domain of the mass spectroscopist, with relatively little contribution by icr. The icr spectrometer is a perfectly good mass spectrometer when operated in the low pressure regime, and a low pressure scan is a standard check on compound purity in the course of some icr studies. It is likely that the imminent appearance of routine high resolution icr spectrometers, both cw and pulsed Fourier transform, will bring the use of icr into realistic competition with high resolution conventional instruments, but this has not yet occurred. Much of this chapter, and particularly Section III, is concerned with means of identifying and characterizing ionization products by icr-oriented procedures. A few additional topics focusing on the properties of the parent neutral will be included here.

1. Appearance Potential Measurements

The extremely high theoretical sensitivity of the icr instrument, particularly under conditions where long residence times permit ions to be accumulated in the cell, makes the instrument attractive for ionization efficiency measurements very near threshold and for determining appearance potentials. Buttrill and Magil (1975) adapted the icr spectrometer to a careful appearance potential measurement technique, and were able to measure appearance potential values for $C_3H_5^+$ from propylene and cyclopropane to ± 2 kcal.

2. Fragmentation Patterns

Morton and Beauchamp (1975) looked at the fragmentation of some diethers $RO-(CH_2)_n-OR^+$. Of interest was the observation of the decomposition process (19) which appears to be specific and diagnostic for the possibility of forming the six membered cyclic transition state **21**, and

$$\tag{19}$$

21

is thus a potentially useful chain-length determination for dialkoxyalkanes and related bifunctional molecules.

3. Systematic Spectroscopy

One of the major advantages of an optical spectroscopic analytical technique is the possibility of immediately identifying chromophoric structural groups in the molecule. Dunbar (1976a) pointed out that, for molecules forming an observable parent ion on ionization, the ion photodissociation spectroscopic approach may combine these advantages with the mass-spectrometer virtues of sensitivity and specific mass selection of the species under examination. Four types of chromophore in hydrocarbon radical cations were studied in this work from the point of view of their utility in systematic spectroscopic classification of parent ions. All four, the olefin, diene, benzene, and saturated-carbon-chain groups, are characterized by substantial photodissociation cross sections and by spectral features which are reproducible in wavelength position among series members. The intensities of the photodissociation peaks did not appear to have as much diagnostic value as their positions, although some regularity of intensities might be anticipated among chromophores embedded in larger molecules.

The well-studied benzene chromophore (Dunbar, 1973a,b,d; Dunbar and Fu, 1973; Dymerski et al., 1974; Freiser and Beauchamp, 1975) illustrates that the substituted benzene ions show a universal and characteristic peak near 280 nm, about 28 nm wide, with a cross section of 1 to 5×10^{-18} cm^2, and a peak or pair of peaks of comparable intensity in the visible region, whose shape and width are somewhat variable and are related to the nature of the side chain.

4. Chemical Ionization

While many icr studies are really chemical ionization studies, this term is usually reserved for an experiment in which a reagent gas, often methane, is present at very high pressure (Field, 1972). The primary electron impact ionization takes place largely on reagent-gas molecules, and the high concentration of neutrals permits the possible sequences of ion–molecule reactions to proceed essentially to completion, so that the actual ionic reagent acting on the unknown sample consists of one or a few stable ions which are unreactive with respect to reagent neutrals: in methane the reagent ions are largely CH_5^+ and $C_2H_5^+$, in ammonia NH_4^+, in water H_3O^+ (possibly clustered), and so forth.

The chemical ionization (CI) idea is particularly adapted to the high pressure mass spectrometer; in some ways the icr spectrometer, with its capability for more detailed analysis of the ionic chemistry of the unknown, is a more powerful tool. However, it is perfectly feasible to do CI work in an icr instrument. Pesheck and Buttrill (1974) made a careful study of methane chemical ionization of esters: the results were shown to be quite comparable to conventional CI results, with some understandable differences resulting from the improbability of stabilizing and deactivating association species such as $(M + C_2H_5)^+$ under the low pressure conditions of the icr experiment.

Clow and Futrell (1972a) investigated methane chemical ionization in the icr spectrometer for a number of hydrocarbons, and they, too, found differences from results in conventional instruments, which could be explained on the basis of inefficient collisional stabilizing of internally excited ions under the long mean free times of the icr experiment. For instance, a $C_5H_9^+$ ion is stabilized and observed in high pressure CI analysis of cyclo-C_6H_{12}, but apparently undergoes further fragmentation too fast to be observed under icr conditions. Since the goal of CI work often is to provide a means of ionizing an unknown neutral with minimal fragmentation, the icr spectrometer is thus, other things being equal, an inferior choice for simple CI work. On the other hand, the sharper probe of ion chemistry available through icr double resonance analysis of CI chemistry has been and will be valuable in increasing the understanding of CI results with conventional reagents, and also in finding and evaluating potentially important new CI reagents [e.g., the exploration by Vogt and Beauchamp (1975) of CF_2H^+, and by Williamson and Beauchamp (1975) of NO^+ as potentially useful CI reagents].

B. Characteristic Ionic Reactions

The most ancient concept of qualitative analysis is the addition of a reagent to the unknown sample, with the occurrence (or nonoccurrence) of a characteristic reaction serving to characterize the unknown. It has occurred to many people that analysis of unknown neutral samples in the icr spectrometer [or in the chemical-ionization mass spectrometer (Field, 1972)] using chosen ionic species as reagents and characteristic ion–molecule reactions as the analytical reactions is a natural extension of this idea. Attractive in this program is the uniting of the specificity and microscopic sample-handling capabilities of mass spectrometry with the enormous power of chemical qualitative analysis. While much effort has been directed along these lines and much progress made, the point has not yet been reached where a sufficient systematic data base exists to make this a useful routine method. The intent here is to structure the existing body of data in terms of its relevance to such a qualitative analysis concept. The behavior of various classes of neutral molecules with ionic reagents will be surveyed; then a number of the more promising ionic reagents will be discussed in terms of what is known about their reactivity and selectivity.

1. *Alcohols and Ethers*

The ion–molecule chemistry of the alcohols is rich, with numerous possible reaction channels whose competition depends on several factors, one of the most important of which is the excess energy available in the collision intermediate. Work by Beauchamp and his colleagues has elucidated much of this chemistry (Beauchamp, 1969; Beauchamp *et al.*, 1974; Caserio and Beauchamp, 1972; Ridge and Beauchamp, 1971) and in particular has related much of what occurs to the formation of hydrogen bonds between ionic reactant and neutral alcohol.

(*a*) *Protonated-Oxygen Ionic Reactants* Two types of reactions are found important in reactions of alcohols with ROH_2^+: one is the "ionic dehydration" reaction, typified by *t*-butyl alcohol, (20). This reaction requires

$$(CH_3)_3C-OH + (CH_3)_3COH_2^+ \longrightarrow [(CH_3)_3C-\overset{H}{O}--H^+ - -\overset{H}{O}\underset{H-CH_2}{\overset{}{-}}C(CH_3)_2] \longrightarrow$$

$$(CH_3)_3COH-H^+-OH_2 + CH_2{=}C(CH_3)_2 \quad (20)$$

the presence of a hydrogen β to the alcohol oxygen to occur, and may be an effective probe for β hydrogens in alcohols (and perhaps other *n*-donor bases). The postulated transition state in (20) is supported by this β-hydrogen requirement and by labeling studies. This reaction has been compared with analogous acid-catalyzed dehydration of neutrals (Beauchamp *et al.*, 1974). While the dehydration reaction is apparently common in secondary and tertiary alcohols, it has not been observed in primary alcohols, perhaps because of excessive activation energy, and this is a distinction of possible utility.

The other principal reaction type is "condensation" with loss of water, exemplified by methanol, reaction (21).

$$CH_3OH + CH_3OH_2^+ \longrightarrow CH_3O\overset{H}{\cdots\cdots}H^+ \longrightarrow CH_3-\underset{CH_3}{\overset{}{OH^+}} + H_2O \quad (21)$$

$$H_3C-\underset{H}{\overset{}{O}}$$

In *t*-butyl alcohol with $(CH_3)_2COH^+$ as the ionic reagent, the condensation and dehydration reactions can compete with each other, and also with proton transfer and with $C_4H_9^+$ production. With protonated parent as the reagent ion, only dehydration and condensation occur (along with collisional stabilization of the intermediate). These competitions, and the extent of fragmentation of the product ions, are governed by thermochemical considerations, so that while the condensation and dehydration processes are characteristic of alcohols, the nature of the ultimately observed products is subject more than one would like to thermochemical rather than structural variables. This is true for many ion–molecule reaction systems, and one important goal is to find more structure-sensitive rather than energy-sensitive reagent ions.

(b) *Acetylating Ionic Reagents* Bursey and colleagues have explored extensively the acetylating reagent ions obtained by ionizing biacetyl at moderately high pressure: these are the biacetyl ion itself ($CH_3COCOCH_3^{+\cdot}$), acetyl ion (CH_3CO^+), and triacetyl ion ($(CH_3CO)_3^+$). Two, $(CH_3CO)_2^{+\cdot}$ and $(CH_3CO)_3^+$, are effective and selective acetylating reagents for oxygen-containing molecules (Kao *et al.*, 1976; Bursey *et al.*, 1970), as for example,

$$(CH_3CO)_2^{+\cdot} + C_2H_5OH \rightarrow C_2H_5OH(CH_3CO)^+ + CH_3CO\cdot \qquad (22)$$

(c) *Ethers and Diethers* In some contrast to alcohols, the principle reaction of R_2O with R_2OH^+ is the formation, and stabilization, of the proton-bridged dimer R_2O---H^+---OR_2 (Morton and Beauchamp, 1972). Since the reaction producing this species has no neutral product to carry off excess energy, the proton-bridged dimer must be stabilized by collision, and is hence observed only at high pressure ($\sim 10^{-5}$ Torr and above). While the lower reactivity of ethers versus alcohols probably has diagnostic importance, classes of reagent ions showing differential reactivity with these two groups still need to be mapped out.

Intramolecular H-bonding in diethers leads to an interesting phenomenon of potential utility for their analysis (Morton and Beauchamp, 1972): if intramolecular H-bonding is possible to give a species **22** for the protonated

22

parent, further reaction to form the proton-bridged dimer is suppressed; the degree of suppression of the reaction depends on the favorability of the intramolecular chelation; when $R = (CH_2)_n$ in **22**, the chelation is most effective, and formation of proton-bridged dimer most effectively suppressed, for large n, with $n = 4$ giving complete product suppression up to 10^{-3} Torr. This is a potentially useful probe of the separation of two ether groups in a molecule, and may well be extended to other functional groups having strong H-bonding ability.

(*d*) *Esterification* In a number of cases, an esterification reaction occurs between a protonated acid and an alcohol (Tiedemann and Riveros, 1974), (23). The reaction appears quite general, and may well be a probe for the

$$CH_3C\overset{\displaystyle O}{\underset{\displaystyle OH_2^+}{\diagdown}} + ROH \longrightarrow CH_3C\overset{\displaystyle O}{\underset{\displaystyle \underset{\displaystyle H}{OR^+}}{\diagdown}} + H_2O \qquad (23)$$

alcohol functionality. Only with protonated formic acid does this reaction not proceed smoothly.

2. Alkyl Halides

A variety of reactions are observed in the alkyl halides and related molecules (Asubiojo *et al.*, 1975; Beauchamp *et al.*, 1972; Sullivan and Beauchamp, 1976; Dawson *et al.*, 1973; Blint *et al.*, 1974; Beauchamp and Park, 1976; Beauchamp, 1976a,b). Proton transfer from protonating reagent ions is often observed, and condensation with cationic reagents with HX elimination, analogous to (21), is also possible, as for instance (Beauchamp *et al.*, 1972), reaction (24). Most characteristic, however, are the anion reactions of these molecules.

$$C_2H_5ClH^+ + C_2H_5Cl \to (C_2H_5)_2Cl^+ + HCl \qquad (24)$$

(*a*) *Base-Induced* HF *Elimination* When an anion attacks a fluorine-containing molecule containing acidic hydrogens, a number of outcomes are possible, depending to a great extent on thermochemical factors (Ridge and Beauchamp, 1974a,b; Sullivan and Beauchamp, 1976): proton transfer to the anion is likely if it is an exoergic process, and if a great deal of energy is available further fragmentation to F^- may occur; the most interesting and characteristic reaction is the energetically favorable HF elimination (25).

$$CH_3O^- + CH_3CF_3 \longrightarrow CH_3O^- \cdots\cdots H \quad F \longrightarrow CH_3OHF^- + CH_2CF_2 \quad (25)$$
$$\underset{CH_2 - CF_2}{}$$

For mild reagent ions no further fragmentation occurs, and this reaction is often favored. A great deal is now known about the thermochemical factors involved in determining the course of reaction of Y^- with RX.

(*b*) *Backside Displacements* Reactions analogous to solution S_2N displacements are observed in the case of methyl chloride (Brauman *et al.*, 1974), reaction (26), where Y = F, Cl, or CH_3S and X = Cl or Br.

$$Y^- + CH_3X \to CH_3Y + X^- \qquad (26)$$

While the factors leading to displacement of the type (26) in preference to proton transfer or HX elimination in reactions such as (25) are not yet clear, it is evident that in the absence of acidic hydrogen β to the halogen, (26) is the normal and characteristic reaction of the halogenated molecules. One might hope that the occurrence of one or the other of these reactivity patterns may be diagnostic for acidic β hydrogens in fluorides; however, in chlorides and bromides, displacement of X^- from the halide predominates regardless of the presence of β-hydrogens (Ridge and Beauchamp, 1974b), except for the interesting case of 2-chloroethanol, where assistance by the oxygen may be responsible for the elimination reaction (27).

$$\underset{\text{CH}_3\text{O}^- + \text{CH}_2-\text{CH}_2}{\overset{\overset{\text{OH} \quad \text{Cl}}{|\qquad|}}{}} \longrightarrow \text{CH}_3\text{OHCl}^- + \underset{\text{CH}_2-\text{CH}_2}{\overset{\text{O}}{\overset{/\,\backslash}{}}} \qquad (27)$$

(c) *Nucleophilic Displacements* Reaction (21) is an example of a large class of cation reactions which have been termed nucleophilic displacements. They occur so widely that their analytical specificity is limited; they are not particularly characteristic of alkyl halides, but are mentioned here because so many examples have been reported involving halogen-containing species. The reaction involves displacement of a molecular fragment from an ionic substrate following nucleophilic attack by a basic neutral molecule, reaction (28). Other examples include reactions (29) and (30). The attacking nucleophilic neutral may be any of a large variety of types, including halides, amines,

$$\text{RX} + \text{R}'\text{Y}^+ \to \text{RXR}'^+ + \text{Y} \qquad (28)$$

$$\text{H}_2\text{O} + \text{CH}_3\text{ClH}^+ \to \text{CH}_3\text{OH}_2^+ + \text{HCl} \qquad (29)$$

$$\text{CH}_3\text{OH} + \text{H}_2\text{B} - \text{BH}_3^+ \to \text{CH}_3\text{OHBH}_2^+ + \text{BH}_3 \qquad (30)$$

alcohols, xenon, and N_2. Two rules have been proposed for predicting when this reaction will occur (Beauchamp, 1977): (1) it must be exoergic, and (2) there must not be a favorable proton transfer from the ionic substrate to the nucleophile. While it is obviously central in the understanding of reaction processes in Lewis-basic neutrals, it is not clear that this reaction type has promising characteristics for general analytical purposes.

3. Aldehydes and Ketones

The carbonyl compounds react in many of the ways expected for *n*-donor bases (van der Hart and Van Sprang, 1977; Karpas and Klein, 1976; Isolani *et al.*, 1973; Hass *et al.*, 1976). Protonation of the oxygen is common, as is attack by electrophiles such as various acylating reagent ions (including the ionization products of the compound itself). There are a number of reactions which show a promising degree of selectivity.

(a) *Acylation*   As part of their work on acylation reactions, Bursey *et al.*
(1974) examined reactions of several ketones, for which a typical reaction is
(31). Of particular interest is the finding that reactions of the relatively bulky

$$CH_3COCOCH_3^{+\cdot} + \quad \longrightarrow \quad + CH_3CO \qquad (31)$$

$(CH_3CO)_2^{+\cdot}$ group are quite sensitive to the environment near the carbonyl,
with bulky groups apparently acting to inhibit acylation. For reaction (31),
the rate of acylation declines as R goes from H to n-C_4H_9 through the series
of n-alkyl substituents, and the rate with R = C_4H_9 is slower by a factor
of almost 15 than that with R = H. A similar effect was observed for the
pair of decalone isomers **23a** and **23b**, with the less hindered carbonyl of **23b**
being acylated twice as fast as that of **23a**.

23a (cis)   **23b** (trans)

Even more striking steric effects on reactivity are found using the reagent ion **24** (1,2-dicyclopropylethanedione), which reacts by transfer of a

24

$C_3H_5CO^+$ group (Henion *et al.*, 1975; Bursey *et al.*, 1975). For this reaction,
2-n-butylcyclohexanone reacts 70 times more slowly than cyclohexanone;
with the decalones, **23a** did not react to an observable extent, indicating that
it reacted less than $\frac{1}{3}$ as fast as **23b**. The remarkable selectivity of this acylating
reagent, which correlates in a sensible and predictable way with the crowding
around the carbonyl oxygen, is a very hopeful step in the progress of this
analytical approach.

Tiedemann and Riveros (1973) have also studied the acylation of carbonyl
compounds, using as acylating reagents the parent ions of the compounds
involved. Of particular interest is the finding that the presence of a γ-hydrogen
in the molecule alters the reactivity pattern: for the ketones without γ-
hydrogens, simple acylation occurs, reaction (32), while if a γ-hydrogen is

$$R_2C{=}O + R_2C{=}O^{+\cdot} \rightarrow R_2COCOR + R\cdot \qquad (32)$$

present, a more complex pattern of reactivity is found, including the re-arrangement reaction exemplified by (33), which is apparently analogous

$$CH_3{-}\overset{\overset{\displaystyle O}{\|}}{C}{-}C_3H_7 + CH_3{-}\overset{\overset{\displaystyle O}{\|}}{C}{-}C_3H_7^{+\cdot} \longrightarrow (CH_3\overset{\overset{\displaystyle O}{\|}}{C}C_3H_7)C_3H_6O^{+\cdot} + C_2H_4 \quad (33)$$

to the McLafferty rearrangement fragmentation of the $CH_3COC_3H_7^{+\cdot}$ ion. It is not yet clear whether the γ-hydrogen requirement applies to the ion, the neutral, or both.

(b) *Decarbonylation by* $NiCp^+$ A reaction perfectly illustrating the type of selective reagent ion one would like to use is the decarbonylation reaction (34), (35) discovered by Corderman and Beauchamp (1976a). The reagent

$$\text{Ni(C}_5\text{H}_5)^+ + \text{RCHO} \Big\langle \begin{array}{l} \nearrow \;\; \text{CpNiCO}^+ + \text{RH} \qquad\qquad (34) \\[1em] \searrow \;\; \text{CpNiRH}^+ + \text{CO} \qquad\qquad (35) \end{array}$$

ion is readily generated by ionization of nickelocene, and reaction (34) (with acetaldehyde) is very fast (8×10^{-10} cm^3 molecule^{-1} sec^{-1}). One or the other of the two decarbonylation channels (34) and (35) occurs for all the aldehydes studied, with (34) being observed for the alkyl aldehydes, and (35) for 3-methylacrolein and benzaldehyde. On the other hand, no decarbonyla-tion was observed for H_2CO, CF_3CHO, CH_3COCl, CH_3COBr, Me_2CO, or MeOAc, so that this process appears to be truly selective for aldehydes, and has great promise as a selective probe for the aldehyde functionality.

(c) *Reaction with* O^- Although it is not a selective reagent, O^- reacts with carbonyl compounds to give a variety of products (Harrison and Jennings, 1976), of which the most common are OH^-, $(M{-}H)^{\cdot-}$, $(M{-}H_2)^-$, and $RCOO^-$, where $(M{-}H)^{\cdot-}$ and $(M{-}H_2)^-$ represent the parent ion minus one and two mass units, respectively, and R is one of the ketone alkyl groups. The ratios of these products vary for different ketones: it is clear that O^- is an energetic reagent capable of populating a variety of product channels, determined at least partly by thermochemical factors.

(d) *Reaction with* CX_3^+ The ions CF_3^+, CF_2Cl^+, CCl_2F^+, and CCl_3^+ (Ausloos et al., 1975a,b; Eyler et al., 1974) react with carbonyl compounds in a set of reactions which can be formulated as involving fragmentation from a four-center transition state (36). The result of the reaction is replace-ment of the carbonyl oxygen with X^+. Loss of HX is a common further

$$CX_3^+ + R'COR \longrightarrow [R\underset{\underset{R'}{|}}{\overset{\overset{\displaystyle O\cdots\cdots CX_{2+}}{|}}{C}}\cdots\cdots X \quad] \longrightarrow CX_2O + R'CXR^+ \qquad (36)$$

further fragmentation

fragmentation of the $R'CXR^+$ product. In cases where there is a weak bond in the carbonyl compound, a competing reaction (37) may be observed to some extent.

$$CX_3^+ + R'COR \rightarrow CX_3R + R'CO' \qquad (37)$$

(e) Cl^- *Transfer* A chloride ion may be transferred from Cl_2^- to carbonyl compounds, reaction (38) (Asubiojo *et al.*, 1975; Karpas and Klein,

$$Cl_2^- + RR'CO \rightarrow RR'COCl^- + Cl\cdot \qquad (38)$$

1976). It is found that the rate of this reaction is strongly sensitive to the size of the R and R' groups, and Karpas and Klein (1976) have developed a quantitative set of parameters for various alkyl substituents which appear to allow successful prediction of the rate of (38) for a variety of aldehydes and ketones. This appears to be the first reaction type for which a reasonable quantitative connection has been made between a reaction rate and the environment of the reactive neutral functionality.

4. *Amines*

The chemistry of amines might be expected to show many features in common with other *n*-donor base compounds, and presumably many of the reactions and reagents useful for alcohol and halide compounds will also give useful characterization and selectivity for amines. However, except for their proton-transfer chemistry, the amines are much less heavily mapped out than these other extensively studied functional groups. Aside from the detailed characterization of the amine-group environment by thermo-chemical means (to be described) there do not appear to be any known features of amine reactivity patterns having analytical utility.

5. *Benzene Derivatives*

(a) *Acetylation* The rate at which aromatic compounds are acetylated by the $(CH_3CO)_2^{+\cdot}$ reagent is sensitive to the steric environment around the site of attack, although in general there are also other factors at work and no straightfoward quantitative relationship has yet been found between the nature of neighboring substituents and the rate of acetylation (Chatfield and Bursey, 1975a,b, 1976a,b).

The steric crowding effects seem to be most nearly dominant and most easily understood in the case of ortho substitution, and Bursey and co-workers have studied two cases in which this particularly clear, the phenols and the pyridines.

Acetylation of compounds of the type **25** (Henion *et al.*, 1973) occurs at a reasonable rate if the ortho substituents are H, CH_3, or C_2H_5, but no product is observed if R is $CH(CH_3)_2$ or $C(CH_3)_3$; however, compounds **26**

      25   **26**

are acetylated even when R is $C(CH_3)_3$, so clearly it is steric hindrance around the OH which is decisive in inhibiting acetylation.

The alkyl pyridines **27** are all acetylated by the $(CH_3CO)_2^{+\cdot}$ reagent, but

      27

the rate of reaction decreases with increasing bulk of the R group (except that $R = CH_3$ is a little faster than $R = H$); the rate for $R = C(CH_3)_3$ is 12 times slower than for $R = H$, an effect which again is presumably due largely to steric effects. From the numerous examples of alkyl substitution Bursey has examined, it is clear that a methyl group, even in the ortho position, is rather ineffective in decreasing the rate of acetylation, and strong effects do not become evident until groups at least as bulky as $CH(CH_3)_2$ are involved. It may well be that the expected acceleration of electrophilic ring attack by the alkyl substituent is at work in opposition to the steric effects, and only for very hindered attack sites does steric crowding become dominant.

(*b*) *Other Aromatic-Ring-Attacking Cations*   Since a gas phase cation is inherently a powerful electrophile, one expects that the attack of a cation on an aromatic ring should reflect some features of electrophilic aromatic substitution which are well known in solution, in particular the effect on rate and position of attack arising from various types of ring substituents. Unfortunately it is not yet possible in most cases to identify the site of ring attack, [in at least one case where this was possible (Dunbar, 1973a), no positional selectivity was evident]; however, it has in some cases been possible to examine substituent rate effects and to identify rate variations characteristic of electrophilic attack.

The acetylating reagents $(CH_3CO)_2^{+\cdot}$ and $(CH_3CO)_3^+$ attack a variety of substituted benzenes according to the general reaction (39). The trend of

$$Y\!-\!COCH_3^+ + \underset{}{\bigcirc}\!\!-\!X \longrightarrow \underset{CH_3CO^+}{\bigcirc}\!\!^X + Y \qquad (39)$$

substituent rate effects (see Table I) clearly follows the trend expected, with electron donating substituents tending to enhance the rate of electrophilic attack (Lehman and Bursey, 1976). (The NO_2 substituent is apparently anomalous, and it is generally presumed that the attack here is on the side chain rather than the ring.)

TABLE I

Relative Rates of Aromatic Substitution[a]

X:	NO_2	CF_3	Cl	F	H	C_2H_5	CH_3	OCH_3	OH	NH_2
K_{AC}:	20	[b]	[b]	14	1	3.5	10	15	50	17
K_{NO_2}:	10	0.3	0.25	0.55	1	0[c]	0.3	0[c]	[b]	0[c]

[a] The ring substituents X are ranked in increasing order of electron-donating ability (decreasing Hammett σ constant). K_{AC} refers to the rate of reaction (39) relative to benzene with $Y = CH_3CO\cdot$, K_{NO_2} to the rate of reaction (40) also relative to benzene.
[b] Not reported.
[c] No product observed.

Nitration of aromatic substrates is also observed using the $CH_2O\!-\!NO_2^+$ reagent ion (Dunbar et al., 1972), according to reaction (40). Most surprisingly, the substituent effects in this case (Table I) are counter to the

$$CH_2ONO_2^+ + \underset{}{\bigcirc}\!\!^X \longrightarrow \underset{NO_2^+}{\bigcirc}\!\!^X + CH_2O \qquad (40)$$

expected trend, and clearly suggest that the rate determining step is enhanced by electron-withdrawing substituents, behavior which would normally be associated with a nucleophilic attacking reagent. While this remarkable behavior may not be understood, the suggestion is that the rate-determining step is not the initial attack of the cation, but rather involves a nucleophilic-type attack on the ring by one of the highly electron-rich terminal oxygens of the $-NO_2$ group; one might then imagine rapid, reversible formation of an intermediate such as **28**, with the rate-determining step being formation

28

of an oxygen–ring bond with CH_2O expulsion. [An excellent fuller discussion of this work of Dunbar, Shen, and Olah is given by Lehman and Bursey (1976).]

In addition to the reactions already mentioned, which involve (formally) transfer of CH_3CO^+, CH^+, or NO_2^+ to the aromatic, there are a variety of (presumably) electrophilic ring-attack reactions, some of which involve rearrangements to stable products. A few of these (Lehman and Bursey, 1976) are (41)–(45).

$$ClCH_2^+ + C_6H_6 \rightarrow C_7H_7^+ + HCl \tag{41}$$

(analogous to the Blanc chloromethylation reaction in solution);

$$CH_3^+ + C_6H_6 \rightarrow C_7H_7^+ + H_2 \tag{42}$$

(analogous to Friedel–Crafts alkylation);

$$NO_2^+ + C_6H_6 \rightarrow C_6H_6O^+ + NO \tag{43}$$

(analogous to solution nitration in acid medium);

$$CH_3CNH^+ + C_6H_6 \rightarrow C_7H_8N^+ + H_2 \tag{44}$$

and (Shen et al., 1974)

Substituent effects have not been examined in these, although labeling studies have elucidated the mechanisms of (44) and (45) at least.

(c) *Deuterium Exchange* Freiser et al. (1975) have described a promising novel analytical approach for substituted benzene molecules. When introduced into the trapping-cell spectrometer with a relatively high pressure of D_2O under ionizing conditions, the aromatic neutral protonates rapidly; then the protonated aromatic sequentially exchanges all its labile protons with the excess D_2O until finally an ionic species is reached in which all exchangeable hydrogens have been replaced by deuterium. To illustrate for mesitylene, the sequence (46) occurs quite rapidly. At long times in the trapping cell the $C_6D_4Me_3^+$ becomes the predominant ion, and its composition

$$C_6H_3Me_3 \xrightarrow[\text{source}]{\text{deuteron}} C_6H_3Me_3D^+ \xrightarrow{D_2O} C_6H_2D_2Me_3^+ \xrightarrow{D_2O}$$

$$C_6HD_3Me_3^+ \xrightarrow{D_2O} C_6D_4Me_3^+ \longrightarrow \text{no further reaction} \quad (46)$$

immediately indicates the presence in mesitylene of exactly three exchangeable hydrogens. The evidence gathered by Freiser et al. suggests that only ring protons are exchangeable, but that even ring protons are not necessarily labile. Illustration of the analytical potential of this experiment is provided by the case of difluorobenzenes: ortho- and para-difluorobenzenes rapidly exchange all four ring protons, whereas the meta isomer exchanges only one of its protons. With more clarification of the rules governing exchangeability of protons, this may be a generally useful approach.

6.   Cyclohexane and Norbornane Derivatives

Quite promising are a series of experiments probing subtle differences in geometry in these systems using acylating reactions (Bursey et al., 1975). For the pair of isomers **29** and **30** (*trans*- and *cis*-4-*t*-butylcyclohexyl

29   **30**

acetates), various acetylating reagents show a degree of differential reactivity. The most pronounced difference was observed for the reagent ion $C_2H_5COCOCH_3COCH_3^+$, with which the trans isomer **29** reacted 2.6 times as fast as the cis isomer **30**.

A similar phenomenon is observed for the relative rates of acetylation by $(CH_3CO)_3^+$ of endo and exo isomer pairs of several alcohols and acetates having the norbornane skeleton (Kao et al., 1976). For instance, *exo*-norborneol **31** is acetylated about four times as fast as the endo isomer **32**;

31   **32**   **33**

the reaction with **31** is close to the predicted collision rate, suggesting no important steric hindrance in this case. If the steric hindrance in **32** is reduced by forming the related compound **33** in which the 6-hydrogen is substantially removed from the vicinity of the —OH, then the acetylation reaction no longer shows inhibition, and **33** and its exo isomer react at the same much higher rate, close to the encounter-controlled rate.

7. Acid Anhydrides

Bowie et al. (Bowie and Williams, 1974a,b, 1975a,b; Bowie, 1975; Wilson and Bowie, 1975) have studied anion reactions in a number of acid anhydrides. Characteristic of these systems is formation of collision-stabilized adducts, of which a typical example in acetic anhydride is (47). Near zero

$$CH_3CO_2^- + (CH_3CO)_2O \rightarrow CH_3CO_2(CH_3CO)_2O^- \qquad (47)$$

electron energy various maleic and phthalic anhydrides also form molecular negative ions, which associate with a parent neutral to form collision-stabilized negative ions at mass $2M$. Of particular interest is the finding that if the molecular ion of some of these anhydrides, for example **34**, is formed at an appropriate electron energy between 1 and 5 eV, a molecular anion is formed which does *not* form the $2M$ association product; Bowie et al. attribute this to formation of a long-lived excited electronic state of the anion.

34

8. Alkenes

π-Bonded compounds are highly reactive, and have been of interest to ion–molecule chemists since the earliest days. The analytical chemist would like to (1) establish the presence of one or more double bonds, (2) characterize their nearby environment, (3) establish their relative locations, if there are multiple π bonds, and (4) establish their cis or trans configuration. One is still a long way from achieving these goals. Even the apparently simple matter of finding reagents to distinguish π-donor from n-donor sites for electrophilic ion attack, as the way of answering (1), is little advanced. The spectroscopic methods described have particular promise as approaches to (2) and especially (3), and seem likely to be enthusiastically pursued. Chemical reactivity methods are at least suggestive of ways to approach all these points, and in particular (4) seems to be a question profitably attacked only by selective-reagent methods, neither spectroscopic methods nor other mass spectrometric techniques being apparently very effective in distinguishing cis from trans.

(a) *Use of* $C_4H_6^{+\cdot}$ The butadiene parent ion is a particularly promising reagent for probing double-bonded molecules, presumably because if conditions in the attacked neutral are favorable a concerted Diels–Alder-type of cyclic reaction mechanism drastically affects the rate and course of the reaction. Gross *et al.* (1972) looked at reactions of seven isomeric C_5H_{10} compounds with $C_4H_6^{+\cdot}$, **35–41**. All seven were clearly differentiated by the

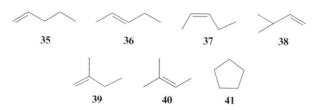

nature and relative abundances of products. The particularly important challenge of distinguishing **36** from **37** was met nicely, as indicated in Table II. In this case, the ratio of m/e 95: m/e 96 is most characteristic, going from 8:1 for cis to 3.5:1 for trans.

TABLE II

Products of Reaction of $C_4H_6^{+\cdot}$ with
2-Butene Isomers

	m/e:	81	82	95	96	109
cis		13	28	70	9	16
trans		7	29	40	11	13.6

(b) *Reaction with Double-Bond-Containing Ions* Many reactions of olefins with olefin ions are rationalized in terms of an addition mechanism (48).

$$R_1CH{=}CHR_2^{+\cdot}$$
$$R_3CH{=}CHR_4$$
$$\longrightarrow \begin{bmatrix} R_1CH{-}\overset{+}{C}HR_2 \\ | \\ R_3CH{-}\overset{\cdot}{C}HR_4 \end{bmatrix} \longrightarrow \text{products} \qquad (48)$$

42

The site of addition is apparently determined by the thermodynamic stability of the carbonium ion and radical sites in the addition complex **42**, with tertiary sites favored over secondary, secondary over primary. As is common for decomposition of radical ions of hydrocarbons, the complex **42** decomposes by a variety of paths of comparable activation energy, so that, while the reaction mechanism is understood in reasonable detail, the deduction of reactant structure from product distributions in (48) may not be straightforward in complex cases (Henis, 1970a,b; Bowers *et al.* 1968).

(c) *Diels–Alder Type Additions* The attack of $C_4H_6^{+\cdot}$ on olefins described yields products which are for the most part rationalized by postulating a cyclic intermediate of Diels–Alder type, reaction (49) (Lehman and

$$\text{products} \qquad (49)$$

43

Bursey, 1976, p. 102). Presumably the rules of orbital symmetry favoring the Diels–Alder process in neutrals also carry some weight here, although one expects them to be less important because

(1) this is a $3 + 2$ not a $4 + 2$ cycloaddition, and
(2) there is so much energy available in the complex due to the attractive interaction between the ion and the neutral that even if **43** is the lowest-energy transition state, other transition state structures are probably also accessible.

However, ionic processes mimicing the Diels–Alder reaction do seem to be favored, and several others have been reported.

The reaction sequence (50) is observed, and the postulated cyclic intermediate

$$+ C_6H_6 \qquad (50)$$

structure is supported by labeling experiments (the product structure is presumptive) (M. L. Gross, quoted by Lehman and Bursey, 1976, p. 103).

This reaction is similar to the reaction of styrene with its own parent ion, (51) (in which the product structure is again presumptive) (Wilkins and Gross, 1971).

$$\text{(structure)} + \phi \longrightarrow \text{(structure)}^{+\cdot} \longrightarrow$$

$$\text{(naphthalene)}^{+\cdot} + \phi H \qquad (51)$$

(d) *Reactions with* $CH_2{=}CH{-}O{-}CH_3^{+\cdot}$ The parent ion of methyl vinyl ether undergoes a reaction which may be quite general with olefins (Ferrer-Correia *et al.*, 1975), and which can be formulated, (52). The formal

$$
\begin{array}{ccc}
\text{R—CH=CH—R'} & & \text{RCH} \quad \text{R'CH} \\
+ & \longrightarrow & \| \quad + \quad \| \\
\text{CH}_2{=}\text{CH}_{+\cdot} & & \text{CH}_2 \quad \text{CH}_{+\cdot} \\
\quad\diagdown & & \qquad\quad\diagdown \\
\quad\text{OCH}_3 & & \qquad\quad\text{OCH}_3
\end{array}
\qquad (52)
$$

result is the replacement of CH_2 with CHR' in the reagent ion. The interesting feature here is that the indicated cleavage of the olefin substrate is apparently clean and specific, and provides a precise method of locating the position of a double bond. This is as close as any known reaction comes to the ideal analytical reaction, in that it apparently provides a single, unambiguous piece of structural information, the position of the double bond along a chain.

(e) *Methylmercury* The CH_3Hg^+ cation obtained from $(CH_3)_2Hg$ forms a collision stabilized association product with a variety of unsaturated compounds (Bach *et al.*, 1978), as for example (52a). Of analytical interest is the

$$
\text{CH}_3\text{Hg}^+ + \text{CH}_2{=}\text{CH}_2 \longrightarrow
\begin{array}{c}
\text{H}_2\text{C}{=\!=\!=}\text{CH}_2 \\
\diagdown\,\overset{+}{}\,\diagup \\
\text{Hg} \\
| \\
\text{CH}_3
\end{array}
\qquad (52a)
$$

fact that the rate of formation of product is dependent on the double-bond geometry, with the cis isomer of 2-butene being favored by 5:3 over the trans isomer.

(f) $O^{\overline{\cdot}}$ Goode and Jennings (1974) have studied reactions of $O^{\overline{\cdot}}$ with olefins. Of interest is reaction (52b). The fact that the two hydrogens are

$$O^{\overline{\cdot}} + \text{RCH}{=}\text{CH}_2 \rightarrow \text{H}_2\text{O} + \text{RCHC}^{\overline{\cdot}} \qquad (52b)$$

lost from the same carbon was established by labeling, and if it is found in further exploration that these two hydrogens are required, this reaction could be a useful test for terminal olefins.

9. Transition-Metal Compounds

The ion chemistry of these compounds is parallel to their solution chemistry: common reaction types include those which result in formation of binuclear and polynuclear species, those which involve expansion of the coordination sphere, and also those which involve ligand displacement. Reactions in which a mononuclear ionic complex reacts with neutral compound to give binuclear (and higher) condensation product ions are known for both cationic (Foster and Beauchamp, 1971, 1975c) and anionic (Dunbar et al., 1973b) reactions of metal carbonyls, including $Fe(CO)_5$, $Cr(CO)_6$, and $Ni(CO)_4$, and other metal-containing molecules (Foster and Beauchamp, 1975d; Corderman and Beauchamp, 1976b). More significant from an analytical standpoint are the ligand substitution processes: various neutral molecules can displace ligands from $Fe(CO)_n{}^+$, a typical observed sequence being (53) (Foster and Beauchamp, 1971). Some nucleophilic ligands studied

$$Fe(CO)_3{}^+ \xrightarrow[-CO]{H_2O} Fe(CO)_2(H_2O)^+ \xrightarrow[-CO]{H_2O} Fe(CO)(H_2O)_2{}^+ \tag{53}$$

include HCl, CH_3F, H_2O, NH_3, and C_6H_6. From reactions such as these, some progress has been made toward establishing relative metal–ligand binding energies for various ligands (Foster and Beauchamp, 1971, 1975c). Other ligand displacement reactions involve attack by an anion or a cation, reactions (54) and (55). Expansion of the coordination sphere through simple association reactions is also known, as in reaction (56), to produce a "mixed-sandwich" ion which is isoelectronic with ferrocene (Hemberger and Dunbar, 1977).

$$Fe(CO)_5 \xrightarrow{F^-} Fe(CO)_3F^- + 2CO \tag{54}$$

$$Fe(CO)_5 \xrightarrow{(CH_3)_2F^+} FeCH_3(CO)_5{}^+ + CH_3F \tag{55}$$

$$(C_5H_5)Fe^+ + C_6H_6 \longrightarrow (C_5H_5)(C_6H_6)Fe^+ \tag{56}$$

10. Other Compound Types

Many other classes of compounds have received attention in the icr literature, although detailed descriptions of their chemistry are not warranted here. Some, with representative references, are:

Cyanides: Staley et al. (1976), Bowie and Williams (1974b), Harrison and Jennings (1976).

Fluorocarbons: In addition to the work already described under alkyl halides (Section 2), see Anicich and Bowers (1974b,c), Su and Bowers (1973a,b), Drewery et al. (1976).

Cyclopropanes: Gross and Lin (1974), Luippold and Beauchamp (1976).

C. Reagent Ions

A number of reagent ions which have received attention in the icr literature are listed here. We have tried to include those which seem to have or may be found to have some utility in terms of selectivity, as well as those which demand attention simply because of their frequent occurrence in the literature.

(1) $C_4H_6^{+\cdot}$ Exceptionally promising for characterizing olefins (Gross et al., 1972).

(2) $(CH_3CO)_2^{+\cdot}$ and $(CH_3CO)_3^{+}$ Acylating agents attacking a wide variety of n- and π-donor bases, with important effects of steric crowding at the attacking site on the rate of acylation.

(3) $(C_3H_5CO)_2^{+\cdot}$ An acylating reagent similar to $(CH_3CO)_2^{+\cdot}$ in reaction patterns, but tending to show more selectivity because of the greater bulk of the ion.

(4) $CH_2{=}CHO{-}CH_3^{+}$ A reagent which apparently cleaves double-bonded compounds at the double bond, with potential as a double-bond location probe (Ferrer-Correia et al., 1975). It has also been used for $C_3H_6O^{+}$ isomer distinction (Drewery and Jennings, 1976).

(5) $Ni(C_5H_5)^{+}$ A selective decarbonylation reagent for aldehydes; also dehydrogenates, and dehydrohalogenates, fluorides, and chlorides (Corderman and Beauchamp, 1976a).

(6) CX_3^{+} $(X_3 = $ any combination of F and Cl)   With aldehydes and ketones, it undergoes a characteristic four-center attack resulting in formal replacement of O with X^{+}. With acids, esters, and acetic anhydride, a variety of products result, apparently following attack on the acyl oxygen (Ausloos et al., 1975a,b).

(7) CH_3^{+} and other carbonium ions   Attack on Lewis bases proceeds through formation of a strong covalent bond, and the large available internal energy brings about one of a variety of processes, including displacement, rearrangement, and disproportionation, with the outcome apparently being thermochemically controlled (Beauchamp, 1977).

(8) Dimethyl sulfoxide ion $((CH_3)_2SO)^{+\cdot}$ has been examined as a source of CH_3SO^{+} transfer analogous to acetylation reactions [reagents (2) and (3)]. While methylsulfurylation is observed, it competes poorly with other processes, chiefly proton transfer (Nixon et al., 1978).

(9) CF_2H^{+} This ion is an efficient protonating reagent for neutrals with proton affinity ≥ 172 kcal. It is suggested that its lack of clustering tendency might make it preferable to water as a protonating chemical ionization reagent (Vogt and Beauchamp, 1975).

(10) CH_5^{+}, H_3^{+} These ions, standard chemical ionization protonating agents, protonate almost all large molecules with considerable release of

energy. CH_5^+ shows differences in reactivity between propylene and cyclopropane (Fiaux *et al.*, 1974). H_3^+ has been used to distinguish C_2H_4O isomers (Bowers and Kemper, 1971).

(11) CH_3Hg^+ This ion attacks double bonds, showing increasing reactivity with increasing substitution on the double bond, and reacts faster with *cis*- than *trans*-2-butene (Bach *et al.*, 1978, 1975, 1972).

(12) Li^+ and other alkali ions Li^+, transferred to the unknown species from an appropriate Li^+ donor, is a general chemical ionization reagent which will ionize molecules with higher Li^+ affinity than the donor (Hodges and Beauchamp, 1976; Staley and Beauchamp, 1975b; Wieting *et al.*, 1975).

(13) NO^+ This ion reacts with a variety of bases, the reaction patterns varying but being related to the Lewis-acid character of the ion. NO^+ transfers readily between basic functionalities, and might be a useful general chemical ionization reagent (Williamson and Beauchamp, 1975).

(14) CO^+ This ion extracts F^- from a variety of perfluoroalkanes (Bowers, 1976).

(15) O^- This is an energetic anion reagent which exhibits a set of rather characteristic reactions with alkenes, alkynes, and carbonyl compounds, usually involving cleavage of one or two hydrogens from the unsaturated substrate (Harrison and Jennings, 1976; Goode and Jennings, 1974; Dawson and Jennings, 1976).

(16) Cl_2^- This anion transfers Cl^- to carbonyl compounds, with substantial substituent effects on the rate (Asubiojo *et al.*, 1975; Karpas and Klein, 1976).

(17) SF_6^- This ion reacts with several compounds containing acidic hydrogen, including hydrogen halides and formic acid, although not H_2O or ROH. Various products result, and the charge may end up on SF_5^- or on products resulting from F^- transfer (Foster and Beauchamp, 1975a,b).

(18) CH_3O^- and RO^- Attack of RO^- on aromatic fluorides results (formally) in replacement of F by O^-, as does the attack on CH_2CF_2 (Riveros and Takashima, 1976). CH_3O^- attacks acid β hydrogens in fluoroalkanes to induce HF elimination, and reacts with chlorides and bromides to give Cl^- and Br^- (Ridge and Beauchamp, 1974b).

D. Quantitative Techniques

Several analytically relevant icr capabilities depend on accurate quantitative measurement. One such approach is the measurement of reaction rates of ion–molecule reactions: such measurements have been implicit or explicit in much of the foregoing and need no further discussion. Two other quantitative approaches are the determination of bond strengths, and the measurement of collision frequencies via icr line widths.

1. *Bond Strengths*

The first bond strength to receive careful attention was the proton-to-molecule bond, measured as the proton affinity (PA), which is defined as the enthalpy of reaction (57), where M is the neutral molecule (or anion)

$$MH^+ \to M + H^+ \tag{57}$$

whose proton affinity is in question. Since the methods for measuring other bond strengths in ions are similar to those for proton affinities, only the measurement of proton affinity will be described. Two basic approaches serve to measure this quantity.

(*a*) *Bracketing* The observation of an exothermic reaction, e.g. (58), indicates that PA(N) ≥ PA(M). Since a large number of proton affinity

$$MH^+ + N \to M + NH^+ \tag{58}$$

values are already known (Milne and Lacy, 1974), the proton affinity of an unknown molecule can be found by bracketing it between two known values, if reagents can be found which transfer a proton both to and from the unknown. This widely used approach can often determine a proton affinity to ± 1 kcal.

(*b*) *Equilibrium* If molecules M and N are introduced into a trapped-ion cell along with a proton source (usually one of the parent ions will act as a proton source), and a sufficient length of time is allowed to pass before the detection pulse, it is often possible to establish the equilibrium (59) if the

$$MN^+ + N \rightleftharpoons M + NH^+ \tag{59}$$

proton affinities of M and N are not too different (Bowers *et al.*, 1971; McIver, 1975). The approach to and establishment of equilibrium for a proton-transfer reaction is shown in Fig. 6. From a knowledge of the neutral molecule concentration and the measurement of the relative amounts of MH^+ and NH^+ at equilibrium, the equilibrium constant K_{eq} can be determined, and from it $\Delta G = -RT \ln K_{eq}$. In many cases ΔS is assumed negligible, and ΔG is taken to be equal to ΔH, the PA difference of M and N; sometimes ΔS is calculated from theoretical considerations, and sometimes K_{eq} is measured as a function of temperature to give both ΔH and ΔS. The ΔG of reaction (57) has been referred to as the gas phase basicity of M. The ΔG values obtained are highly accurate, and proton affinity values coming from equilibrium measurements are typically accurate to ± 0.1 kcal or so relative to some standard molecule.

The proton–molecule bond strength is determined both by the intrinsic basicity of the site of protonation, and by the charge-stabilizing (solvating) ability of the rest of the molecule: it is the latter effect which is of most

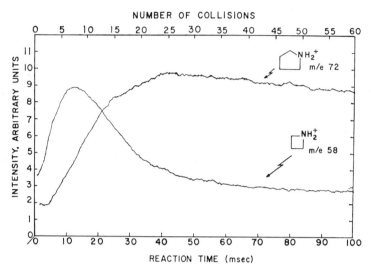

Fig. 6 Establishment of equilibrium in a mixture of $(CH_2)_3NH$ and $(CH_2)_4NH$ (6.4:1 mixture) at 10^{-5} Torr pressure using a pulsed-icr technique. During the initial 10 msec both protonated-molecule peaks increase as the primary parent radical cations transfer protons to the neutrals to give the $(P + 1)$ species, after which the competition for protons between the two neutral species results in establishment of equilibrium within 70 msec. [Reprinted by permission from M. T. Bowers, D. H. Aue, H. M. Webb, and R. T. McIver, *J. Am. Chem. Soc.*, **93**, 4314 (1971). Copyright by the American Chemical Society.]

interest, since the "solvation" of the protonation site is strongly affected by the number, the nature, and the location of surrounding groups, and thus gives a very powerful probe of molecular properties. An excellent illustration of this is the proton affinities of amines, for which a few isomeric series are given in Table III. The differences, while small and clearly requiring the use of the equilibrium technique for measurement, are nevertheless distinct and significant, and show that the distinguishing of even small changes in the

TABLE III

Proton Affinities of Some Amines[a]

Amine	PA (kcal/mole)	Amine	PA (kcal/mole)
n-PrNH$_2$	222.8	s-BuNH$_2$	224.8
i-PrNH$_2$	223.7	t-BuNH$_2$	225.8
n-BuNH$_2$	223.3	MeEtNH	227.3
i-BuNH$_2$	223.8	Me$_3$N	228.6

[a] Aue *et al.* (1976a).

distribution of alkyl substituents near the basic nitrogen can readily be discerned.

The effect of internal-chelating solvation of the proton is much greater, as shown in a study of diamines (Aue *et al.*, 1973). The effect of a second nitrogen in enhancing the solvation of the proton is illustrated by comparing the proton affinities of *n*-pentylamine (223.1 kcal) with that of 1,4-diaminobutane (243.3 kcal). The replacement of the terminal methylene with a chelating NH_2 group is worth 20 kcal of enhanced stabilization of the proton. This approach will be able to probe very delicately the geometrical relation of other basic groups relative to the site of protonation.

Another useful quantity related to the proton affinity is the hydrogen affinity (HA), defined as the enthalpy of reaction (60). This is seen to correspond to the proton affinity except for retention of the charge on the molecule;

$$MH^+ \rightarrow M^{+\cdot} + H\cdot \qquad (60)$$

the consequence of this is that charge solvation effects are minimal in determining the hydrogen affinity, and $HA(M^+)$ is almost a pure measure of the intrinsic M—H bond strength. This is supported by the finding that HA(M) is constant for homologous series (Arnett *et al.*, 1972; Caserio and Beauchamp, 1972). The hydrogen affinity is related to the proton affinity by (61), where

$$HA(M^+) = PA(M) + IP(M) - IP(H) \qquad (61)$$

IP(X) is the ionization potential of X, and thus requires knowledge of the ionization potential of the molecule, not always readily available. In cases where it can be determined, however, $HA(M^+)$ accurately reflects the nature of the site of protonation: for instance, primary amines have a hydrogen affinity of ~ 107 kcal, secondary amines ~ 97, tertiary amines ~ 90, alcohols ~ 115, alkyl chlorides ~ 107, alkyl bromides ~ 93, alkyl iodides ~ 77, and so forth.

Another feature probed by hydrogen affinity measurements is nonbonding electronic interactions between different parts of a molecule. This is suggested by the comparison of the hydrogen affinities of the two closely similar molecules, **44** and **45**, which have hydrogen affinities of 97.4 and 83.3 kcal, respectively (Staley and Beauchamp, 1974a). The difference is attributed to

nonbonded interactions of the two nitrogens in **45**. This is a potentially useful probe of the structural relationship of lone pair (and π-) groups.

There are numerous other thermochemical measurements which can be made by similar techniques, and whose analytical application is similar.

Some are the acidity of a molecule, defined by (62), the methyl cation affinity, defined by (63), and the lithium ion affinity, defined by (64).

$$MH \rightarrow M^- + H^+ \tag{62}$$

$$MCH_3^+ \rightarrow M + CH_3^+ \tag{63}$$

$$MLi^+ \rightarrow M + Li^+ \tag{64}$$

2. Line Broadening and Collision Rates

Being a resonance phenomenon, icr is governed by the same kinds of lineshape rules which apply to other types of resonant phenomena: the lineshapes are theoretically Lorentzian, and the linewidth is determined by the relaxation or damping mechanism. In the icr case at high pressure, the relevant damping mechanism is the collision of ions with neutrals (Beauchamp, 1967b), so that the icr lineshape (power absorption as a function of oscillator frequency) contains information about collision processes. The analysis shows that the parameter actually determined by measurement of the icr linewidth for a given ion (assuming that there are no chemical reactions involved) is the diffusion cross section σ_d which is the same parameter measured in ion-diffusion experiments (the longitudinal diffusion cross section). This cross section is equivalent to and readily converted to two other common parameters, the ion mobility μ, and the momentum transfer rate constant K. The reader is referred to the literature for fuller discussion of the relation of ion-collision theory to icr results (Dymerski et al., 1974a). Of importance here is the fact that σ_d is influenced by two important properties of the neutral molecules involved in the collision processes, namely the dipole moment and the van der Waals radius. The icr linewidth measurements thus provide an approach to measurement of these properties.

If the ion is considered to be a point charge and the neutral molecule a small sphere of polarizable material, the ion–neutral collision dynamics are governed entirely by the charge-induced-dipole attractive force, and lead to a momentum transfer rate constant independent of ion velocity, given by

$$K = 9.97 \times 10^{-10}(M_r\alpha)^{1/2}/m \tag{65}$$

where M_r is the reduced mass of the ion–neutral pair, α the neutral polarizability, and m the ion mass. [It has been suggested (Dymerski and Dunbar, 1972; Dymerski et al., 1974a) that because molecular rotation may be restricted during the collision, it may be more appropriate to use the maximum component α' of the polarizability in place of α in (65), and this does give substantially improved agreement between theory and experiment.]

(a) Hard-Sphere Correction   Under appropriate conditions, the finite size of the ion and the molecule can contribute significant hard-sphere repulsive terms to the ion–molecule interaction potential; addition of a hard-sphere repulsion modifies the theory in a well-understood way (McDaniel, 1964), and Eq. (65) becomes

$$K = 5.09 \times 10^{-10}(M_r\alpha')^{1/2}/mA \tag{66}$$

where A is a tabulated hard-sphere correction (Hass, 1926). The icr lineshapes for Cl^- and $Cr(CO)_5^-$ ions in a variety of nonpolar neutrals show good agreement with (66) (Dymerski *et al.*, 1974a), and the determination of A for a given ion-neutral pair potentially gives information about the size of the molecule.

(*b*) *Dipole Effects*   If the neutral molecule has a permanent dipole moment, an additional attractive potential modifies the collision event, and the momentum transfer rate constant may be approximated by

$$K = 9 \times 10^{-10}(M_r^{1/2}/m)[\alpha'^{1/2} + CD/(2E)^{1/2}]$$

where D is the molecular dipole moment, and C is a constant less than unity which adjusts for the fact that the dipole moment is not always lined up pointing at the ion, but is to some extent averaged out by rotation. Bowers and Su (1975) and Barker and Ridge (1976) have developed detailed theoretical expressions to take the dipole interactions into account, with the expression of Ridge being apparently most accurate. A series of comparisons of theory and experiment for dipolar molecules with both nonreactive and reactive ions [see Barker and Ridge (1976) for a summary] have given generally good agreement with theory, and this approach may have considerable capability for measuring the dipole moment of unknown neutral molecules via comparison of the icr linewidth with theory, leaving D as the adjusted parameter.

E.   Other Methods

1.   Trapped-Electron Spectra

A number of molecules have the ability to capture near-zero-energy electrons with very high efficiency, as for instance the dissociative electron capture in SF_6, reaction (67) (Foster and Beauchamp, 1975a). Such molecules,

$$SF_6 \xrightarrow{\;e^-\;} F^- + SF_5 \tag{67}$$

among which are also included CCl_4 and other chlorides, act as effective scavengers for slow-moving electrons in the icr cell; the anion formed can be readily detected by the icr detector. Ridge and Beauchamp (1969) demonstrated the ingenious use of this phenomenon to obtain inelastic electron scattering spectra of neutral molecules by introducing the neutral of interest into the icr spectrometer along with a trace of a scavenger such as SF_6 and monitoring the F^- signal while varying the electron beam energy. Whenever the electron energy matches an excitation process taking the neutral into an excited state, it is possible for electrons to transfer all their kinetic energy into internal energy of the neutral and to be inelastically scattered at very low velocity. These scattered electrons are scavenged and observed as an increase in F^- signal, the sequence of events being those shown in (68).

$$e^-(\text{energy } E_0) + M \longrightarrow e^- (\text{zero-energy}) + M^* (\text{internal energy } E_0) \xrightarrow{SF_6}$$

$$F + SF_5^- + M^* \tag{68}$$

A trace of SF_5^- signal intensity versus electron energy thus traces out an in elastic electron scattering spectrum (or trapped-electron, TE, spectrum) for the neutral in question. This method displays the energies of excited states of the molecule as does optical spectroscopy, but is not governed by the same selection rules, and is thus a useful complementary technique to other spectroscopic methods (Mosher *et al.*, 1975, 1974; Staley *et al.*, 1975).

2. *Penning Ionization*

The Penning ionization process involves ionization of the sample molecules by highly excited metastable molecules of a reagent such as N_2, e.g. reaction (69). Huntress and Beauchamp (1969) described the adaptation of

$$N_2^* + C_6H_6 \rightarrow N_2 + C_6H_6^+ + e^- \tag{69}$$

the icr spectrometer to observation of reactions such as (69). This interesting technique has not been much pursued by icr workers.

3. *Neutral-Product Collection*

Lieder and Brauman (1975, 1974) have shown that the neutral products can be accumulated in a closed-off icr system, and subsequently analyzed. Among several systems examined, the most interesting was the displacement of bromide from *cis*- and *trans*-4-bromocyclohexanols, reaction (70).

$$Cl^- + C_6H_{10}BrOH \rightarrow C_6H_{10}ClOH + Br^- \tag{70}$$

The reaction was carried out for an extended period of time in a mixture of CH_3Cl (the Cl^- source) and the bromocyclohexanol, under conditions suitable for abundant Cl^- production. The spectrometer was then switched to a positive-ion mode, and the chlorocyclohexanol product molecules were analyzed from their icr spectra. Careful control and calibration studies were done, and it was found that the *cis*-4-bromocyclohexanol yielded a *trans*-4-

chlorocyclohexanol product and conversely, indicating inversion of configuration in the nucleophilic displacement. Although technically difficult, this technique has great potential for yielding reaction-mechanism information which cannot be deduced from product-ion identification alone.

References

Literature coverage since 1973 is extensive and includes a number of significant citations not explicitly mentioned in the text. Lehman and Bursey (1976) have comprehensively reviewed the literature through 1973.

Anders, L. R. (1966). Ph.D. Thesis, Harvard Univ.
Anders, L. R. (1969). *J. Phys. Chem.* **73**, 469.
Anders, L. R., Beauchamp, J. L., Dunbar, R. C., and Baldeschwieler, J. D. (1966). *J. Chem. Phys.* **45**, 1062.
Anicich, V. G., and Bowers, M. T. (1973). *Int. J. Mass Spectrom. Ion Phys.* **12**, 231.
Anicich, V. G., and Bowers, M. T. (1974a). *J. Am. Chem. Soc.* **96**, 1279.
Anicich, V. G., and Bowers, M. T. (1974b). *Int. J. Mass Spectrom. Ion Phys.* **13**, 351.
Anicich, V. G., and Bowers, M. T. (1974c). *Int. J. Mass Spectrom. Ion Phys.* **13**, 359.
Aoyagi, K. (1974). *Bull. Chem. Soc. Jpn.* **47**, 519.
Aoyagi, K. (1976). *Bull. Chem. Soc. Jpn.* **49**, 26.
Arnett, E. M. (1973). *Accounts Chem. Res.* **6**, 404.
Arnett, E. M., and Abboud, J. L. M. (1975). *J. Am. Chem. Soc.* **97**, 3865.
Arnett, E. M., and Petro, C. (1976). *J. Am. Chem. Soc.* **98**, 1468.
Arnett, E. M., and Wolf, J. F. (1975). *J. Am. Chem. Soc.* **97**, 3262.
Arnett, E. M., Jones, F. M., III, Taagepera, M., Henderson, W. G., Beauchamp, J. L., Holtz, D., and Taft, R. W. (1972). *J. Am. Chem. Soc.* **94**, 4727.
Arnett, E. M., Small, L. E., McIver, R. T., Jr., and Miller, J. S. (1974). *J. Am. Chem. Soc.* **96**, 5638.
Arnett, E. M., Johnston, D. E., and Small, L. E., (1975). *J. Am. Chem. Soc.* **97**, 5598.
Asubiojo, O. J., Blair, L. K., and Brauman, J. I. (1975). *J. Am. Chem. Soc.* **97**, 6685.
Aue, D. H., Webb, H. M., and Bowers, M. T. (1973). *J. Am. Chem. Soc.* **95**, 2699.
Aue, D. H., Webb, H. M., and Bowers, M. T. (1975a). *J. Am. Chem. Soc.* **97**, 4137.
Aue, D. H., Webb, H. M., and Bowers, M. T. (1975b). *J. Am. Chem. Soc.* **97**, 4136.
Aue, D. H., Webb, H. M., and Bowers, M. T. (1976a). *J. Am. Chem. Soc.* **98**, 311.
Aue, D. H., Webb, H. M., and Bowers, M. T. (1976b). *J. Am. Chem. Soc.* **98**, 318.
Aue, D. H., Webb, H. M., Bowers, M. T., Liotta, C. L., and Hopkins, H. P. (1976c). *J. Am. Chem. Soc.* **98**, 854.
Ausloos, P., Lias, S. G., and Eyler, J. R. (1975a). *Int. J. Mass Spectrom. Ion Phys.* **18**, 261.
Ausloos, P., Eyler, J. R., and Lias, S. G. (1975b). *Chem. Phys. Lett.* **30**, 21.
Bach, R. D., Gauglhofer, J., and Kevan, L. (1972). *J. Am. Chem. Soc.* **94**, 6860.
Bach, R. D., Patane, J., and Kevan, L. (1975). *J. Org. Chem.* **40**, 257.
Bach, R. D., Weibel, A. T., Patane, J., and Kevan, L. (1978). *J. Am. Chem. Soc.* (to be published).
Baldeschwieler, J. D. (1968). *Science* **159**, 263.
Baldeschwieler, J. D., and Woodgate, S. S. (1971). *Accounts Chem. Res.* **4**, 114.
Barker, R. A., and Ridge, D. P. (1976). *J. Chem. Phys.* **64**, 4411.
Beauchamp, J. L. (1967a). Ph.D. Thesis, Harvard Univ.
Beauchamp, J. L. (1967b). *J. Chem. Phys.* **46**, 1231.
Beauchamp, J. L. (1969). *J. Am. Chem. Soc.* **91**, 5925.

Beauchamp, J. L. (1971). *Ann. Rev. Phys. Chem.* **22**, 527.

Beauchamp, J. L. (1974). *Adv. Mass Spectrom.* **8**, 717.

Beauchamp, J. L. (1975). *In* "Interactions Between Ions and Molecules" (P. Ausloos, ed.). Plenum Press, New York.

Beauchamp, J. L. (1976a). *J. Chem. Phys.* **64**, 718.

Beauchamp, J. L. (1976b). *J. Chem. Phys.* **64**, 929.

Beauchamp, J. L. (1978). *Accounts Chem. Res.* (to be published).

Beauchamp, J. L., and Armstrong, J. T. (1969). *Rev. Sci. Instrum.* **40**, 123.

Beauchamp, J. L., and Park, J. Y. (1976). *J. Phys. Chem.* **80**, 575.

Beauchamp, J. L., Holtz, D., Woodgate, S. D., and Patt, S. L. (1972). *J. Am. Chem. Soc.* **94**, 2798.

Beauchamp, J. L., Caserio, M. C., and McMahon, T. B. (1974). *J. Am. Chem. Soc.* **96**, 6243.

Blair, L. K., Isolani, P. C., and Riveros, J. M. (1973). *J. Am. Chem. Soc.* **95**, 1057.

Blint, R. J., McMahon, T. B., and Beauchamp, J. L. (1974). *J. Am. Chem. Soc.* **96**, 1269.

Bloom, M., and Riggin, M. (1974). *Can. J. Phys.* **52**, 436.

Bowers, M. T. (1976). Thermal energy ion-molecule interactions, *in* "High Power Gas Lasers" (G. Bekefi, ed.). Wiley, New York.

Bowers, M. T., and Chau, M. (1976). *J. Phys. Chem.* **80**, 1739.

Bowers, M. T., and Kemper, P. R. (1971). *J. Am. Chem. Soc.* **93**, 5352.

Bowers, M. T., and Su, T. (1975). *In* "Interactions Between Ions and Molecules" (P. Ausloos, ed.). Plenum Press, New York.

Bowers, M. T., Ellenan, D. D., and Beauchamp, J. L. (1968). *J. Phys. Chem.* **72**, 3599.

Bowers, M. T., Aue, D. H., Webb, H. M., and McIver, R. T. (1971). *J. Am. Chem. Soc.* **93**, 4314.

Bowers, M. T., Chesnavich, W. J., and Huntress, W. T. (1973). *Int. J. Mass Spectrom. Ion Phys.* **12**, 357.

Bowers, M. T., Kemper, P. R., and Laudenslager, J. B. (1974). *J. Chem. Phys.* **61**, 4394.

Bowers, M. T., Chau, M., and Kemper, P. R. (1975). *J. Chem. Phys.* **63**, 3656.

Bowie, J. H. (1975). *Aust. J. Chem.* **28**, 559.

Bowie, J. H., and Williams, B. D. (1974a). *Aust. J. Chem.* **27**, 1923.

Bowie, J. H., and Williams, B. D. (1974b). *Aust. J. Chem.* **27**, 769.

Bowie, J. H., and Williams, B. D. (1975a). *Int. J. Mass Spectrom. Ion Phys.* **17**, 395.

Bowie, J. H., and Williams, B. D. (1975b). *Org. Mass Spectrom.* **10**, 141.

Brauman, J. I., and Blair, L. K. (1973). *In* "Determination of Organic Structures by Physical Methods" (F. C. Nachod and J. J. Zuckerman, eds.) Vol. 5. Academic Press, New York.

Brauman, J. I., and Smyth, K. C. (1969). *J. Am. Chem. Soc.* **91**, 7778.

Brauman, J. I., Lieder, C. A., and White, M. J. (1973). *J. Am. Chem. Soc.* **95**, 927.

Brauman, J. I., Olmstead, W. N., and Lieder, C. A. (1974). *J. Am. Chem. Soc.* **96**, 4030.

Briscese, S. M. J., and Riveros, J. M. (1975). *J. Am. Chem. Soc.* **97**, 230.

Briscese, S. M. J., Blair, L. K., and Riveros, J. M. (1973). *An. Acad. Brasil. Cienc.* **45**, 359.

Bursey, M. M., Elwood, T. A., Hoffman, M. K., Lehman, T. A., and Tesarek, J. M. (1970). *Anal. Chem.* **42**, 1370.

Bursey, M. M., Kao, J., Henion, J. D., Parker, C. E., and Huang, T. X. (1974). *Anal. Chem.* **46**, 1709.

Bursey, M. M., Hass, J. R., and Stern, R. L. (1975). *Anal. Chem.* **47**, 1452.

Buttrill, S. E., Jr., and Magil, G. C. (1975). *Int. J. Mass Spectrom. Ion Phys.* **17**, 287.

Caserio, M. C., and Beauchamp, J. L. (1972). *J. Am. Chem. Soc.* **94**, 2638.

Chatfield, D. A., and Bursey, M. M. (1975a). *Int. J. Mass Spectrom. Ion Phys.* **18**, 239.

Chatfield, D. A., and Bursey, M. M. (1975b). *J. Am. Chem. Soc.* **97**, 3600.

Chatfield, D. A., and Bursey, M. M. (1976a). *J. Chem. Soc. Faraday Trans. I* **72**, 417.

Chatfield, D. A., and Bursey, M. M. (1976b). *Int. J. Mass Spectrom. Ion Phys.* **20**, 101.

Chesnavich, W. J., Su, T., and Bowers, M. T. (1976). *J. Chem. Phys.* **65**, 90.
Clow, R. P., and Futrell, J. H. (1972a). *J. Am. Chem. Soc.* **94**, 3748.
Clow, R. P., and Futrell, J. H. (1972b). *Int. J. Mass Spectrom. Ion Phys.* **8**, 119.
Comisarow, M. B. (1971). *J. Chem. Phys.* **55**, 205.
Comisarow, M. B., and Marshall, A. G. (1974a). *Chem. Phys. Lett.* **26**, 489.
Comisarow, M. B., and Marshall, A. G. (1974b). *Chem. Phys. Lett.* **25**, 282.
Comisarow, M. B., and Marshall, A. G. (1974c). *Can. J. Chem.* **52**, 1997.
Comisarow, M. B., and Marshall, A. G. (1975). *J. Chem. Phys.* **62**, 293.
Comisarow, M. B., and Marshall, A. G. (1976). *J. Chem. Phys.* **64**, 110.
Corderman, R. R., and Beauchamp, J. L. (1976a). *J. Am. Chem. Soc.* **98**, 5700.
Corderman, R. R., and Beauchamp, J. L. (1976b). *Inorg. Chem.* **15**, 665.
Corderman, R. R., and Beauchamp, J. L. (1976c). *J. Am. Chem. Soc.* **98**, 3998.
Dawson, J. H. J., and Jennings, K. R. (1976). *J. Chem. Soc. Faraday Trans. II* **72**, 700.
Dawson, J. H. J., Henderson, W. G., O'Malley, R. M., Jennings, K. R. (1973). *Int. J. Mass Spectrom. Ion Phys.* **11**, 61, 89, 99, 111.
Diekman, J., MacLeod, J., Djerassi, C., and Baldeschwieler, J. D. (1969). *J. Am. Chem. Soc.* **91**, 2069.
Drewery, C. J., and Jennings, K. R. (1976). *Int. J. Mass Spectrom. Ion Phys.* **19**, 287.
Drewery, C. J., Goode, G. C., and Jennings, K. R. (1972). *In* "MTP International Review of Science, Mass Spectrometry, and Physical Chemistry," (A. D. Buckingham and A. MacColl, eds.), Ser. 1, Vol. 5. Butterworth, London.
Drewery, C. J., Goode, G. C., and Jennings, K. R. (1976). *Int. J. Mass Spectrom. Ion Phys.* **20**, 403.
Dunbar, R. C. (1968). *J. Am. Chem. Soc.* **90**, 5676.
Dunbar, R. C. (1971a). *J. Chem. Phys.* **54**, 711.
Dunbar, R. C. (1971b). *J. Am. Chem. Soc.* **93**, 4354.
Dunbar, R. C. (1973a). *J. Am. Chem. Soc.* **95**, 472.
Dunbar, R. C. (1973b). *J. Am. Chem. Soc.* **95**, 6191.
Dunbar, R. C. (1974). The study of gas-phase ions by ion cyclotron resonance, *in* "Chemical Reactivity and Reaction Paths" (G. Klopman, ed.). Wiley (Interscience), New York.
Dunbar, R. C. (1975a). *Spectrochim. Acta* **31A**, 797.
Dunbar, R. C. (1975b). *In* "Interactions Between Ions and Molecules" (P. Ausloos, ed.), Plenum Press, New York.
Dunbar, R. C. (1975c). *J. Am. Chem. Soc.* **97**, 1382.
Dunbar, R. C. (1975d). *Chem. Phys. Lett.* **32**, 508.
Dunbar, R. C. (1976a). *Anal. Chem.* **48**, 723.
Dunbar, R. C. (1976b). *J. Am. Chem. Soc.* **98**, 4671.
Dunbar, R. C., and Fu, E. (1973). *J. Am. Chem. Soc.* **95**, 2716.
Dunbar, R. C., and Hutchinson, B. B. (1974). *J. Am. Chem. Soc.* **96**, 3816.
Dunbar, R. C., and Klein, R. (1976). *J. Am. Chem. Soc.* **98**, 7994.
Dunbar, R. C., and Kramer, J. M. (1973). *J. Chem. Phys.* **58**, 1266.
Dunbar, R. C., Shen, J., and Olah, G. A. (1972). *J. Am. Chem. Soc.* **94**, 6862.
Dunbar, R. C., Shen, J., Melby, E., and Olah, G. A. (1973a). *J. Am. Chem. Soc.* **95**, 7200.
Dunbar, R. C., Ennever, J. F., and Fackler, J. P., Jr. (1973b). *Inorg. Chem.* **12**, 2734.
Dunbar, R. C., Bursey, M. M., and Chatfield, D. A. (1974). *Int. J. Mass Spectrom. Ion Phys.* **13**, 195.
Dymerski, P. P., and Dunbar, R. C. (1972). *J. Chem. Phys.* **57**, 4049.
Dymerski, P. P., and Dunbar, R. C. (1974a). *Rev. Sci. Instrum.* **45**, 124.
Dymerski, P. P., and Dunbar, R. C. (1974b). *Rev. Sci. Instrum.* **45**, 1293.
Dymerski, P. P., Dunbar, R. C., and Dugan, J. V. (1974a). *J. Chem. Phys.* **61**, 298.

Dymerski, P. P., Fu, E., and Dunbar, R. C. (1974b). *J. Am. Chem. Soc.* **96**, 4109.

Eadon, G., Diekman, J., and Djerassi, C. (1969). *J. Am. Chem. Soc.* **91**, 3986.

Eadon, G., Diekman, J., and Djerassi, C. (1970). *J. Am. Chem. Soc.* **92**, 6205.

Eyler, J. R. (1974). *Rev. Sci. Instrum.* **45**, 1154.

Eyler, J. R. (1976). *J. Am. Chem. Soc.* **98**, 6831.

Eyler, J. R., and Atkinson, G. H. (1974). *Chem. Phys. Lett.* **28**, 217.

Eyler, J. R., Ausloos, P., and Lias, S. G. (1974). *J. Am. Chem. Soc.* **96**, 3673.

Faigle, J. F. G., Isolani, P. C., and Riveros, J. M. (1976). *J. Am. Chem. Soc.* **98**, 2049.

Ferrer-Correia, A. J. V., Jennings, K. R., and Sen Sharma, D. K. (1975). *J. Chem. Soc. Chem. Commun.* 973.

Fiaux, A., Smith, D. L., and Futrell, J. H. (1974). *Int. J. Mass Spectrom. Ion Phys.* **15**, 9.

Field, F. H. (1972). Chemical ionization mass spectrometry, *in* "Ion-Molecule Reactions" (J. L. Franklin, ed.), p. 261. Plenum Press, New York.

Foster, M. S., and Beauchamp, J. L. (1971). *J. Am. Chem. Soc.* **93**, 4924.

Foster, M. S., and Beauchamp, J. L. (1975a). *Chem. Phys. Lett.* **31**, 479.

Foster, M. S., and Beauchamp, J. L. (1975b). *Chem. Phys. Lett.* **31**, 482.

Foster, M. S., and Beauchamp, J. L. (1975c). *J. Am. Chem. Soc.* **97**, 4808.

Foster, M. S., and Beauchamp, J. L. (1975d). *J. Am. Chem. Soc.* **97**, 4814.

Foster, M. S., and Beauchamp, J. L. (1975e). *Inorg. Chem.* **14**, 1229.

Freiser, B. S., and Beauchamp, J. L. (1974). *J. Am. Chem. Soc.* **96**, 6260.

Freiser, B. S., and Beauchamp, J. L. (1975). *Chem. Phys. Lett.* **35**, 35.

Freiser, B. S., and Beauchamp, J. L. (1976a). *J. Am. Chem. Soc.* **98**, 3136.

Freiser, B. S., and Beauchamp, J. L. (1976b). *J. Am. Chem. Soc.* **98**, 265.

Freiser, B. S., McMahon, T. B., and Beauchamp, J. L. (1973). *Int. J. Mass. Spectrom. Ion Phys.* **12**, 249.

Freiser, B. S., Woodin, R. L., and Beauchamp, J. L. (1975). *J. Am. Chem. Soc.* **97**, 6893.

Freiser, B. S., Staley, R. H., and Beauchamp, J. L. (1976). *Chem. Phys. Lett.* **39**, 49.

Fu, E. W. (1976). Ph.D. Thesis, Case Western Reserve Univ.

Fu, E. W., Dymerski, P. P., and Dunbar, R. C. (1976). *J. Am. Chem. Soc.* **98**, 337.

Fujiwara, S. (1970). *Bull. Chem. Soc. Jpn.* **43**, 561.

Futrell, J. H. (1971). *In* "Dynamic Mass Spectrometry" (D. Price, ed.), Vol. 2. Heyden, New York.

Gauglhofer, J., and Kevan, L. (1972). *Chem. Phys. Lett.* **16**, 492.

Goode, G. C., and Jennings, K. R. (1974). *Adv. Mass Spectrom.* **6**, 797.

Goode, G. C., Ferrer-Correia, A. J., and Jennings, K. R. (1970). *Int. J. Mass Spectrom. Ion Phys.* **5**, 229.

Gray, G. A. (1971). *Adv. Chem. Phys.* **19**, 141.

Gross, M. L. (1972). *J. Am. Chem. Soc.* **94**, 3744.

Gross, M. L., and Lin, P. H. (1974). *Org. Mass Spectrom.* **9**, 1194.

Gross, M. L., and McLafferty, F. W. (1971). *J. Am. Chem. Soc.* **93**, 1267.

Gross, M. L., Lin, P. H., and Franklin, S. J. (1972). *Anal. Chem.* **44**, 974.

Harrison, A. G., and Jennings, K. R. (1976). *J. Chem. Soc. Faraday Trans. I* **72**, 1601.

Hass, J. R., Bursey, M. M., Kingston, D. G., and Tannenbaum, H. P. (1972). *J. Am. Chem. Soc.* **94**, 5095.

Hass, J. R., Cooks, R. G., Elder, J. F., Jr., Bursey, M. M., and Kingston, D. G. I. (1976). *Org. Mass Spectrom.* **11**, 697.

Hassé, H. R. (1926). *Phil. Mag.* (Ser. 7) **1**, 139.

Hemberger, P. H., and Dunbar, R. C. (1977). *Inorg. Chem.* **16**, 1246.

Henderson, W. G., Holtz, D., Beauchamp, J. L., and Taft, R. W. (1972). *J. Am. Chem. Soc.* **94**, 4728.

Henion, J. D., Sammons, M. C., Parker, C. E., and Bursey, M. M. (1973). *Tetra. Lett.* **49**, 4925

Henion, J. D., Kao, J., Nixon, W. B., and Bursey, M. M. (1975). *Anal. Chem.* **47**, 689.

Henis, J. M. S. (1968). *J. Am. Chem. Soc.* **90**, 844.

Henis, J. M. S. (1970a). *J. Chem. Phys.* **52**, 282.

Henis, J. M. S. (1970b). *J. Chem. Phys.* **52**, 292.

Henis, J. M. S. (1972). *In* "Ion-Molecule Reactions" (J. L. Franklin, ed.), Vol. 2. Plenum, New York.

Hodges, R. V., and Beauchamp, J. L. (1975). *J. Inorg. Chem.* **14**, 2887.

Hodges, R. V., and Beauchamp, J. L. (1976). *Anal. Chem.* **48**, 825.

Hoffman, M. K., and Bursey, M. M. (1971). *Tet. Lett.* 2539.

Huizer, A. H., and van der Hart, W. J. (1977). *Mol. Phys.* **33**, 897.

Huntress, W. T., and Beauchamp, J. L. (1969). *Int. J. Mass Spectrom. Ion Phys.* **3**, 149.

Huntress, W. T., and Bowers, M. T. (1973). *Int. J. Mass Spectrom. Ion Phys.* **12**, 1.

Isolani, P. C., and Riveros, J. M. (1975). *Chem. Phys. Lett.* **33**, 362.

Isolani, P. C., Riveros, J. M., and Tiedemann, P. W. (1973). *J. Chem. Soc. Faraday Trans. II* 1023.

Kao, J., Simmonton, C. A., III, and Bursey, M. M. (1976). *Org. Mass Spectrom.* **11**, 140.

Karpas, A., and Klein, F. S. (1976). Presented at the *Int. Mass Spectrom. Conf., 7th, Florence* September.

Kemper, P. R., and Bowers, M. T. (1973). *J. Chem. Phys.* **59**, 4915.

Kemper, P. R., and Bowers, M. T. (1974). *Adv. Mass Spectrom.* **6**, 809.

Kemper, P. R., and Bowers, M. T. (1975). *Chem. Phys. Lett.* **36**, 183.

Kemper, P. R., and Bowers, M. T. (1977). *Rev. Sci. Instrum.* **48**, 1477.

Kramer, J. M., and Dunbar, R. C. (1973). *J. Chem. Phys.* **59**, 3092.

Kramer, J. M., and Dunbar, R. C. (1974). *J. Chem. Phys.* **60**, 5122.

Laudenslager, J. B., Huntress, W. T., and Bowers, M. T. (1974). *J. Chem. Phys.* **61**, 4600.

LeBreton, P. R., Williamson, A. D., Beauchamp, J. L., and Huntress, W. T. (1975). *J. Chem. Phys.* **62**, 1623.

Ledford, E. B., Jr., and McIver, R. T., Jr. (1978). (to be published).

Lehman, T. A., and Bursey, M. M. (1976). "Ion Cyclotron Resonance Spectrometry." Wiley, New York.

Lew, H., and Heiber, I. (1973). *J. Chem. Phys.* **58**, 1246.

Lias, G. S., Eyler, J. R., and Ausloos, P. (1976). *Int. J. Mass Spectrom. Ion Phys.* **19**, 219.

Lieder, C. A., and Brauman, J. I. (1974). *J. Am. Chem. Soc.* **96**, 4028.

Lieder, C. A., and Brauman, J. I. (1975). *Int. J. Mass Spectrom. Ion Phys.* **16**, 307.

Luippold, D. A., and Beauchamp, J. L. (1976). *J. Phys. Chem.* **80**, 795.

Marshall, A. G. (1971). *J. Chem. Phys.* **55**, 1343.

McAdoo, D. J., McLafferty, F. W., and Bente, P. F., III (1972). *J. Am. Chem. Soc.* **94**, 2027.

McDaniel, E. W. (1964). "Collision Phenomena in Ionized Gases." Wiley, New York.

McIver, R. T., Jr. (1970). *Rev. Sci. Instrum.* **41**, 555.

McIver, R. T., Jr. (1973). *Rev. Sci. Instrum.* **44**, 1071.

McIver, R. T., Jr. (1975). *Org. Mass Spectrom.* **10**, 396.

McIver, R. T., Jr., and Baranyi, A. D. (1974). *Int. J. Mass Spectrom. Ion Phys.* **14**, 449.

McIver, R. T., and Dunbar, R. C. (1971). *Int. J. Mass Spectrom. Ion Phys.* **7**, 471.

McIver, R. T., Jr., and Miller, J. S. (1974). *J. Am. Chem. Soc.* **96**, 4323.

McIver, R. T., Jr., and Silvers, J. H. (1973). *J. Am. Chem. Soc.* **95**, 8462.

McIver, R. T., Jr., Ledford, E. B., and Miller, J. S. (1975). *Anal. Chem.* **47**, 692.

McMahon, T. B., and Beauchamp, J. L. (1972). *Rev. Sci. Instrum.* **43**, 509.

Memel, J., and Nibbering, N. M. M. (1973). *J. Chem. Soc. Perkins II* 2089.

Milne, G. W. A., and Lacey, M. J. (1974). *Crit. Rev. Anal. Chem.* **4**, 45.

Morton, T. H., and Beauchamp, J. L. (1972). *J. Am. Chem. Soc.* **94**, 3672.
Morton, T. H., and Beauchamp, J. L. (1975). *J. Am. Chem. Soc.* **97**, 2355.
Mosher, O., Foster, M. S., Flicker, W. M., Kuppermann, A., and Beauchamp, J. L. (1974). *Chem. Phys. Lett.* **29**, 236.
Mosher, O. A., Foster, M. S., Flicker, W. M., Beauchamp, J. L., and Kuppermann, A. (1975). *J. Chem. Phys.* **62**, 3424.
Murphy, M. K., and Beauchamp, J. L. (1976). *J. Am. Chem. Soc.* **98**, 1433.
Nibbering, N. M. M. (1973). *Tetrahedron* **29**, 385.
Nixon, W. B., Bursey, M. M., and Henion, J. D. (1978). *J. Am. Chem. Soc.* (submitted).
Odom, R. W., Smith, D. L., and Futrell, J. H. (1974). *Chem. Phys. Lett.* **24**, 227.
Odom, R. W., Smith, D. L., and Futrell, J. H. (1975). *J. Phys. B: At. Mol. Phys.* **8**, 1349.
Orth, R. G., and Dunbar, R. C. (1977). *J. Chem. Phys.* **66**, 1616.
Orth, R. G., Dunbar, R. C., and Riggin, M. (1977). *Chem. Phys.* **19**, 279.
Pearson, P. K., Schaefer, H. F., III, Richardson, J. H., Stephenson, L. M., and Brauman, J. I. (1974). *J. Am. Chem. Soc.* **96**, 6778.
Pesheck, C. V., and Buttrill, S. E., Jr. (1974). *J. Am. Chem. Soc.* **96**, 6027.
Pitt, C. G., Bursey, M. M., and Chatfield, D. A. (1976). *J. Chem. Soc. Perkin Trans. II* 434.
Pound, R. V., and Knight, W. D. (1950). *Rev. Sci. Instrum.* **21**, 219.
Reed, K. J., and Brauman, J. I. (1974). *J. Chem. Phys.* **61**, 4830.
Reed, K. J., and Brauman, J. I. (1975). *J. Am. Chem. Soc.* **97**, 1625.
Reed, K. J., Zimmerman, A. H., Andersen, H. C., and Brauman, J. I. (1976). *J. Chem. Phys.* **64**, 1368.
Richardson, J. H., and Stephenson, M. (1975). *J. Am. Chem. Soc.* **97**, 2967.
Richardson, J. H., Stephenson, L. M., and Brauman, J. I. (1973). *J. Chem. Phys.* **59**, 5068.
Richardson, J. H., Stephenson, L. M., and Brauman, J. I. (1974a). *Chem. Phys. Lett.* **25**, 318.
Richardson, J. H., Stephenson, L. M., and Brauman, J. I. (1974b). *Chem. Phys. Lett.* **25**, 321.
Richardson, J. H., Stephenson, L. M., and Brauman, J. I. (1974c). *J. Am. Chem. Soc.* **96**, 3671.
Richardson, J. H., Stephenson, L. M., and Brauman, J. I. (1975a). *Chem. Phys. Lett.* **30**, 17.
Richardson, J. H., Stephenson, L. M., and Brauman, J. I. (1975b). *J. Chem. Phys.* **62**, 1580.
Richardson, J. H., Stephenson, L. M., and Brauman, J. I. (1975c). *J. Am. Chem. Soc.* **97**, 1160.
Richardson, J. H., Stephenson, L. M., and Brauman, J. I. (1975d). *J. Chem. Phys.* **63**, 74.
Ridge, D. P., and Beauchamp, J. L. (1969). *J. Chem. Phys.* **51**, 470.
Ridge, D. P., and Beauchamp, J. L. (1971). *J. Am. Chem. Soc.* **93**, 5925.
Ridge, D. P., and Beauchamp, J. L. (1974a). *J. Am. Chem. Soc.* **96**, 637.
Ridge, D. P., and Beauchamp, J. L. (1974b). *J. Am. Chem. Soc.* **96**, 3595.
Riggin, M. T., and Dunbar, R. C. (1975). *Chem. Phys. Lett.* **31**, 539.
Riggin, M., and Woods, I. B. (1974). *Can. J. Phys.* **52**, 456.
Riggin, M. T., Orth, R., and Dunbar, R. C. (1976). *J. Chem. Phys.* **65**, 3365.
Riveros, J. M. (1974). *Adv. Mass Spectrom.* **6**, 277.
Riveros, J. M., and Takashima, K. (1976). *Can. J. Chem.* **54**, 1839.
Riveros, J. M., and Tiedemann, P. W. (1972). *An. Acad. Brasil Ciene* **44**, 413.
Riveros, J. M., Breda, A. C., and Blair, L. K. (1973a). *J. Am. Chem. Soc.* **95**, 4066.
Riveros, J. M., Tiedemann, P. W., and Breda, A. C. (1973b). *Chem. Phys. Lett.* **20**, 345.
Robinson, F. N. H. (1959). *J. Sci. Instrum.* **36**, 481.
Shen, J., Dunbar, R. C., and Olah, G. A. (1974). *J. Am. Chem. Soc.* **96**, 6227.
Smith, D. L., and Futrell, J. H. (1974a). *Int. J. Mass Spectrom. Ion Phys.* **14**, 171.
Smith, D. L., and Futrell, J. H. (1974b). *Chem. Phys. Lett.* **24**, 611.
Smith, D. L., and Futrell, J. H. (1975). *J. Phys. B: At. Mol. Phys.* **8**, 803.
Smith, D. L., and Futrell, J. H. (1976). *Chem. Phys. Lett.* **40**, 229.
Smith, R. D., and Futrell, J. H. (1974). *Chem. Phys. Lett.* **27**, 493.

Smith, R. D. and Futrell, J. H. (1975a). *Int. J. Mass Spectrom. Ion Phys.* **17**, 233.
Smith, R. D., and Futrell, J. H. (1975b). *Chem. Phys. Lett.* **36**, 545.
Smith, R. D., and Futrell, J. H. (1976a). *Int. J. Mass. Spectrom. Ion Phys.* **19**, 201.
Smith, R. D., and Futrell, J. H. (1976b). *Org. Mass Spectrom.* **11**, 309.
Smith, R. D., and Futrell, J. H. (1976c). *Int. J. Mass Spectrom. Ion Phys.* **20**, 71.
Smith, R. D., and Futrell, J. H. (1976d). *Org. Mass Spectrom.* **11**, 445.
Smith, R. D., and Futrell, J. H. (1976e). *Int. J. Mass Spectrom. Ion Phys.* **20**, 43.
Smith, R. D., and Futrell, J. H. (1976f). *Int. J. Mass Spectrom. Ion Phys.* **20**, 59.
Smith, R. D., and Futrell, J. H. (1976g). *Int. J. Mass Spectrom. Ion Phys.* **20**, 33.
Smith, R. D., Smith, D. L., and Futrell, J. H. (1975). *Chem. Phys. Lett.* **32**, 513.
Smith, R. D., Smith, D. L., and Futrell, J. H. (1976a). *Int. J. Mass Spectrom. Ion Phys.* **19**, 369.
Smith, R. D., Smith, D. L., and Futrell, J. H. (1976b). *Int. J. Mass Spectrom. Ion Phys.* **19**, 395.
Smyth, K. C., and Brauman, J. I. (1972). *J. Chem. Phys.* **56**, 1132, 4620, 5993.
Sommer, H. S., Thomas, H. A., and Hipple, J. A. (1949). *Phys. Rev.* **76**, 1877.
Sommer, H. S., Thomas, H. A., and Hipple, J. A. (1951). *Phys. Rev.* **82**, 697.
Staley, R. H., and Beauchamp, J. L. (1974a). *J. Am. Chem. Soc.* **96**, 1604.
Staley, R. H., and Beauchamp, J. L. (1974b). *J. Am. Chem. Soc.* **96**, 6252.
Staley, R. H., and Beauchamp, J. L. (1975a). *J. Chem. Phys.* **62**, 1998.
Staley, R. H., and Beauchamp, J. L. (1975b). *J. Am. Chem. Soc.* **97**, 5920.
Staley, R. H., Corderman, R. R., Foster, M. S., and Beauchamp, J. L. (1974). *J. Am. Chem. Soc.* **96**, 1260.
Staley, R. H., Harding, L. B., Goddard, W. A., III, and Beauchamp, J. L. (1975). *Chem. Phys. Lett.* **36**, 589.
Staley, R. H., Kleckner, J. E., and Beauchamp, J. L. (1976). *J. Am. Chem. Soc.* **98**, 2081.
Su, T., and Bowers, M. T. (1973a). *J. Am. Chem. Soc.* **95**, 1370.
Su, T., and Bowers, M. T. (1973b). *J. Chem. Phys.* **58**, 3027.
Su, T., and Bowers, M. T. (1973c). *J. Am. Chem. Soc.* **95**, 7609.
Su, T., and Bowers, M. T. (1973d). *J. Am. Chem. Soc.* **95**, 7611.
Su, T., and Bowers, M. T. (1974). *J. Chem. Phys.* **60**, 4897.
Su, T., and Kevan, L. (1973a). *J. Phys. Chem.* **77**, 148.
Su, T., and Kevan, L. (1973b). *Int. J. Mass Spectrom. Ion Phys.* **11**, 57.
Sullivan, S. A., and Beauchamp, J. L. (1976). *J. Am. Chem. Soc.* **98**, 1160.
Teng, H., and Dunbar, R. C. (1978). *J. Am. Chem. Soc.* **100**, 2279 (1978).
Tiedemann, P. W., and Riveros, J. M. (1973). *J. Am. Chem. Soc.* **95**, 3140.
Tiedemann, P. W., and Riveros, J. M. (1974). *J. Am. Chem. Soc.* **96**, 185.
van der Hart, W. J. (1973). *Chem. Phys. Lett.* **23**, 93.
van der Hart, W. J., and Van Sprang, H. A. (1975). *Chem. Phys. Lett.* **36**, 215.
van der Hart, W. J., and Van Sprang, H. A. (1977). *J. Am. Chem. Soc.* **99**, 32.
Vogt, J., and Beauchamp, J. L. (1975). *J. Am. Chem. Soc.* **97**, 6682.
Wieting, R. D., Staley, R. H., and Beauchamp, J. L. (1975). *J. Am. Chem. Soc.* **97**, 924.
Wilkins, C. L., and Gross, M. L. (1971). *J. Am. Chem. Soc.* **93**, 895.
Williamson, A. D., and Beauchamp, J. L. (1975). *J. Am. Chem. Soc.* **97**, 5714.
Wilson, J. C., and Bowie, J. H. (1975). *Aust. J. Chem.* **28**, 1993.
Winkler, J., and McLafferty, F. W. (1973). *J. Am. Chem. Soc.* **95**, 7533.
Wobschall, D. (1965). *Rev. Sci. Instrum.* **36**, 466.
Wobschall, D., Graham, J. R., and Malone, D. P. (1963). *Phys. Rev.* **131**, 1565.
Woods, I. B., Riggin, M., Knott, T. F., and Bloom, M. (1973). *Int. J. Mass Spectrom. Ion Phys.* **12**, 341.

PHYSICAL METHODS IN MODERN CHEMICAL ANALYSIS, VOL. 2

Refractive Index Measurement

Thomas M. Niemczyk

Department of Chemistry
University of New Mexico
Albuquerque, New Mexico

I. Introduction and Theory

A. Definitions

1. Snell's Law

The refractive index of a substance is the ratio of the velocity of light in vacuum (or more practically, in air) to the velocity of light in the substance. The indices thus describe the extent by which various substances cause a decrease in the speed of light compared to that in a vacuum. The velocity

337

reduction is dependent on the wavelength of the light passing through the substance and on the temperature, pressure, chemical composition, homogeneity, purity, etc. of the substance. Thus the refractive index is an important parameter in many measurements concerned with the testing or control of products. For example, the purity of a substance may be estimated on the basis of a refractive index determination. Refractive index measurements have long been used in many situations and many instruments have been developed that can be used to make the measurements simple and reliable. It is the purpose of this chapter to review the means by which refractive index measurements can be made and discuss some of the applications.

The fundamental basis of refractive index measurements is the phenomenon of refraction. When light of a given wavelength passes from one isotropic medium, j in Fig. 1, into a second optically denser medium J, it undergoes a change in wave velocity ($v_j \rightarrow v_J$) and its direction also changes unless the ray is perpendicular (N) to the surface boundary. The relation between the angle of incidence i and the angle of refraction r is expressed in Snell's law of refraction (Jenkins and White, 1957).

> The refracted ray lies in the plane of incidence, and the sine of the angle of refraction bears a constant ratio to the sine of the angle of incidence.

The second part of this law states

$$\sin i / \sin r = \text{const} \tag{1}$$

The constant in Eq. (1) is defined as the refractive index n of the medium J relative to the medium j and is equal to the characteristic ratio v_j / v_J. By experimentally measuring the angles i and r one can determine the values of the refractive index for various transparent media.

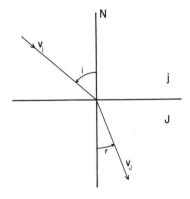

Fig. 1 Refraction of a beam of light upon incidence, at angle i, on a plane interface between medium j and a second optically denser medium J.

In the refraction at a boundary between two isotropic substances having indices of refraction n and n' Snell's law may be written in the symmetrical form

$$n \sin r = n' \sin i \tag{2}$$

The ratio n'/n is called the relative index of the second medium with respect to the first. If measurements are to be comparable, it is necessary to specify the reference medium as well as to control any other variables which affect the velocity of light in the sample itself. In most cases it is convenient to use air as the first medium, j in Fig. 1, for making refractive index measurements. The value of n generally quoted in the literature for liquids and solids refer to air at ambient conditions or sometimes to air at specified conditions such as 20°C. From Eq. (2) it is easy to see that the index of any substance j, relative to another J, may be converted to refer to another standard, air at STP (0°C, 760 mm, dry), by multiplication of the measured index by the index of air at STP relative to air at laboratory conditions,

$$n_{j/\text{lab air}} \times n_{\text{lab air/STP}} = n_{j/\text{STP}} \tag{3}$$

provided the index $n_{\text{lab air/STP}}$ is known.

If measurements are made using ambient air as the "standard" a considerable amount of error can be introduced. The ambient air may go through temperature and pressure fluctuations and the moisture and CO_2 content may vary. For precise measurements of refractive index, much care in controlling the conditions must be exercised, but in most work only the temperature of the air need be taken into account. Tilton (1935) has discussed the effects such as temperature and pressure on the reference medium. He shows that a sample of high refractive index, 1.8 (assume constant with temperature), when measured on a warm day and again on a cold day (20°C difference) would be different by 4×10^{-5} merely because the reference air is denser. Similarly, if the barometer reads 30 to 45 mm higher when a refractive index determination is repeated, the second result will appear lower by 2×10^{-5}. The degree of humidity in the ambient air is of much less importance than is a precise knowledge of the temperature or pressure.

2. Absolute Index of Refraction

The refractive index of a gaseous sample is generally referred to as the reference medium. Whenever a vacuum is used as the reference standard, the refractive index is called an absolute index of refraction, n_{vac}. In practice, the difference between n_{vac} and n is very small (0.03%). It follows from Eq. (3)

and the definition of n_{vac} that

$$n_{vac} = (n_{vac} \text{ of air}) \times n \tag{4}$$

For ordinary laboratory conditions using yellow light $n_{vac\,air} = 1.00027$.

B. Dependence of Refractive Index on Temperature and Pressure

The refractive index of a substance can be determined accurately only if the substance is homogeneous on a scale comparable with the dimension of a wavelength of the light used to make the measurement. Solids should be free from stress effects and from striations and visible defects. In a liquid sample suspended material can be tolerated, but the sample should contain no soluble impurities. Given a sample that fits these requirements the variable most likely to have an effect on the accuracy of the measurement is the temperature. The refractive indices of all liquids decrease fairly substantially with increasing temperature.

1. Density Variations

The effect of temperature on the measured refractive index is the result of three factors. Best known is the fact that as the temperature increases, the density of both the air and the liquid decreases; hence, the light beam encounters fewer molecules per unit volume in passing through these media. The change in refractive index due to temperature change of the reference air has been previously discussed. The change in the absolute refractive index n_{vac} of a medium can be related to the temperature change in density d by the equation

$$\partial n_{vac}/\partial t = f(n_{vac})\,dd/dt \tag{5}$$

where $f(n_{vac})$ is a positive function, variously approximated as, for example $(n_{vac} - 1)$, $(n_{vac}^2 - 1)$, or $(n_{vac}^2 - 1)/(n_{vac}^2 + 2)$, this change in index is usually negative and not related to wavelength.

2. Spectral Variations

The third factor involved in the effect of temperature on the measured refractive index is the alteration in the spectral characteristics of the molecular structures involved. This is mainly due to changes in the resonant frequencies of certain vibrators in the media. Increase in temperature tends to shift ultraviolet absorption bands toward longer wavelengths and this leads to an increased refractive index. The reasons for these effects will be treated in more detail in Section I,C,3.

3. *Control and Specification of Temperature*

Published temperature coefficients of refractive indices of liquids show that all such values are negative, demonstrating that the expansion of the liquid predominates over the other effects. For a large number of transparent and semitransparent liquids (including numerous oils) the value of $\Delta n_{vac}/\Delta t$ is approximately expressed as -4×10^{-4}, although a number of organic liquids have values ranging from -4×10^{-4} to -6×10^{-4}. Water and aqueous solutions, which have a value for $\Delta n_{vac}/\Delta t$ of -1×10^{-4}, are important exceptions. The small temperature coefficient of water and aqueous solutions is principally a consequence of the relatively small expansion coefficient of the liquid. There are some optically dense liquids that have unusually large negative temperature coefficients of index, such as -7×10^{-4} and -8×10^{-4} for methylene iodide and carbon disulfide, respectively.

The values of $\Delta n_{vac}/\Delta t$ for solids are much less uniform from substance to substance but as a rule are considerably smaller than for liquids and may be positive or negative. For many glasses an increase in temperature causes a small but not negligible increase in n; e.g., $\Delta n_{vac}/\Delta t = +1.8 \times 10^{-6}$ per degree for one of the Pulfrich refractometer prisms at room temperature. The temperature coefficient of the index is a parameter of prime importance in characterizing thermal distortion effects in ir transmitting materials. Thus, there have been considerable data published in recent years concerning this type material illustrating the diversity of data for $\Delta n_{vac}/\Delta t$ for solids. For example, $\Delta n_{vac}/\Delta t$ for alkaline earth fluorides is about $-1.5 \times 10^{-5}/°C$ as reported by Lipson *et al.* (1976), and $\Delta n_{vac}/\Delta t$ for germanium and silicon range from 1×10^{-4} to $4 \times 10^{-4}/°C$ as reported by Icenogle *et al.* (1976).

It is evident that temperature is an important parameter in refractive index measurement and should be carefully controlled when work is to be precise. In addition the temperature should be specified; this is done by writing $t°C$ as a superscript, e.g., n^{25}. Ward and Kurtz (1938) recommend the use of 20°C as the standard temperature for hydrocarbons and indeed much work with organic liquids has been done at this temperature; however, in conformity with other physicochemical data, 25°C might seem more appropriate. For fifth decimal place accuracy the requirements of temperature control are $\pm 0.02°C$ for liquids of average thermo-optical sensitivity. This kind of control generally requires the use of thermostats and circulating pumps that are available commercially.

The effect of pressure on refractive index measurements is much less severe than that of temperature. According to Timmermans and Hennaut-Rowland (1932) and Timmermans and Delcourt (1934), an increase in pressure of one atmosphere causes the refractive index of most liquids to increase by about 3×10^{-5}. Thus, normal fluctuations of barometric pressure produce a negligible change in n_{liq}. The effect on solids is even smaller.

C. *Dependence of Refractive Index on Wavelength*

1. *Dispersion*

The well-known phenomenon of the dispersion of white light into a spectrum by passage through a prism is a graphic illustration of the fact that light of different wavelengths is refracted to different angles by a given medium, and hence the fact that the index of refraction of the medium is a function of the wavelength. If measurements of n versus λ are made for various media and the values of n plotted against wavelength, a curve like one of those in Fig. 2 is obtained. The curve for any material transparent over the wavelength range illustrated will differ in detail but will have the same general shape. These curves are representative of normal dispersion, for which the following facts are to be noted:

(1)   As the wavelength decreases the index of refraction increases.

(2)   The rate of increase is greater at shorter wavelengths.

(3)   For different substances at a given wavelength the curve is usually steeper the larger the index of refraction.

(4)   The curve of one substance will generally not overlap that of another by a shift in scale.

From these facts concerning normal dispersion it is easily seen that the wavelengths used in the measurement of the refractive index of a substance

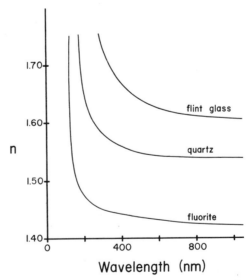

Fig. 2 Dispersion curves for different materials that are commonly used for lenses and prisms.

should be specified. The possible errors involved when comparing the index for a medium determined at two different wavelengths is dependent upon the magnitude of the index, thus more care is needed for media of higher refractive index.

2. Dispersion Equations

Many attempts have been made to develop equations that relate normal dispersion to wavelength. The first successful attempt was made by Cauchy in 1836. His equation may be written

$$n = A + (B/\lambda^2) + (C/\lambda^4) + \cdots \tag{6}$$

The number of terms to be taken in the equation depends on the range of wavelengths to be considered and A, B, and C are constants which are characteristic of a particular substance. Often only the first two terms, $A + B/\lambda^2$, are used, and it is found that over short wavelength ranges this formula is accurate to a few units of the fourth decimal place.

If, however, the data are of any value in the fifth decimal place or a wider wavelength range, e.g., the entire visible region, is to be covered, a more elaborate dispersion formula is advisable. Several have been developed including that of Conrady (1903)

$$n = n_0 + a/\lambda + b/\lambda^{3.5} \tag{7}$$

This formula has proved very successful for optical glass and has been suggested worthy of trial for other media, with a possible modification of the exponent of λ in the last term. Other equations of a similar nature have been tested with similar results. Numerous tests of the Hartmann equation

$$n = n_0 + c/(\lambda - \lambda_0)^a \tag{8}$$

indicate that it is remarkably reliable to the fifth decimal of index for glass, plastics, and other media. Other equations include that of Wright (1921)

$$n_y - n_x = a(n_F - n_C) - b \tag{9}$$

where F and C denote the F and C lines in the sodium spectrum and x and y denote any two spectral lines, and that of Waldmann (1938)

$$(n_D - n_C) = K(n_F - n_C) \tag{10}$$

in which K is said to be constant within 10% for a wide range of organic compounds excepting those of high dispersion.

3. Anomalous Dispersion

If measurements of refractive index of a transparent substance such as quartz are extended into the infrared region of the spectrum, the dispersion curve begins to show marked deviations from the Cauchy or any other

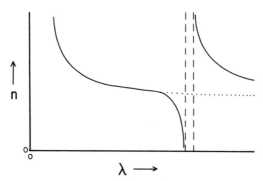

Fig. 3 Dispersion curve for a material that demonstrates a region of anamalous dispersion. The region between the dashed lines represents an absorption band. The dotted line shows the values one would obtain from the Cauchy equation if used in this region.

previously mentioned equation. The deviation is always of the type illustrated in Fig. 3. The dotted line represents the Cauchy equation and it can be seen that the values for n decrease more rapidly than predicted as the wavelength approaches a region where the material has an absorption band (the region in Fig. 3 set off by dashed lines), such as might be the case for quartz in the infrared. This is an absorption band and it can be seen that on the long wavelength side of the band the index is very high, decreasing with wavelength in the same manner as seen for quartz in the far uv. The Cauchy equation can be applied in this region with different constants. Discontinuity in a dispersion curve as it crosses an absorption band gives rise to the term anomalous dispersion. The term "anomalous" seems somewhat inappropriate since all materials show like behavior in regions of absorption bands.

The first attempt at a dispersion equation of more general applicability than the Cauchy or similar equations was obtained by postulating a mechanism by which the medium could affect the velocity of the light wave. It was assumed that the medium contains particles bound by elastic forces, so that they are capable of vibrating with a certain definite frequency, v_0. Passage of light waves through the medium is then assumed to exert a periodic force on the particles, which causes them to vibrate. If the frequency v of the light does not agree with v_0 the vibrations will be of very small amplitude. As the frequency approaches v_0 the response is greater and a large amplitude is reached at $v = v_0$. These vibrations will in turn react on the light wave and alter its velocity. A mathematical investigation of this mechanism was made in 1871 by Sellmeier, who obtained the equation

$$n^2 - 1 = \sum_j \frac{K_j \lambda^2}{\lambda^2 - \lambda_j^2} \tag{11}$$

There are two constants in this equation, K_j and λ_j, both related to a specific absorption j, the latter by the equation $\lambda_j v_j = C$. This equation accounts for the region of anomolous dispersion as well as giving a more accurate representation of n in regions far from absorption bands than does a Cauchy or a similar equation.

It can be readily seen from the above discussion that it is necessary to specify the wavelength to which a given value of the refractive index of a substance refers. For example, the refractive index of benzene is 1.49759 when measured with red light of $\lambda = 656.3$ nm, but it is 1.52487 with violet light of $\lambda = 434.1$ nm. The kind or the wavelength of light used should be designated by a subscript, as in n_D or $n_{589.3}$. When this subscript is lacking it has been customary to assume the value refers to the sodium D line (589.3 nm). This line actually consists of two lines, D_1, (589.0 nm) and D_2 (589.6 nm) of nearly equal intensity. The difference between n_{D_1} and n_{D_2} can actually be considerable for substances of high dispersion, for example, $n_{D_1} - n_{D_2}$ for benzene is 2×10^{-5}. Guild (1917) discusses the effect of the doublet on the critical boundary in the Pulfrich refractometer. With colored materials it is sometimes necessary to determine n, or the dispersion, at other wavelengths. There are many choices of convenient light sources now available. Sources that can be used in refractive index measurements are discussed in Section II,A.

D. The Effect of Concentration on the Refractive Index of Solutions

1. Determination of Concentration

The refractive index is often used as a measure of concentration. The most reliable method for a measurement of this type is to interpolate from an empirical calibration curve. In some cases there exists a linear relationship between n and concentration, but more generally the relationship is represented by a curved line. The relationship is obviously strongly influenced by the units in which concentration is expressed, with molarity M the most likely to yield a straight line. In many applications of refractometry, n is used to determine a weight percent p of a component, say sugar, but a plot of n versus p is far less likely to be linear. This is illustrated by the data of Koenig-Gressmann (1938) on acetone (n_1)–carbon tetrachloride (n_2) solutions. The relationship $M_1 = k(\Delta n)$, where $\Delta n = n - n_1$, $k = 1000 \, d_1/\mathrm{MW}(n_1 - n_2)$, $d_1 = $ density, and MW $= $ molecular weight of pure acetone, shows only a slight deviation from linearity over the entire concentration range. A plot of p versus Δn shows considerable nonlinearity.

In the analysis of aqueous protein solutions, a term called the specific refractive increment k' is often used. This term is defined as $k' = (n - n_0)/C_w$ and is seen to be constant over a wide range of concentration C_w, when C_w is expressed in grams per milliliter. These relationships have been discussed by Doty and Geiduschek (1953) and Perlmann and Longsworth (1948). For example, for human serum albumin at pH 4.85 and 0.5°C, the difference $n - n_0$ between solution and solvent is 0.1887 ± 0.0001 per unit C_w, from 1.8% to 7.7% protein. Most proteins have k' values within about 0.010 of 0.185 for yellow light, depending on pH, salt concentration, and temperature.

It has been shown by Arshid and co-workers (1955, 1956a,b) that plots of n^2 versus mole fraction often closely fit straight lines. In cases of strong interactions between the components in a solution the plot of n^2 versus mole fraction consists of two straight lines which intersect at the mole ratio corresponding to the interaction complex. These workers have reported extensive studies, using this technique to follow intermolecular complex formation between organic substances in aqueous and nonaqueous solutions.

2. Relationship to Density

The relation between refractive index and density d has been the subject of investigation since the time of Newton, who considered $(n^2 - 1)/d$ worthy of attention because it was approximately constant for several different substances.

Gladstone and Dale investigated the changes in n and d as the temperatures of liquids were varied, and, instead of the Newton constant, they found better constancy for $(n - 1)/d$, and this ratio became known as specific refractivity.

Lorenz of Copenhagen and Lorentz of Leyden independently derived the expression $(n^2 - 1)/d(n^2 + 2) = C$ for a certain class of liquids of which water was said to be typical. This ratio, called specific refraction, has enjoyed wide popularity and has been applied, probably in many cases, under circumstances for which it was never intended. The constant 2 in the denominator is only an approximation in special cases for a more generally applicable variable X. It is not surprising to find that Eykman, after much work on organic liquids, advocated $(n^2 - 1)/d(n + 0.4) = C$ as more suitable for the particular kinds of media and the particular variability in which he was interested. This Eykman relation has been favored by Kurtz and Ward (1937) in certain work on petroleum products, and also by Gibson (1938) in his work on benzene at high pressures.

When Zehnder (1888) and Röntgen and Zehnder (1891) applied pressure to liquids, they concluded that the $(n - 1)/d$ relation was the more closely followed. When Pockels (1902) analyzed his data on unidirectional pressure applied to glass, he favored the Newton constant, $(n^2 - 1)/d$. Tool and Tilton (1926) at the National Bureau of Standards, when investigating the effect of annealing temperatures on the refractive indices and densities of glasses, found that Newton's relation was approximated. Young and Finn (1940), in their work at the National Bureau of Standards, have favored the use of $(n - 1)/d$ when dealing with variations caused by small changes in the chemical composition of similar glasses.

In the field of gases much testing of the various relations has been attempted, especially with variations in pressure, but some of these relationships become so nearly alike when the density of oscillators is low and n approaches unity, as in gases, that it is difficult to find experimental data that are sufficiently accurate for discriminatory use in such tests.

It should be remembered that the index of refraction and the density are influenced by many variables, such as temperature, pressure, chemical composition, isomeric rearrangements, physicochemical rearrangements (heat treatment), and change of state. Thus, when specific refractions are being used for various purposes, and atomic factors derived and when, subsequently, the additivity of molecular refractions is the question, it is suggested that the arbitrary nature of the cases be considered before relying altogether on the various explanations that are sometimes offered for lack of expected perfection of these relationships. They are of value only as they are proved useful in investigation and analysis and not as a basis for negative arguments. Present theories indicate that no single, simple and widely applicable relation exists between refractive index and density.

3. Mixtures

If two fluids mix without significant volume change, as usually happens with gases but not very generally with liquids, except possibly for dilute solutions, the index of a mixture of them may be estimated from their volumes v_1, and v_2, as $n = (n_1 v_1 + n_2 v_2)$ or, similarly, with $(n - 1)$ in place of n, $(n_1 - 1)$ in place of n_1, etc.

If there is reason to suppose that Gladstone and Dale's constant is applicable, one writes

$$\frac{100(n - 1)}{d} = \frac{p_1(n_1 - 1)}{d_1} + \frac{(100 - p_1)(n_2 - 1)}{d_2} \tag{12}$$

where p is percent by weight. A similar equation can be written for Newton's constant, $(n^2 - 1)/d$, or for the Lorentz–Lorenz constant, for example, depending on the individual investigator's experience or preference.

In cases where one knows the contraction or expansion of volume

$$c = (v_1 + v_2 - V)/(v_1 + v_2) \tag{13}$$

where V is the solution volume, one may write

$$\frac{100(n - 1)}{d} \frac{(1 - ac)}{(1 - c)} = \frac{p_1(n_1 - 1)}{d_1} + \frac{(100 - p_1)(n_2 - 1)}{d_2} \tag{14}$$

where a varies with temperature and wavelength.

To solve Eq. (12) for n, an experimental value for the density should be known. For instance, for a mixture of water and acetone (65.61% acetone) at 25°C, $d_{exp} = 0.8809$, and, since $d_1 = 0.7862$ for acetone and $d_2 = 0.9977$

for water, one obtains from Eq. (12) $n_{calc} = 1.3620$, which agrees well with $n_{exp} = 1.3634$. Note that this value is greater than either the refractive index of pure water or of pure acetone which indicates that a fairly strong inter-action between water and acetone takes place. One should, whenever possible, rely only on interpolations from known data rather than resort to computation by any rule for mixtures.

E.   Precision of Refractive Index Measurements

1.   General Considerations

From the previous discussion it can be seen that factors such as variations in temperature or the presence of impurities in a sample can lead to substantial imprecision in refractive index measurements. Bauer et al. (1959) even point out that one should proceed with caution when using refractive indices found in standard reference works or in the journal literature. They cite, for example, five indices published for o-xylene that cover a range from 1.5046 to 1.5071.

In certain cases one can estimate the degree of purification necessary for measurements of a given accuracy from the following formula:

$$n - n_0 \approx 0.01 p_1 (n_1 - n_0) \tag{15}$$

where n, n_0, and n_1 refer to the mixture, the pure substance, and the impurity, respectively, and p_1 is the weight percent of impurity. This equation is based on Eq. (12) and the assumption that the density of the mixture is equal to that of the major component. This is generally not an unreasonable assumption since in the purification of a substance the most difficult impurities to remove will be those with chemical properties similar to those of the solvent.

One impurity that can have a relatively large effect on a refractive index measurement is dissolved air. Dissolved air may lower the refractive index of liquids by about 0.01% at saturation under one atmosphere. Changes in the humidity of the dissolved air may influence the fifth decimal place.

The precision and accuracy of the refractive index determination also depends on the method used to make the measurement. Absolute methods require a primary standard, generally air, and a direct measurement of the angle of refraction. Several prismatic standards are well known and it is possible to measure indices in this way with reproducibility within $\pm 3 \times 10^{-6}$, for a given laboratory, especially on samples of isotropic crystals, homogeneous glass, and distilled water. Absolute methods require great care and considerable experience if precise work is to be realized.

More commonly refractive index measurements are made by comparing a measurement for a sample with a measurement made on a standard material for which the refractive index is well known. The comparison standard used in such a measurement is a material that is transparent, homogeneous, and has been carefully measured with respect to air. If the sample is compared to the standard there are always some difficult-to-control sources of error that can be assigned more liberal tolerances than is the case in absolute measurements because, to the extent the sample resembles the standard, they may be affected alike during the comparison measurement. Thus the result for the sample is automatically free from the effect of certain conditions that prevailed during the comparison, and the result holds good, to a large extent, for the conditions that were more carefully controlled when the standard was compared with the primary standard.

Most refractometers in use are comparison instruments. In general they are more reliable the shorter their total range in the scale of refractive index. The instruments used are capable of relatively high precision and in general are not the limiting factor in the data obtained in a refractive index measurement. The ease of obtaining high precision with modern refractometers has caused some researchers to make unwarranted assumptions regarding the accuracy which is being reached. Tilton and Taylor (1950) stated it is impossible to obtain an accuracy of better than ± 2 (or 3) $\times$ 10^{-5} in refractometry of solid samples with commercial instruments. On the average the errors made on liquids are larger. Probably there are many different residual error sources that have the possibility of contributing to lower precision.

Because of this it is recommended that the substitution process be used when it is desired that the refractive index measurement be highly accurate and precise. That is, the unknown and a standard sample very like the unknown should be compared, either simultaneously or in rapid succession, at as near the same test conditions as possible. If the principle of substitution can be strictly followed, the accuracy attained will necessarily closely approximate the precision with which the whole comparison operation can be done.

2. *Reference Standards*

If a laboratory is to do excellent work by comparison methods over a wide range in index, it should be provided with a number of refractometers, each covering a short interval in the index scale. In order to work by substitution methods it must also be supplied with a series of standard refractive index samples for which the indices are known over an appropriate range of temperatures.

Since some refractometers do not always give the same results for liquids and for solids for identical index, it is necessary to have some liquid as well as solid standards. This is particularly unfortunate because the questions of original purity, contamination, sensitivity to temperature, and expendability makes the use of liquid standards much less convenient. However, for all serious problems concerning liquids that can be handled by immersion refractometers, it will be satisfactory to use solids, if obtainable, as the substitution standards.

Tilton and Tool (1929) point out that solid standards in the form of 60° prisms (with polished triangular ends for use on refractometers) can be certified in the fifth decimal place by the National Bureau of Standards. Glass can be commercially obtained for the purpose only with n from 1.46 to 1.9, but lithium fluoride ($n = 1.39$) and fluorite ($n = 1.43$) are satisfactory also. Equivalent solid standards for any index can be made, as suggested by Richter (1930) and such prisms are furnished in some instances by Zeiss for adjusting refractometers.

Among standard liquids, distilled water takes first place. Its indices for sodium light are given for every 0.1° from 0 to 60°C and at every 0.5° for 13 different wavelengths (Tilton and Taylor, 1938, pp. 449, 446–447). Excerpts are given here as Tables I and II. Other standards are toluene, 1.49693, 2,2,4–trimethylpentane, 1.39145, and methylcyclohexane, 1.42312 (for D lines at 20°C), which are available from the National Bureau of Standards as certified samples 211a, 217a, and 218a, respectively, with fifth decimal indices at 20, 25, and 30°C for each of seven wavelengths, as listed in Table III.

TABLE I

Sodium-Lines Index of Refraction of Distilled Water

t (°C)	n_D	t (°C)	n_D	t (°C)	n_D	t (°C)	n_D
0	1.333 949	15	387	30	1.331 940	45	1.329 852
1	947	16	315	31	819	46	693
2	940	17	238	32	695	47	533
3	927	18	158	33	569	48	370
4	908	19	075	34	440	49	204
5	884	20	1.332 988	35	308	50	037
6	855	21	897	36	173	51	1.328 867
7	821	22	803	37	036	52	696
8	782	23	706	38	1.330 897	53	522
9	738	24	606	39	754	54	346
10	691	25	503	40	610	55	168
11	638	26	396	41	463	56	1.327 989
12	582	27	287	42	314	57	807
13	521	28	174	43	162	58	623
14	456	29	059	44	008	59	437
						60	249

TABLE II

Index of Refraction of Distilled Water for Various Spectral Lines[a]

t (°C)	706.52 Helium	667.81 Helium	656.28 Hydrogen	589.26 Sodium	587.56 Helium	576.96 Mercury	546.07 Mercury	501.57 Helium	486.13 Hydrogen	471.31 Helium	447.15 Helium	435.85 Mercury	404.66 Mercury
0	1.330 948	1.331 816	1.332 094	1.333 949	1.334 003	1.334 345	1.335 440	1.337 339	1.338 113	1.338 925	1.340 425	1.341 214	1.343 756
5	889	755	032	884	1.333 937	279	372	269	042	854	352	141	681
10	704	567	1.331 843	691	744	085	176	070	1.337 842	653	149	1.340 938	476
15	410	269	545	387	440	1.333 781	1.334 869	1.336 760	531	341	1.339 835	623	158
20	020	1.330 876	151	1.332 988	041	380	466	353	123	1.337 931	423	210	1.342 742
25	1.329 545	398	1.330 672	503	1.332 556	1.332 894	1.333 977	1.335 860	1.336 628	435	1.338 925	1.339 710	239
30	1.328 993	1.329 843	116	1.331 940	1.331 993	331	411	289	055	1.336 860	347	131	1.341 656
35	371	218	1.329 489	308	360	1.331 697	1.332 774	1.334 647	1.335 411	215	1.337 699	1.338 481	001
40	1.327 685	1.328 528	1.328 798	1.330 610	1.330 662	1.330 998	071	1.333 940	1.334 702	1.335 504	1.336 984	1.337 765	1.340 280
45	1.326 939	1.327 778	047	1.329 852	1.329 904	238	1.331 308	171	1.333 932	1.334 731	208	1.336 987	1.339 497
50	136	1.326 971	1.327 239	037	089	1.329 422	1.330 489	1.332 346	104	1.333 902	1.335 375	152	1.338 655
55	1.325 280	111	1.326 378	1.328 168	1.328 220	1.328 552	1.329 615	1.331 467	1.332 223	018	1.334 487	1.335 261	1.337 758
60	1.324 373	1.325 200	1.325 466	1.327 249	1.327 301	1.327 631	1.328 690	1.330 536	1.331 290	1.332 082	1.333 547	1.334 319	1.336 809

[a] These values were computed by means of a general interpolation formula.

Thomas M. Niemczyk

TABLE III

Indices of Refraction of 2,2,4-Trimethylpentane, Methylcyclohexane, and Toluene

Wavelength (nm)	Designation of line	2,2,4-Trimethylpentane			Methylcyclohexane			Toluene		
		20°C	25°C	30°C	20°C	25°C	30°C	20°C	25°C	30°C
667.81	Helium	1.389 17	1.386 71	1.384 25	1.420 64	1.418 12	1.415 61	1.491 80	1.489 03	1.486 19
656.28	Hydrogen, C	1.389 45	1.386 98	1.384 52	1.420 95	1.418 43	1.415 92	1.492 43	1.489 66	1.486 82
589.26	Sodium, D_1, D_2	1.391 46	1.388 99	1.386 51	1.423 13	1.420 59	1.418 07	1.496 93	1.494 13	1.491 26
546.07	Mercury, e	1.393 17	1.390 68	1.388 21	1.424 99	1.422 44	1.419 91	1.500 86	1.498 03	1.495 14
501.57	Helium	1.395 45	1.392 95	1.390 46	1.427 44	1.424 91	1.422 34	1.506 20	1.503 34	1.500 41
486.13	Hydrogen, F	1.396 41	1.393 90	1.391 39	1.428 47	1.425 91	1.423 35	1.508 47	1.505 59	1.502 65
435.83	Mercury, g	1.400 30	1.397 78	1.395 24	1.432 71	1.430 11	1.427 54	1.518 00	1.515 06	1.512 06

The National Bureau of Standards sells samples of certified refractive index for use in the calibration of refractometers. Currently, samples of 2,2,4-trimethylpentane (SRM 217b and SRM 1816) and *n*-heptane (SRM 1815) are available. The reference materials available are changed from time to time, so the interested researcher should refer to a current edition of NBS Special Publication 260 to check availability and secure ordering information.

F. Relationship of Refractive Index to Electronic Structure

1. Molar Refraction

The equation of Lorentz and Lorenz has been extended to define a term called the Lorentz–Lorenz molar refraction given by the following equation

$$R = [(n^2 - 1)/(n^2 + 2)]M/d \qquad (16)$$

where M is the molecular weight. It has often been used in the analysis of mixtures for, in many cases, even small deviations of R from additivity allow one to draw important conclusions concerning inter- and intramolecular forces and the electronic structure of molecules. For example, Grosse (1946) has demonstrated the extent of fluorination of naphthenes can be determined based on the fact that the specific Lorentz–Lorenz refraction is practically a constant for fluoronaphthenes ($R = 0.098$ c^3/gm) and for naphthenes ($R = 0.33$ c^3/gm), independent of molecular weight, and considerably different for the two types of compounds. This method has also been used to study paraffin–fluoroparaffin mixtures and the fluorination of chlorocarbons.

Much effort has been extended to measure the Lorentz–Lorenz molar refraction, especially by Fajans and his students (Fajans, 1924, 1941; Fajans and Johnson, 1942; Fajans *et al.*, 1931). From this and other work several conclusions can be drawn about the way in which exact values of R can be obtained. In the measurement of liquids two experimental precautions are important in order to obtain reliable values: (i) The measurement of density and of refractive index should both be made on the same sample in order to minimize the influence of impurities. (ii) The temperature of measurement of n and of d should not differ appreciably. The temperature difference permissible for a given accuracy can be calculated from the temperature coefficient of n and of d (Section I,B,1) according to Eq. (5). By differentiating Eq. (16) one obtains an equation for evaluating the error ΔR in the molar refraction caused by a given experimental error, Δn in n or Δd in d:

$$\frac{\Delta R}{R} = \frac{6n}{(n^2 + 2)(n^2 - 1)} \Delta n - \frac{\Delta d}{d} \qquad (17)$$

The determination of R for crystals is much more difficult than that for liquids. Two methods are used which allow an estimation of R: (i) Measurements of n (and d) on the molten substance; and (ii) evaluation of

R_{app}, the apparent molar refraction, of the substance by measuring the n and d of its solution of a known concentration in a suitable solvent. The refractions of mixtures of organic substances show, in general, such small deviations from additivity that the value of R_{app} should in most cases be equal to R_{cryst} within a few hundredths of a cubic centimeter.

2. Additive Refractive Increments

For over a hundred years scientists have been trying to understand the relationship between R and the electronic properties and structure of a given molecule. For a long time the term "characteristic of the molecule" was understood to mean that the value of R is independent of temperature, pressure, and what state it was in. "Indicative of the molecular structure" was understood to mean that R of a molecule can be additively composed of constant empirical contributions of atoms and atomic groups present in the molecule; since these increments proved to be dependent on the kind of binding between the components, the molar refraction has been considered to be at the same time additive and constitutive.

Based on Newton's equation for specific refraction, Berthelot (1856) expressed the refractive power of substances as a molar quantity according to the equation

$$R_B = (n^2 - 1)M/d \tag{18}$$

He showed that the CH_2 group makes an approximately constant contribution to R_B. This sort of work was extended by others who developed similar expressions culminating, perhaps, in the work of Lorenz and Lorentz who derived their equation from the electromagnetic theory of light. The refractive index and therefore R, depends on wavelength so this must be taken into account. Also, R is not quite independent of temperature as was suggested, so often the molar refraction is given as R_D^{20}, referring to 20°C and the sodium D line.

Although the methods of constant additive refractometric increments were criticized early by Swietoslawski (1920), they have served for many years as a convenient and effective method of testing or verifying structural formulas for the organic chemist. Although some relatively recent work by Vogel et al. (1954) has been based on additive refractometric increments, these methods have largely been displaced by more modern spectrometric techniques.

The use of molar refraction as a quantity composed of additive constants assigned to bonded atoms and groups, or to bonds between them, leads to innumerable deviations from additivity. The various purely empirical methods of registering these deviations or accounting for them are based mostly on the classical structural formulas of organic chemistry. Therefore this approach cannot be applied logically to organic ions, nor can it account for the influence of intramolecular polarity on molar refraction demonstrated by the behavior of inorganic substances.

A more general theory is based on the correlation of molar refraction with electronic polarizability and intramolecular forces. The Lorentz–Lorenz molar refraction is an exact measure of the electronic polarizability of free atoms and mononuclear ions, and when these particles form molecules, crystals, or solutions, the accompanying deviations from additivity of molar refraction indicate that the overall strength with which the electronic systems are bound by the positive ion has changed. Then, the analysis of the deviations of molar refraction from constancy or additivity is relatively simple for interactions which do not involve discontinuous changes of electron configurations. For a much more detailed discussion of molar refraction and electronic structure see Section 7 in the work of Bauer *et al.* (1959).

II.   Apparatus and Experimental Procedures

A.   *Light Sources*

Much refractometry is done today using white light, but the use of monochromatic-light refractometers is rapidly increasing. The increase in the use of monochromatic light can be directly related to the development of intense, stable, monochromatic sources. There are sources available that cover the entire visible spectrum. Some of the lines used in refractometry are listed in Table IV.

A large step forward in monochromatic sources was the development of metal vapor arc lamps. These lamps contain internal electrodes and a fill gas and are sometimes called Osram lamps or Philips lamps after the manufacturer. They are especially suited to metals with high vapor pressure such as the alkali metals and certain others such as cadmium or mercury. These lamps can be designed to operate conveniently on 110 V alternating current. Perhaps the most popular for use in refractometry is the sodium arc lamp. The radiation from a sodium arc lamp consists mainly of the very intense yellow D_1 and D_2 doublet; the other sodium lines have low intensity and the spectrum of the fill gas can easily be filtered out. A low pressure mercury arc lamp produces four intense lines between yellow and violet and is recommended for dispersion measurements. Unfortunately mercury gives no intense lines at the red end of the spectrum.

Various types of hydrogen or noble gas discharges have also been used in refractometry. Various Geissler-type tubes have been used but these must

TABLE IV

Spectral Lines Used in Refractometry

Source	Symbol	Color	λ in air (nm)	λ^2 (μ^2)
Potassium arc	A′	Red	767.86	0.589 61
Helium tube	He_{r1}	Red	706.519	0.499 169
Helium tube	He_{r2}	Red	667.815	0.445 977
Hydrogen tube	H_α or C	Red	656.279	0.430 702
Cadmium	Cd_r	Red	643.847	0.414 539
He–Ne laser	—	Red	632.820	0.400 461
Sodium arc	D_1	Yellow	589.593	0.347 620
Sodium arc	D_m or D	Yellow	589.3	0.3473
Sodium arc	D_2	Yellow	588.996	0.346 917
Helium tube	D_3	Yellow	587.562	0.345 229
Mercury arc	Hg_{y1}	Yellow	579.066	0.335 317
Mercury arc	Hg_{y2}	Yellow	576.960	0.332 883
Mercury arc	Hg_g or e	Green	546.074	0.298 197
Cadmium	Cd_g	Green	508.582	0.258 656
Helium tube	He_g	Green	501.568	0.251 570
Helium tube	He_{bg}	Blue-green	492.193	0.242 254
Hydrogen tube	H_β or F	Blue	486.133	0.236 325
Cadmium	Cd_{b1}	Blue	479.991	0.230 391
Helium tube	He_b or C	Blue	471.314	0.222 137
Cadmium	Cd_{b2}	Blue	467.815	0.218 851
Helium tube	He_v	Blue-violet	447.148	0.199 941
Mercury arc	Hg_{bv} or g	Blue-violet	435.834	0.189 951
Hydrogen tube	H_γ or G^1	Violet	434.047	0.188 396
Mercury arc	h_1	Violet	407.783	0.166 287
Mercury arc	h_2	Violet	404.656	0.163 747

be operated at a high potential (500–1500 V). Helium tubes have been often used because they emit well-separated spectral lines distributed over the entire visible range. They found wide use with Pulfrich refractometers, because the instrument itself served to separate the spectral lines. The helium tube has the great advantage that the yellow helium D_3 line has practically the same wavelength (587.6 nm) as the commonly used sodium D line. This makes corrections from n_{D_3} to n_D easy and accurate.

The hydrogen lamp has often been used in the past because of the simplicity of the spectrum and the readily available fill gas in pure form. Hydrogen lamps generally suffer from the fact that some of the lines fade out as the lamp heats up, while none of the lines are as intense as those from the best sodium or mercury lamps. From the work of Campanile and

Lantz (1954) it appears that the best hydrogen source for critical angle refractometry is probably the Shell hydrogen lamp. A high intensity over a wide area is achieved by water cooling and intermittent operation. A palladium membrane is attached to the lamp through which the hydrogen can be periodically refilled.

The greatest development in light sources over the last two decades is the laser. The features of complete monochromaticity, good collimation and high intensity, make it a potentially useful source in refractive index measurements. Although a variety of lasers are available with lines throughout the visible region the helium–neon laser promises to have the most impact in refractometry because of the availability of very inexpensive models. Models with outputs on the order of 1–5 mW at 632.8 nm are available from many sources. There are two reasons for the conversion to a laser source: (i) the reduction of single slit diffraction effects in the instrument due to the natural parallelicity of the laser rays, and (ii) the laser intensity is high enough to allow for the use of high f-number optics. Block (1971) has discussed a technique of reproducibly introducing laser light into a refractometer.

B.   Filters

1.   Absorption Filters

To separate the lines from a line source or isolate certain regions of a continuous source some sort of monochromator or filter must be used. In general a monochromator is too expensive and cumbersome to be routinely used in refractometry so the major wavelength isolation device used is a filter. There are two types of filters available, absorption filters and interference filters. Absorption filters are sometimes called color filters and are available as bandpass, highpass, or lowpass filters. Commonly available absorption filters have relatively wide spectral bandwidth, generally in the the range of 35–40 nm, measured at a transmittance of one-half the maximum. Often it is necessary to combine filters to provide a suitably narrow bandwidth. Figure 4 illustrates, in an idealized way, how this can be accomplished. Substance A transmits below 600 nm and absorbs above this region. Substance B transmits above 580 nm and absorbs below. The combination of these two substances into a single filter produces a filter with maximum transmission at about 590 nm with relatively sharp cutoff both above and below this wavelength. A filter with characteristics such as those in Fig. 4 would be well adapted to isolate the sodium D lines. Colored or absorption filters are inexpensive and available from a variety of sources.

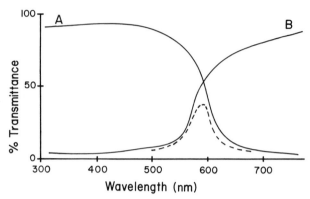

Fig. 4 Combination of two filters, A, which absorbs above 600 nm, and B, which absorbs below 580 nm, to produce a bandpass filter. The transmittance of the two filter combination is represented by the dashed line.

2. Interference Filters

In the last decade a wide range of interference filters have become available. These filters are bandpass filters with bandwidths at half transmittance of 10 nm or lower. The peak transmittance can often be as high as 50%. The filter itself is composed of several parallel, half-reflecting surfaces carefully spaced by a transparent material. In Fig. 5 two such surfaces, x and y, are shown to illustrate the operation of an interference filter. The half-reflecting surfaces are separated by one-half of the desired transmission wavelength. As light of wavefront AB enters the filter, a fraction is reflected from surface y to surface x. If the spacing is such that distance CD equals one-half the wavelength, $\lambda/2$, and DE also equals one-half the wavelength, $\lambda/2$, then the reflected ray CE + DE will be in phase with the ray entering directly at E. Therefore, light of wavelength λ will undergo constructive

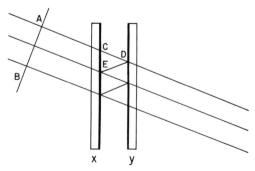

Fig. 5 Light path through an interference filter.

interference, or be reinforced, and emerge from the filter. Wavelengths other than λ, except harmonics, should undergo destructive interference on emergence from the filter and thus be greatly diminished in intensity. In Fig. 5 the entering beam is drawn at an oblique angle to illustrate the principle of the filter. Interference filters are designed to be used with the incidence beam normal to the surface of the filter. If radiant energy enters the filter at some angle other than normal, the spectral band passed by the filter shifts toward shorter wavelength.

Frequencies harmonically related to the fundamental frequency of the interference filter also will be reinforced. A filter of maximum transmittance at 600 nm also will transmit at 300 and 200 nm, corresponding to a path length for CD + DE of 2λ and 3λ, respectively. If these harmonically related frequencies are a source of difficulty, they can be easily removed by combining the interference filter with a color filter with peak transmittance at the same wavelength as the fundamental of the interference filter. Such a combination will provide the better spectral isolation of the interference filter with removal of the harmonically related transmission peaks. Interference filters are available from several sources and can be found with most any peak transmission wavelength desired, especially those corresponding to wavelengths such as the mercury, helium, or sodium lines.

C. The Critical Angle Phenomenon

The most widely used refractometers depend on the critical angle phenomenon. The sample is put in contact with a medium of known higher index in the form of a glass block or refracting prism that constitutes the most critical part of the instrument. Consider Fig. 6 where a sample S is

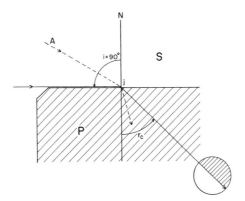

Fig. 6 Representation of the critical boundary inside a prism.

placed on the surface of a prism P having a greater index of refraction than S. The sample has refractive index $n = n_P \sin r / \sin i$ according to Eq. (2). The sample and prism are then illuminated with a monochromatic source whose rays are incident as shown. If some of the rays at the interface are incident at $i = 90°$, grazing incidence, the corresponding refracted rays, along r_c, can be precisely located because no rays are refracted into P at an angle greater than the angle r_c, which is called the critical angle. Rays that are incident on the surface at angles less than 90°, such as ray A in Fig. 6, will be refracted into P at angles less than r_c. Thus a sharp dividing line called the critical boundary between a light and a dark region is formed inside the prism as shown in Fig. 6. Angle r_c is sometimes called the critical angle of reflection because a ray traveling in the reverse direction of the arrow in Fig. 6 and striking point j at an angle slightly greater than r_c would not pass into the sample S but would be totally reflected.

In practice the conditions represented by Fig. 6 must be somewhat modified. The value of r_c is not obtained directly. The rays pass from the prism into air causing another change in direction and are then viewed with a telescope. Thus, as shown for the critical ray in Fig. 7, the observed apparent angle of refraction r_c' is smaller than r_c. The relationship between r_c' and r_c is given in the equation

$$\sin(90° - r_c') = (n_s/n_{air}) \sin(90° - r_c) = n_s \sin(90° - r_c) \tag{19}$$

Moreover, in actual refractometers, the whole surface of the prism instead of a point may be illuminated with light at or near grazing incidence. The telescope, if it consists of perfectly corrected lenses, collects all rays traveling parallel to each other, i.e., all rays with the same r, into one line in the focal plane. Thus the critical boundary shows up in the telescope of most instruments as a sharp, nearly straight line of demarcation between a dark field

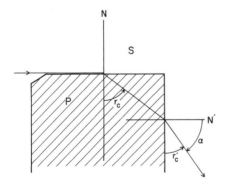

Fig. 7 Emergent critical rays. Note the angle measured, α or r_c', is not the actual critical angle.

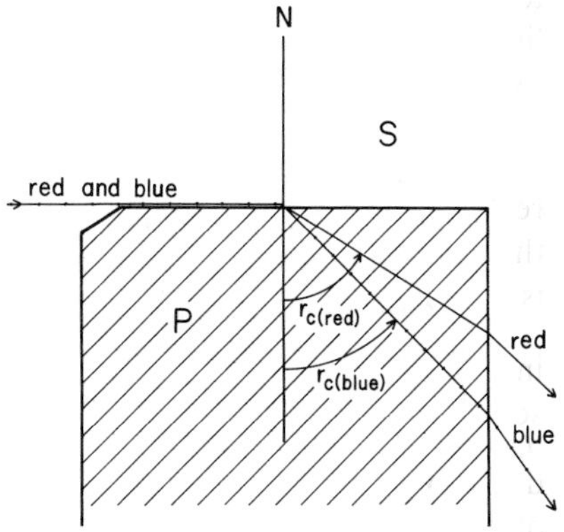

Fig. 8 Illustration of the difference in critical angles when measured with light of different wavelengths.

and a brightly illuminated field having the color of the monochromatic light used.

Most refractometers in use today use a light source. As discussed in Section I,C, the refractive index and thus the critical angle of refraction depends on the wavelength. As shown in Fig. 8, a red ray will show a larger critical angle than will a blue ray. The considerations needed for white light refractometers will be discussed later.

Table V summarizes the accuracy and range of several types of critical angle refractometers.

TABLE V

Accuracy and Range of Critical-Angle Refractometers

Refractometer	Maximum accuracy, N_D	Typical ranges of n
Pulfrich	$\pm 1 \times 10^{-4}$	1.33–1.61
		1.47–1.74
		1.64–1.86
Abbe	$\pm 1 \times 10^{-4}$	1.30–1.70
		1.45–1.84
Precision Abbe	$\pm 2 \times 10^{-5}$	1.20–1.50
		1.40–1.70
		1.33–1.64
Dipping	$\pm 4 \times 10^{-5}$	1.32–1.54[a]
Hand held	$\pm 1 \times 10^{-3}$	1.33–1.44
		1.39–1.50
Hand held (temp. comp.)	$\pm 1 \times 10^{-4}$	1.33–1.37

[a] Several prism changes are necessary to cover this range.

D. The Pulfrich Refractometer

1. Construction and Optical Principles

The Pulfrich refractometer dates from 1887 and very little has been done
to change the basic design since that time. The Pulfrich refractometer is
limited to the use of a monochromatic light source, but solids and liquids
over a wide range of indices can be investigated. Although values can be
determined to the fifth decimal place, absolute values obtained with a Pulfrich
refractometer are generally not reliable in the fifth place, mostly because of
the lack of precise and rugged construction techniques. The refractive index
is not read directly but must be deduced from a set of observations. Once the
instrument has been set up, a value of n can be obtained in about five minutes,
so the inability to directly read n is not a great limitation. Because the optical
design allows for the full illumination of the refracting prism and the shielding
of grazing rays, the critical boundary is particularly sharp in this type
refractometer.

The instrument measures the angle of emergence α ($\alpha = 90° - r_c'$, Fig. 7),
of the critical ray passing from a $90°$ prism P, of index n_P into air. From
$n_P = \sin \alpha / \sin(90° - r_c)$ and $n/n_P = \sin r_c / \sin 90° = \sin r_c / 1$,

$$n = (n_P^2 - \sin^2 \alpha)^{1/2} \tag{20}$$

The basics of the construction of a Pulfrich refractometer are shown in Fig. 9.
Light from source L is focused by a condenser C to provide rays that just
graze the horizontal surface of the beveled right-angle prism P. Because of
the confinement of the rays to near grazing incidence, sources with spectral
lines at well-spaced intervals can be used without a monochromator or
filter. In the case of a multiline source one obtains, depending on the dispersion of the sample and the prism, a separate sharp critical boundary for
each λ. The critical ray emerging from the prism P is reflected $90°$ by a prism

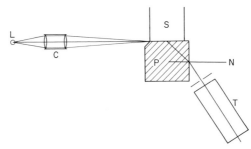

Fig. 9   Schematic view of a Pulfrich refractometer. L, light source; C, condenser; S, sample;
P, refracting prism; N, normal to prism surface; T, telescope.

so that the telescope T in Fig. 9 should actually be coming directly out from the paper. This is done so that the observer will always be able to make his observations in the plane facing the graduated circle.

The telescope is permanently fixed to a large rotatable circle which is graduated in half-degrees of arc. To make an observation the critical boundary is lined up with the cross hairs in the telescope eyepiece. For fine adjustment of the angle a micrometer scan is provided so the cross hairs can be accurately set at the critical boundary line. The final position is read from the graduated circle with the aid of a magnifying lens and a vernier.

The angle that is measured when the cross hairs are set to coincide with the critical boundary is defined as the apparent critical angle of emergence, θ. The true critical angle of emergence is

$$\alpha = \theta - \theta_z \tag{21}$$

where θ_z is the zero point correction. The zero point correction is the graduated circle reading when the telescope axis is perpendicular to the vertical prism surface (line N, Fig. 9). The deviation from zero, θ_z, of the graduated circle reading when the telescope is truly at $90°$ with respect to the prism may be as much as $0.5°$. θ_z must be checked from time to time. However, when a number of similar measurements are made one can usually assume θ_z to remain constant, unless there is a specific reason to suspect a change. Things that may cause a change in θ_z are substitution of one prism for another, if sample vessels have been attached to or detached from the prism, or if there has been a large temperature change. In general, θ_z should be remeasured whenever the instrument has been subjected to any mechanical strain. The changes in θ_z are usually irregular, and during normal use amount to several minutes of arc, or a change in the fourth decimal of n.

With intense spectral lines the apparent critical angle can be reproduced to within $\pm 1 \times 10^{-5}$ in n. For a measurement of the absolute value of n an accuracy of α of $\pm 1.0'$ is sufficient since inherent instrumental errors limit absolute accuracy to $\pm 1 \times 10^{-4}$. It is advisable to check the instrument by running a standard such as water when getting ready to make an unknown measurement.

Each Pulfrich refractometer is supplied with a set of tables for the interpolation of n_λ^{20} from values of α. Generally there are tables for several spectral lines, but it must be remembered that a particular table will only apply directly to one kind of prism and to one temperature, usually $20°$. For wavelengths, λ', not in the table, n_λ^{20} is obtained from α_{λ}' by means of Eq. (20). This requires a knowledge of the refractive index of the prism at λ' and $20°C$, $n_{\lambda'(P)}^{20}$. If this value is not known the best method for its determination is, perhaps, to measure the difference $\Delta\alpha$ between the emergent critical angles, α_λ and α_{λ}', for some standard substance, generally water, having accurately known values of n at the two wavelengths. The required value for the prism is then calculated by using α_λ' and $n_{\lambda'(H_2O)}^{20}$ in Eq. (20).

2. Conditions for a Sharp Boundary

When the telescope is in the proper position the field of view should be a sharp critical boundary with the upper half of the field dark and the lower half bright. When the source is a multiline discharge lamp some or all of the wavelengths should appear as bright colored bands, sharp on the upper edge. There generally is a shutter in a Pulfrich refractometer that can be used to cut off part of the light beam. It does so by blocking a fraction of the upper part of the beam so that as it is moved farther and farther into the light beam the colored bands become narrower, the whole field of view becomes less intense, and the contrast of the critical boundary is improved. There is never any doubt which is the critical boundary since its position will remain constant regardless of the shutter position whereas the noncritical edges of the band will move as the shutter is moved into the light beam. Thus, the position of the shutter is often a major factor in the determination of the sharpness of the critical boundary.

There are several other factors that may cause blurring of the critical boundary. The sample must be properly illuminated. The condensor should be adjusted so that a sharp image of the source can be observed on a piece of tissue paper just in front of the prism. To obtain maximum intensity, the entire surface of the prism should be illuminated with slightly converging rays. Consequently, the image of the source should be at least as wide as the prism and should be approximately bisected by the horizontal surface of the prism. These conditions may not be obtained with a simple adjustment of the condenser, a new relative position for the source and the condenser may be needed.

The position of the telescope relative to the prism block may also need adjustment, especially if very precise work is to be done. The central ray of the illuminating beam should pass directly through the center of the telescope objective. Proper illumination of the telescope objective is achieved when its aperture is completely or at least symmetrically filled with light.

Thermal drifts during the measurement can lead to a fuzzy boundary. This is due to poor temperature control or not letting the instrument come to thermal equilibrium. Temperature control will be discussed in the next section.

Dirt on the refractometer prism or vessel, or turbidity in the sample cause scattering of the incident light. The intensity of the light bands is decreased and the critical boundary contrast can be diminished due to stray light. Obstructions that cause scattering may also prevent rays at grazing incidence from entering the prism. The shielding effect can cause appreciable errors.

Overlapping bands corresponding to neighboring wavelengths from the source can mask the critical boundary. This problem can generally be solved by use of a suitable filter in front of the source. Finally, the source intensity may be too low. Lines from a source that are inherently weak will be difficult to use.

3. Temperature Control

The importance of temperature control on refractive index measurements has been previously pointed out. The degree to which temperature control is needed is dependent upon the desired accuracy of the measurement. For absolute measurement of n_D of liquids to $\pm 1 \times 10^{-4}$ under conditions in which the room temperature is fairly constant, the instrument may be used without special thermal control if the temperature of the sample is measured directly to $\pm 0.2°C$ at the time of measurement. If the room temperature deviates from 20°C, or whatever standard temperature is desired, and information concerning dn/dt is available, the room temperature measurement can be interpolated to the standard temperature. Generally it is just as easy to provide temperature control of $\pm 0.2°C$ via a circulating water system.

The Pulfrich refractometer is capable of determining Δn to $\pm 1 \times 10^{-5}$. To achieve measurements with this precision it is necessary to control the temperature fluctuations of liquid samples to $\pm 0.02°C$. This requirement is generally taken care of by circulation of thermostated water through the prism jacket and through a hollow cylinder which is lowered into the open refractometer vessel until it touches the sample. The cylinder also contains a 0.1°C thermometer. If the working temperature deviates considerably ($> 5°C$) from the room temperature, gradients can be set up in the prism because of its large exposed vertical surface. The best way to eliminate this problem is to enclose the entire instrument in a thermostated box. This has the additional advantage of defining the temperature of the reference medium, air.

Whenever measurements are made considerably above or below the standard temperature it is necessary to take into account the effect of the change in index of the prism block, dn_P/dt. Table VI gives the temperature coefficients for various types of refractometer blocks.

Tilton (1943c) gives a method for the correction of the prism index to temperatures other than standard. The correction is given by

$$k = (n_P/n')(dn_P/dt)(t - 20°) \tag{22}$$

where t is in degrees centigrade. For example, using a prism for which $n_{D_3}^{20} = 1.62$ to measure the refractive index of p-nitrotoluene at 55°C, one finds a

TABLE VI

Temperature Coefficients[a] of Relative (ΔN) and Absolute ($\Delta \bar{N}$) Index of Refraction (Units of Fifth Decimal) for Refractometer Blocks and Similar Glasses at Room Temperatures

Highest index measurable	Block N_D	v[b]	$\dfrac{t+t_0}{2}$	$\lambda = 656.3$ nm ΔN	$\lambda = 656.3$ nm $\Delta \bar{N}$	$\lambda = 589.3$ nm ΔN	$\lambda = 589.3$ nm $\Delta \bar{N}$	$\lambda = 486.1$ nm ΔN	$\lambda = 486.1$ nm $\Delta \bar{N}$	$\lambda = 434.0$ nm ΔN	$\lambda = 434.0$ nm $\Delta \bar{N}$	$\lambda = 404.7$ nm ΔN	$\lambda = 404.7$ nm $\Delta \bar{N}$
	1.5045	64.7	30°	0.29	0.16	0.32	0.19	0.38	0.25	0.44	0.31		
	1.5175	64.2	35°	0.16	0.03	0.17	0.04	0.22	0.09	0.25	0.12		
1.50	1.5220	58.5	28°	0.17	0.04	0.19	0.06	0.24	0.11	0.27	0.14	0.28	0.15
	1.5202	59.6	30°	0.34	0.21	0.35	0.22	0.40	0.27	0.44	0.31	0.30	0.17
(Ib) 1.613	1.6220 R,Z[c]	35.9	55°	—	0.24	—	0.28	0.—	0.39	—	0.50	0.48	0.35
	1.6227	55.6	33°	0.33	0.19	0.34	0.20	0.38	0.24	0.44	0.30		
	1.6561	33.2	35°	0.44	0.30	0.46	0.32	0.60	0.46	0.75	0.61	0.48	0.34
1.70	1.7167	29.4	28°	0.66	0.51	0.74	0.59	0.94	0.79	1.14	0.99	0.92	0.78
(IIb) 1.713	1.7474 R,Z	27.8	57°	—	0.70	—	0.78	—	1.05	—	1.31	1.33	1.17
(IIc) 1.746	1.7548	27.6	30°	0.77	0.62	0.85	0.70	1.12	0.96	1.36	1.20		
	1.7537	27.6	30°	0.70	0.55	0.81	0.66	1.09	0.93	1.34	1.18		
1.75	1.7619	27.1	26°	0.77	0.62	0.86	0.71	1.17	1.01	1.43	1.27		
(IIIb) 1.899	1.9068 R,Z	21.7	?	—	1.03	—	1.21	—	1.71	—	2.26		
(IIIc) 1.910	1.9180	21.0	30°	1.12	0.95	1.30	1.13	1.83	1.66	2.44	2.27		

[a] With the exceptions noted, the coefficients were determined with the apparatus described in Tilton (1936).

[b] Abbe's value, $v = (N_D - 1)/(N_F - N_C)$, is the reciprocal of the relative mean dispersion.

[c] R,Z These coefficients of $\Delta \bar{N}$ as listed by Roth and Eisenlor (1911), are almost identical with those recommended by Carl Zeiss for their Pulfrich refractometer. For block I these values are those found by Pulfrich (1892), for Jena glass O.544 over the range 11°–99°C; similarly for block II, they list Pulfrich's values for glass O.165 over the range 14°–100°C. Their values for block III agree with interpolations between those found by Reed (1898), for glasses O.163 and S.57.

value, $n'_{D_3} = 1.5383$. The value 1.5383 is too low because $n^{55}_{D_{3(P)}}$ is greater than $n^{20}_{D_{3(P)}}$, that is $n^{55}_{D_{33(P)}} = n^{20}_{D_{3(P)}} + k$. Knowing that $dn_P/dt = 0.28 \times 10^{-5}$ for this prism for the D_3 line, and substituting into Eq. (22), we have $k = +(1.62/1.54)(0.28 \times 10^{-5})(55 - 20) = +1.0 \times 10^{-4}$. So $n^{55}_{D_3} = 1.5383 + 0.00010 = 1.5384$ relative to air at 55°C. The manufacturer generally supplies values of the coefficient $dn_{\lambda(P)}/dt$. It must be remembered that the temperature coefficients of a prism depend strongly on the wavelength as well as the type of glass. Then the temperature correction is especially important for dispersion measurements. The change in dn_P/dt is generally about 0.1% per degree, therefore, it is permissible to assume dn_P/dt is constant in Eq. (22) provided $t - 20°C$ is less than about 50°C. If the correction covers a larger temperature span than this it is good practice to measure the effect of dn_P/dt by making measurements at various temperatures using a standard substance having an accurately known temperature coefficient of dispersion.

4. *Liquid Samples*

In general the Pulfrich refractometer is not designed to work with very small amounts of liquid. For best results a layer of liquid about 0.5-mm thick ($\simeq 0.2$ ml) should cover the entire horizontal prism surface. If a smaller area of the prism surface is covered there will be a corresponding decrease in the intensity of the critical boundary. When less than the entire prism surface is covered the symmetry of illumination becomes more critical. For routine work about 4 ml of sample is generally used. About half of this amount is needed to rinse the sample vessel until a constant value of n is obtained.

The standard type sample cell consists of a simple cylindrical tube, ground at the base to fit the prism surface. This container serves to confine the liquid sample to the horizontal prism surface and provides a vertical window for the incident light rays. When this arrangement is used, the vessel may be filled by a pipette, but the liquid should be removed with a piece of lens paper to avoid scratching the prism surface. A modified sample vessel has been used by Geffchen and Kruis (1928) that allows rapid filling and draining while excluding the atmosphere. The vessel has facilities for the flow of thermostated water through the cell so that temperature control can be maintained.

Another type of liquid cell that is sometimes employed is a double chamber cell designed for the comparison of two liquids. These cells have a black glass partition perpendicular to the prism faces and mounted so that it bisects the polished horizontal surface. The vessel is generally cemented to the prism to prevent loss by evaporation and to increase mechanical strength. Care should be taken in the selection of a cement. It should not be soluble in the solvents to be studied since this could lead to substantial error in n. To test for this, the liquid under investigation should be placed in the cell and the instrument thermostated. If the cement dissolves in the liquid, a change in θ over a period of a few hours should be noted. On the other hand, the cement must be dissolved by some solvent. The sample cell used should never be removed by force since this can result in a change in the zero setting of the instrument or, worse, the prism surface might be damaged.

5. *Solid Samples*

The material to be studied must be homogeneous and have two mutually perpendicular faces, one of which must be plane and clear (or capable of being polished). The other face merely acts as a window for the incident light so that it does not have to be so regular. The two surfaces should intersect in a sharp line so that when viewed from the light source a good plane of contact is observed. It is imperative that the source be placed so that, optically, it is in an extension of the plane of contact between sample and refractometer block without intervention of any object that can shield the grazing incident rays. False edges can be readily detected by their variable behavior during relative movements of the instrument and the source. This is one of the most useful and most frequently necessary tests in accurate refractometry.

The clear plane face of the sample is brought into contact with the horizontal refractometer prism face by using a very small drop of an appropriate liquid on the clean prism surface and carefully pressing the solid into place so that a very thin, bubble-free film separates the two surfaces.

The glass refracting block must be cemented in its metal housing such that no portion of the metal or cement rises above the plane of the polished glass surface. Otherwise an unfavorable condition exists for the use of solids. Even though the solid sample is small, and parallelism with the block is obtained through the layer of contact liquid, the metal or cement may shield the sample in such a manner that truly grazing incidence does not exist in the sample. Whenever two polished surfaces of a sample intersect at an angle less than 90°, and the surfaces of the glass block and its metal housing are approximately coplanar, then it is impossible to obtain truly grazing incidence unless the sample can extend approximately as far toward the source as does the housing. The breadth of the source is usually sufficient to take care of sample-free intersection angles somewhat greater than 90°, but a beveled edge of the intersecting polished faces of the sample is almost necessarily fatal. Also, if a slight excess of contact liquid is used, it may emerge toward the source and by refraction or reflection prevent the introduction of light at truly grazing incidence.

The use of a finely ground edge toward the source is safer than a polished surface in these and in related cases, but there is sometimes a serious loss of light which may impair the precision of the index setting if the light is of low intensity. In all cases false edges ascribable to geometrical shielding give rise to lower than normal index readings.

Tilton (1942b, 1943d) has pointed out that if the film between the two surfaces forms a wedge, it may cause an appreciable error by altering the effective angle of the refractometer prism. An error as great as 2×10^{-4} can arise from this source. For accurate refractometry it is certainly imperative to test and adjust the liquid contact layer to parallelism by the use of interference fringes. The interference fringes arise when monochromatic light is passed through or reflected from the two broad surfaces of the wedge. The fringes consist of equally spaced light and dark bands parallel to the edge of the wedge. The safe procedure is to view the fringes in the exit pupil of the telescope with the aid of an auxiliary eyepiece or a simple hand magnifier while adjusting and readjusting the sample with gentle pressure, thus obtaining the parallelism precisely where it is required. A fringe parallel to the refracting edge of the block may appreciably affect refraction in the plane of measurement, but a few fringes normal to the refracting edge produce a negligible effect on measurements. If fringes are viewed through the sample, their permissible number increases as the difference in index between the sample and the contact liquid decreases, but the number seldom exceeds 1 or 2 per centimeter of contact for an error of 1×10^{-5} in the measured index of the sample. For viewing in the exit pupil the tolerance is always $\frac{1}{3}$ fringe per centimeter for yellow light. Thus, for accurate dispersion measurements it would be necessary to adjust the wedge until not more than one fringe could be seen in the exit pupil. For measurements of n_D to the fourth place as many as six or eight fringes can be tolerated.

The choice of a suitable contact liquid for given samples can also be a factor in determining the required tolerance of parallelism. The liquid used should have an index greater than the sample, but only slightly so. On the other hand, the index difference between sample and contact liquid must not be too small lest internal reflections be seriously changed at some interfaces, and contrast decreased either for the fringes or for the critical border line, or lest partial overlapping occur for the critically refracted rays corresponding to the sample and contact liquids. It is suggested that one should use a light oil of index, say $n = 1.45$, when measuring lithium fluoride, $n = 1.39$; that ethylene bromide, $n = 1.54$, or anise seed oil, $n = 1.55$, be used for low-index crown glass; and the usual monobromonaphthalene, $n = 1.66$, and methylene iodide, $n = 1.74$ be used for higher-index glasses as may be necessary.

6. Modified Pulfrich Refractometer

The last commercially available Pulfrich refractometer was a redesigned version manufactured by Bellingham and Stanley, Ltd. London. Figure 10 shows the key features of their design. The graduated circle used is made of

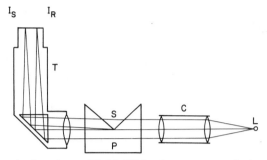

Fig. 10 Schematic view of a modified Pulfrich refractometer employing a differential cell. L, light source; C, collimator; P, differential cell; S, sample; T, telescope, I_S, sample image; I_R, reference image.

glass which minimizes corrosion of the fine scale marks and allows the use of transmitted light for observations. By means of a micrometer screw method the readings can be estimated to single seconds, which corresponds to better than 1×10^{-5} in n. A special differential cell is available that can be used in place of the standard prism. The optical arrangement in Fig. 10 is that for the differential cell. The light from the source is rendered parallel by the collimator where it enters the cell and the sample under study. In its passage through the cell, it is divided into two separate beams, one passing straight through the lower part of the cell and the other being deviated by the sample in the cell. The angular displacement of each beam is measured on the divided circle and the difference is used to determine the refractive index of the sample. This method permits the measurement of refractive index over the wide range of 1.26–1.95.

E. The Abbe Refractometer

1. Construction and Optical Principles

The most commonly used refractometers today are based on the original design of Abbe. Like the Pulfrich, the Abbe refractometer measures the critical angle of refraction (or reflection) relative to a glass prism. The usual type of Abbe refractometer requires a very small sample, 0.05 ml, of liquid and is designed for maximum speed, convenience, and simplicity. Fairly high accuracy is obtainable as well, n_D can be measured to ± 0.0002. Most are designed to work in a specific range, i.e., $n = 1.30–1.71$ or $n = 1.45–1.84$, but some have interchangeable prisms so that a much greater range of measurements is possible. Solids, and with some models even opaque solids, are generally easily handled. The convenience of the Abbe type

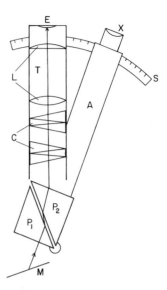

Fig. 11 Schematic view of an Abbe refractometer. S, sector scale; A, alidade; P_1, auxiliary prism; P_2, refracting prism; C, Amici prisms; T, telescope; L, telescope lenses; E, eyepiece; X, magnifier; M, illumination mirror.

refractometers is somewhat affected by limitations when some applications are considered. It is not well suited to the accurate measurement of solutions with a volatile component or of powders. With the exception of the precision Abbe refractometer (Section II,E,7), the usual instrument is not suited to measurement of n for lines other than the sodium D line.

Figure 11 illustrates the mechanical arrangement and path of the critical ray in an Abbe refractometer. As can be seen the light enters the illuminating prism P, from below, being reflected off the mirror M. One to three drops of the liquid sample are placed between prisms P_1 and P_2, which are hinged together at their lower edges for access to the sample. Generally the prisms are partially hollowed to facilitate the circulation of thermostated water. The upper prism P_2 is mounted rigidly on a bearing which has an arm A extending at right angles to the upper face of P_2. The rotatable arm carries a marker and magnifier so that accurate readings can be made on the fixed scale S. The telescope T can be permanently mounted or, in some models, firmly mounted on a bearing of its own so that it can be rotated about the same axis as the movable arm A. The scale is always firmly attached to the telescope so that in the models with the movable telescope it can be positioned in any convenient location. Some instruments are provided with a rack and pinion mounted on S for moving the prism, while in others it is rotated by hand except in the final stages of adjustment, when a slow motion tangent screw is used. The reading of the scale is a measure of the angle r between the normal to the prism surface and the telescope axis. When a properly illuminated sample is placed between the prisms and the telescope cross hairs are on the critical boundary, the resulting angle, r_c, is the emergent critical angle characteristic of the sample, the prism, and the reference medium, air. The critical angle r_c is related to n of the sample by the equation:

$$n = \sin r_c \cos \beta + \sin \beta (n_P^2 - \sin^2 r_c)^{1/2} \qquad (23)$$

where β is the prism angle and n_P its index of refraction. Actually, the scale is calibrated in terms of n_D^{20} or, in some cases, the concentration of an aqueous sugar solution, etc. The standard instruments are calibrated to the nearest 0.001 unit (n_D^{20}), and the magnifier allows the next decimal place to be estimated to ± 0.0001.

Almost all Abbe refractometers are designed to be used with white light. Ordinarily, with white light, the critical boundary is diffuse and has the appearance of a rainbow because of the divergence of critical rays having different wavelengths (dispersion effects). If this is the case it is impossible to measure n accurately. The width of the color band depends on the relative dispersion of sample and prism. The critical boundary can be sharpened and its color removed by means of a compensator C in Fig. 11.

A compensator is a combination of prisms, generally called an Amici prism. A more detailed view of an Amici compensator is given in Fig. 12. The different varieties of glass chosen for an Amici compensator disperse all wavelengths except the yellow sodium D line, which is allowed to pass. When two identical Amici prisms are placed in the path of the critical rays,

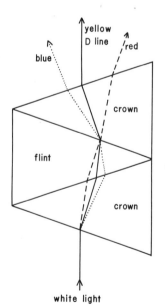

Fig. 12 The dispersion of white light with an Amici compensator.

a dispersion of these rays results which is equal and opposite to that caused by the sample and prism. Then, by collecting the divergent critical rays leaving the prism so that all wavelengths are superimposed on one line in the focal plane of the telescope, the Amici prisms compensate for dispersion of the rays leaving the prism and the resultant critical boundary is practically white. Rays from the sodium D line are not deviated but the dispersion introduced for other wavelengths is proportional to the angle through which one of the Amici prisms is rotated relative to the other. The correct amount of compensating dispersion is obtained by rotating one of the Amici prisms until the boundary is sharp and colorless. There are two positions of the compensator, 180° apart, which give a good boundary. If the sodium D line is used for illumination the compensator is not needed. When a compensator is present in an instrument, monochromatic sources other than the sodium D are not suitable unless an elaborate calibration is made.

The resolution of an Amici compensator is not perfect and its performance should be checked. A matter that can be easily tested is the possible deviation of sodium light by the compensator as shown by Tilton (1943b). This is so important and some of its effects are so well concealed that it should receive explicit attention from every careful user of white light instruments with rotatable compensators. First, the observed index almost always differs for the two possible achromatizing positions. This is caused by lateral deviation of sodium light by the Amici prisms and is very difficult to avoid in the construction of the prisms, but it easily eliminated by using only the averages of readings taken for opposite achromatizations.

Deviation of sodium light in the principal plane of the Amici prisms can be present and may be serious. The effects cannot be eliminated experimentally, and a double-entry table of corrections would be necessary to correct for such a defect. One should observe with sodium light the index readings for a sample while the compensator is rotated. The "high" and "low" readings for a given sample should not differ by more than 2×10^{-4} or 3×10^{-4}.

2. Conditions for Proper Performance

When an instrument is new, is newly set up, or has not been used in a long period it is recommended that the instrument be thoroughly checked before any important measurements are made. The compensator should be checked as described above, and the scale calibration should be checked. A description of simple tests for various common defects of Abbe refractometers is given by Tilton (1942a, 1943a). The scale readings are tested by measurements with a set of standards, such as those mentioned in Section I,E,2. The manufacturer generally supplies a set of glass plates to serve as a convenient set of standards, but it is generally preferable to use these plates as standards only when calibrating the instrument for measurement of solids. The value of n_D obtained for a liquid is usually too low if the instrument is adjusted to give the correct value for a solid of the same n. Discrepancies of as much as 3×10^{-4} for liquids of low n have been reported by Tilton (1942a) and Mair (1932). Discrepancies are attributed to shielding of rays at grazing incidence by the auxiliary prism and/or the presence of Hershel fringes.

Any discrepancy between the known n of the standard and the value n' determined from the instrument being tested can be corrected for in two ways: (i) The standard is placed on the prism and the movable arm is put in the position where the scale reading corresponds to the correct value of n. The telescope objective is adjusted so that the critical boundary is brought into exact coincidence with the cross hair intersection. (ii) The observed difference $\delta n = n - n'$, is applied as a numerical correction to all readings, i.e., $n = n' + \delta n$. The second method is preferred because the error will be different in each region of the scale. Thus, a calibration can be performed in each region and an appropriate δn determined for that region.

3. Temperature Control

Temperature control is required in refractometry not only because of the effect of temperature on the sample, but also because of the effect on the instrument itself. The effects on an Abbe refractometer are not unlike the temperature effects on a Pulfrich refractometer discussed in Section II,D. Both prisms in an Abbe refractometer are mounted in a jacket through which thermostated water can be circulated. A thermometer should be placed in the flowing water leaving the jacket. For most liquids the temperature need be controlled to $\pm 0.2°C$ to realize the full accuracy of the instrument. The thermostating should begin about 30 min. prior to the time the instrument is to be used to ensure temperature equilibrium in the prisms. The first measurements should be checked after a few minutes. If the instrument is used when there are still temperature gradients, the critical boundary will appear blurred. Once equilibrium is reached in the instrument it only takes a few minutes for a new sample to become equilibrated because of the small sample size used with an Abbe refractometer.

If the working temperature is substantially different than the ambient temperature, it is found that it is desirable to have the entire instrument thermostated. If, on the other hand, high temperatures are being used and the instrument has a 2-Amici prism compensator for which one does not have reliably established corrections, then it probably is better to insulate the compensator as completely as possible from the jacketed prism system. Modifications of standard instruments for work at low temperatures (to $-120°C$) have been described by Grosse (1937) and Wiley *et al.* (1950). An instrument for use at high temperatures (100°C) has been described by Black *et al.* (1954).

Generally, an instrument is designed to work at 25°C. If the temperature of the measurement is in the range of $\pm 5°C$ from the standard temperature the instrumental correction is very small. The error, Δn, in determining the relative index of any sample on the usual Abbe-type refractometer at temperature t (not far from t_0) with respect to air at the same temperature is fairly well represented as

$$\Delta n = 0.87(t - t_0)\Delta N \qquad (24)$$

where t_0 is the temperature for which the instrument was calibrated, and ΔN is the temperature coefficient of relative index of the refractometer block (see Table IV).

4. *Liquid Samples*

The auxiliary prism, P_1 in Fig. 11, serves to illuminate and confine the sample. When a drop of liquid is placed on the ground surface and the two prisms are locked together with the clamp, a thin film (0.1 mm) of liquid is held against the upper prism, P_2. When the ground face of P_2 is illuminated by tilting the mirror, some of the light scattered at its mat surface passes into the liquid at grazing incidence and produces a critical boundary. The same light source should be used in an identical manner for calibration and measurements.

There are certain precautions that should be taken whenever refractive index measurements are made. The prism surfaces must be clean. These surfaces scratch easily so great care should always be taken whenever an operation involves exposing the surfaces. They can be cleaned with alcohol and soft lens paper. Following this the surfaces should be rinsed with the solution to be measured. The solution to be measured should be clear or nearly so, otherwise the critical boundary becomes blurred. The handling of colored liquids will be discussed below. After the sample has been placed between the prisms and the prisms have been clamped together one should check to see that the liquid has spread evenly between the prisms and that no bubbles are present in the liquid film. Even more care must be taken with hydroscopic liquids or liquids with volatile components. The two prisms should only be opened to the point necessary for filling and then closed as soon as possible after the sample has been deposited. One of the major problems in working with an Abbe refractometer is the lack of shielding of the sample from the surroundings. Since very small samples are used, gain of water or the partial loss of a component can affect the measurement a great deal. Time should be allowed for the establishment of temperature equilibrium in the sample and the mirror should be adjusted for maximum illumination. A change in the position of the mirror will not change the position of the boundary, if it is the true critical boundary. The movable arm is positioned so that the critical boundary is in the center of the telescope field and the compensator should be adjusted to achromatize the boundary. Readings should be repeated to check reproducibility, which should be $\pm 1 \times 1^{-4}$. The compensator should be moved to the other achromatizing position and the measurement repeated. These measurements will seldom agree, but Tilton (1943a) states that the mean value is free from the instrumental error causing the difference.

There have been some attempts, including that by Altmann (1948), to construct a liquid sample cell that shields the sample better and allows faster sampling by means of a streaming-through arrangement. Also, Young and Rule (1954) have published the design of a plate-like cell constructed from a microscope slide with a cavity for the liquid sample that is especially useful for highly corrosive liquids. This cell mounts on the Abbe prism much like the standard glass plates.

5. Solid Samples

The illuminating prism, P_1 in Fig. 11, is lowered as far as possible so that access to prism P_2 is readily available. The solid sample is then mounted on the lower surface of P_2 in the same manner as described in Section II,D,5 for the mounting of a solid sample on a Pulfrich refractometer. Care should be taken to choose the proper liquid and ensure the surfaces of the prism and the sample are parallel. The illuminating prism is then used to reflect rays into the sample "window" at grazing incidence. The same procedure and precautions are then followed as when measuring a liquid sample.

6. Opaque Samples

One real advantage of some Abbe refractometer models is their ability to work with opaque or colored liquids and solids. The models capable of operation with these samples provide a refracting prism, P_2 in Fig. 11, that has a window for illuminating the lower surface from within. By this means, light can be reflected from the prism surface into the telescope when the surface is covered with an opaque material. Under proper conditions a sharp critical boundary is observed that corresponds to the critical angle of reflection, which is equal to the critical angle of refraction. Generally, the intensity of the field is less than observed with a clear sample illuminated at grazing incidence, but otherwise the observations and measurements are the same.

7. The Precision Abbe Refractometer

The need in industry for convenient refractometers having fifth decimal accuracy has stimulated attempts to improve the Abbe. Bausch and Lomb market a "precision refractometer" that is a modified Abbe designed by by Straat and Forrest (1939) for an absolute accuracy of up to $\pm 2 \times 10^{-5}$, but probably capable of only $\pm 3 \times 10^{-5}$ in the reproducibility of measurements with pure liquids or nonvolatile solutions. An outstanding feature of this model, due to the lack of a compensator, is that it can be easily used to measure n for a variety of wavelengths (dispersion). Three separate interchangeable prism assemblies are available that cover ranges of n of 1.20–1.50, 1.40–1.70, and 1.33–1.64. (A high range prism assembly, 1.55–1.84, is available on special order.)

The improved accuracy is obtained by dispensing with the compensating prism, using a monochromatic source, and using unusually large and precise Abbe prisms. The light source supplied with the instrument is either a sodium arc (589.3 nm) or a mercury lamp (546.1 nm or 435.8 nm), but the tables supplied with the instrument for conversion of scale reading to refractive index also have computations for hydrogen (656.3 nm or 486.1 nm) and helium (667.8 nm or 501.5 nm). The technique of measurement is the same as that described for the standard Abbe with the exceptions that the achromatization step is eliminated and the value of n is determined by converting an arbitrary scale reading to n_λ via a table. With a good critical boundary, a setting of the cross hairs may be reproduced within 0.005 scale divisions. This corresponds to 1×10^{-5} in n at the upper end of the scale and 3×10^{-5} in n at the lower end. For such high precision the temperature of a liquid sample has to be controlled to within $\pm 0.02°C$.

The theoretical accuracy and precision obtainable should not lead one to believe these numbers are easily achievable. There are certain difficulties with the Bausch and Lomb precision refractometer that come to light when a careful analysis is performed. These difficulties can lead to errors and imprecision. In very carefully performed tests of reproducibility and accuracy for sugar solutions, Charles and Meads (1955) found the standard deviation in the error of observation of n to be 6×10^{-5}; while the absolute accuracy varied in the fourth place. Some of the features of the instrument which, if not fully appreciated, can contribute to the observed errors are:

(1)   The most serious source of error and one fundamental to the design of Abbe refractometers is the cutting off of near-grazing rays: those rays incident on the refracting prism at very small angles. The light must enter the liquid at a finite distance from the surface of the refracting prism, thus the angle of incidence at the prism surface is not as large as the ideal case. The rays close to grazing incidence are low in intensity because they originate

from a small area near the edge of the prism. Because of the large proportion of noncritical rays entering the sample the critical boundary becomes somewhat diffuse. As a larger and larger area of the prism is illuminated, the intensity of rays further from the critical angle increases and for a fully illuminated prism the apparent critical boundary is actually a rather broad, shaded region with the position of maximum intensity change at considerably less than the critical angle. When most of the prism is illuminated the width of the shadow corresponds to approximately 1×10^{-4} in n. To minimize this effect a shutter should be placed in front of the source so as to cut out most of the rays far from grazing incidence. This does not enhance the intensity of the critical rays but does increase the ratio of critical to noncritical rays. The edge of the lampshield can be used as a shutter, or a slit can be placed over the illuminating prism. When properly illuminated, the maximum uncertainty in the critical boundary position corresponds to about 1×10^{-5} in n. Forrest (1956) has analyzed the dependence of the shielding error on the length and thickness of the liquid layer. In general this error is not as severe when measuring solid samples because the separation between the sample and refracting prism is very small.

(2) The prism orientation can be shifted by softening of the prism cement by solvent action of the sample or by a lack of stability in the mechanical systems. These defects are mostly present only in older refractometers.

(3) Proper sample preparation and handling will always be a critical step. Evaporation or absorption of water during measurement can cause substantial errors.

(4) The Herschel fringes, discussed by Tilton (1943d), are more prominent in the precision refractometer than in ordinary Abbe refractometers because the face of the illuminating prism of the precision model was polished on many of these instruments. If the observer is aware of these bands and is careful to discriminate against them they should not interfere with the critical boundary determination.

8. *The Dipping Refractometer*

The dipping or immersion refractometer is essentially an Abbe type of short index range without an illuminating prism and without water jackets for temperature control. Figure 13 shows the main components of a dipping refractometer. The refracting prism is dipped or immersed to a depth of 3 or 4 cm in the sample to be measured and the grazing light must be introduced from below, often by means of a mirror. The index and angle of the block or refracting prism are so chosen by the designer, with respect to the small index range to be measured, such that the instrument shall have

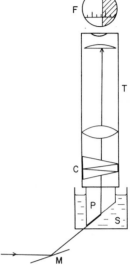

Fig. 13 Schematic diagram of a dipping refractometer. M, illumination mirror; S, sample; P, refracting prism; C, Amici compensator; T, telescope; F, field of view illustrating the critical boundary.

maximum openness of scale compatible with approximately normal emergence of the rays from the second face of the prism. The total range in emergent angle is small, and a small linear scale enclosed in the instrument at the focal plane of the objective is substituted for the sector and scale as used on the Abbe. A table of the refractive index equivalents of the scale readings is provided.

Originally, the Zeiss firm made only one dipping refractometer with a range of 1.325–1.367 (known later as the range for the A prism), and they standardized the model so well that much work during two or three decades was done and published in terms of scale divisions of this dipping refractometer without reference to refractive index. When other makers began marketing dipping refractometers, they did not always adhere exactly to the same table of index equivalents, and thus some care is necessary in order to avoid appreciable error when making intercomparisons of some of the published work.

The index range of dipping refractometers has been greatly extended by interchangeable refracting prisms, so that they can be used for measurements anywhere in the range 1.325–1.544. For each prism at least one standard is provided for adjusting the zero, and after every change of prism on one of these instruments it is necessary to recheck the zero. Otherwise some fine lint or other material may be enclosed between the surfaces that must be precisely seated for correct performance.

For checking these instruments precisely with solid standards it is advisable to mount the tube on a stand so that the polished surface of the dipping prism is nearly horizontal, and interference fringes should be used in the elimination of wedge-effect errors caused by the contact liquid. Care must be taken not to bend the tube. Possibly one should not allow gravity to flex it, if highest repeatability is desired.

The control of temperature for dipping refractometry is ordinarily obtained by placing the unknowns in small beakers and partially immersing the beakers in a thermostated water bath. The instrument is suspended from a wire above the bath. As an alternate procedure, one can sometimes employ the metal cups that are provided with the instrument, especially for use with solutions or mixtures that may evaporate rapidly or differentially. These cups are provided with ground joints and a window through which light may enter at the proper angle. They give good temperature control when immersed in a well-stirred bath, but one cannot be sure about leaks, and there is a cleaning nuisance, if not a problem. At temperatures near room temperature, the repeatability of readings with a dipping refractometer approximates $\pm 1 \times 10^{-5}$. It cannot be assumed, however, that results obtained with a dipping refractometer are accurate because they are precise. Accuracies on the order of $\pm 4 \times 10^{-5}$ in n, corresponding to ± 0.1 scale divisions, are the rule for better dipping refractometers. The dipping refractometer is often used primarily to determine the concentrations of solutions, e.g., the concentration of sugar in aqueous solution. Browne and Zerban (1941) give a good discussion of the use of the dipping refractometer as well as tables for the conversion of scale readings to refractive index. Moilliet (1955) has described the construction of a plastic jacket that permits a dipping refractometer to be used in measurements on a continuously flowing liquid.

9.   Hand Held Models

Perhaps the most popular refractometers on the market today for routine work are hand held Abbe-type refractometers. These refractometers are very similar to the dipping type with a movable cover over the refracting prism that takes the place of the illuminating prism. Like the dipping refractometers, they have a fixed scale, so there is a tradeoff between scale magnification and range. Most of these instruments are fairly narrow range, i.e., 1.33–1.44 n_D, and offer reproducibilities in n of $\pm 1 \times 10^{-3}$, although most are calibrated in specific gravity of urine, protein concentration, sugar concentration, etc.

Temperature control with these instruments can be a problem. The Bausch and Lomb hand held refractometer has a thermometer calibrated in correction units directly attached to the instrument. American Optical has a line of hand held refractometers that offer a new solution to temperature control. These refractometers are equipped with a liquid filled prism whose index changes with temperature, thus compensating for the effects of temperature. These instruments are capable of working accurately over fairly broad temperature ranges. Because significant temperature errors are eliminated, it was found possible to increase the magnifying power of the viewing system resulting in a model covering a range of 1.333 to 1.373 in n with a reproducibility of $\pm 1 \times 10^{-4}$. These instruments are particularly useful for process or quality control, or in field work, as they can be easily carried in one's pocket.

F. Other Critical Angle Methods

The principle of a device called the Wiedemann air plate refractometer is illustrated in Fig. 14. The key element to this technique is a sandwich of two plane-parallel plates separated by a layer of air. This sandwich is immersed in the liquid to be measured and then illuminated with a beam of collimated monochromatic light. The light will emerge from the cell and will be seen by the observer, through a telescope, only if the angle of incidence, i in Fig. 14, is less than or equal to the critical angle. Collimated monochromatic light is used so that the entire beam will be totally reflected almost simultaneously. By turning the plates from extinction on the left to extinction on the right, one measures twice the critical angle, C, for liquid to air, and the index is computed as $n = 1/\sin C$. If the air plate is replaced by a liquid of index lower than n, the instrument becomes a differential refractometer. This method seems worth consideration, especially if accurate index measurements are required at temperatures appreciably higher than those of the room.

Another critical angle refractometer called the Hallwachs double cell refractometer is illustrated in Fig. 15. The center wall separates two media of differing refractive index. The cell is illuminated with collimated, monochromatic light at grazing incidence to the plane-parallel partition. A critical boundary is observed at the exit that depends upon the difference in indices of the media on the two sides of the partition.

The Kohlrausch total reflectometer is somewhat similar in operation to the Wiedeman air plate refractometer. A plate to be measured, or a plate of known index, attached to a graduated circle is immersed in a fluid (standard or sample) of higher refractive index. The apparatus is contained in a cylindrical vessel with a plane-parallel plate affixed to it as a window. The cylinder is illuminated from one side and the plate is rotated until the critical boundary

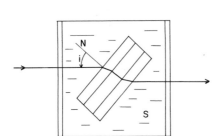

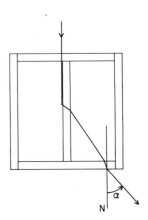

Fig. 14 Wiedemann air plate refractometer. N, normal to the air plates; i, angle of incidence; S, solution being measured.

Fig. 15 Hallwachs double cell refractometer. N, normal to the exit plate; α, critical angle.

observed through the window lies on the cross hairs of a telescope. The illumination is moved to the other side and the measurement is repeated. The angle through which the plate was turned is simply related to the ratio of the refractive indices of the plate and the liquid.

G. Methods Based on Interference Phenomena

1. Interference of Light

If two beams of coherent light are caused to intersect, the region of crossing may show a resultant amplitude and intensity which is very different from the sum contributed by the two beams acting separately. This modification of intensity obtained by the superposition of two or more beams of light is called interference. An interferometer is an arrangement for separating a coherent beam into two distinct paths and then reuniting them to form measurably observable interference fringes. In all cases the fringes are actually formed and located in the region of intersection of the two beams, and they can be seen on a screen except in the special case for which the beams are parallel and the fringes are formed at infinity.

A bright fringe will be formed at the plane of optical symmetry with respect to the two interfering light beams because both were in phase when they were separated and at this plane both have traveled their separate paths in the same time. This phenomenon is called constructive interference. To the right and left of this central bright fringe are regions where the paths for the two beams differ by one-half wavelength, $\lambda/2$, and where, in consequence, the beams differ in phase and dark fringes are formed. This is destructive interference. Furthermore, on each side bright and dark fringes are formed in succession. The distance between like fringes is directly proportional to the wavelength of the light used in the measurement.

With interferometers used for index measurements the optical path lengths for the two separate beams are nearly identical and white light can be used. The central bright fringe in the plane of symmetry will be colorless, but since red light has longer wavelengths than blue light, the lateral bright fringes will be tinged with red on their edges away from the central fringe. Even though white light may be used, the ready availability of intense monochromatic sources cause them to be preferred for interference methods.

When a substance is placed so that one beam of the interferometer must traverse it, and it has a different refractive index than the substance it replaced, perhaps air, it will cause the beam to take a different amount of time to get from the source to the point of intersection. If the substance has a higher index of refraction than the medium through which the second beam must pass, the first will be slowed relative to the second, and the central fringe of the system will be displaced toward the side of the higher index medium. By measuring the shift in position of the interference bands the difference in refractive index between the media in the two paths can be determined. The difference in refractive index, Δn, can be measured very accurately using interference techniques. The ordinary instruments that have been produced commercially have a maximum accuracy of about $\pm 5 \times 10^{-7}$ in Δn for liquids when white light is used. With monochromatic light an accuracy of $\pm 1 \times 10^{-7}$ and, with certain modifications in the instrument, $\pm 1 \times 10^{-8}$ can be obtained. The range which can be covered is inversely proportional to accuracy; it can be adjusted by placing the sample in cells of various lengths. Thus, when an accuracy of $\pm 5 \times 10^{-7}$ is required, the maximum difference which can be measured is about $\Delta n_{max} = 0.0006$, using a 4-cm cell. Increasing the range to $\Delta n_{max} = 0.05$ by using the shortest practical cell (1 mm) causes the error to increase to about $\pm 4 \times 10^{-5}$.

An important advantage of the interferometric technique is that temperature control is not as critical because it is a differential instrument. Also, the range is unlimited as long as reference liquids can be found within the limits $\Delta n = 0.05$. The interferometer is the most convenient, as well as the most accurate, instrument for measuring the apparent molar refraction of solutes in dilute solutions, or $\Delta n = n_{soln} - n_{solv}$. One has to be careful, however, when working in dilute solutions since the error due to impurities is often greater than the measurement error.

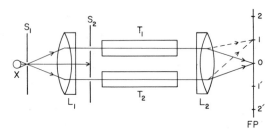

Fig. 16 Schematic diagram of a Rayleigh refractometer. X, light source; S_1 and S_2, slits; T_1 and T_2, sample and reference compartments; L_1, collimating lens; L_2, focusing lens; FP, focal plane.

2. Rayleigh's Refractometer

The principles of operation can be understood by examining the simplified model of Rayleigh's refractometer seen in Fig. 16. Monochromatic light from the source X passes through a slit S_1 and is then rendered parallel by lens L_1. Cells T_1 and T_2 contain the sample and the reference medium. Light beams defined by the slits S_2 pass through each of these cells and then are focused onto the focal plane of the interferometer, FP, by lens L_2. The image formed in the focal plane is a series of alternate light and dark bands, equally spaced and approximately equal in width. The solid lines that intersect at point 0 in the focal plane and the dashed lines that intersect at point 1 illustrate the different paths the light beams can travel. The phase of the two light beams at the intersection in the focal plane, which is determined by the optical path difference between the two beams, determines whether the image at that point is light or dark.

The condition for minimum (zero) light intensity at a point in the focal plane is

$$p = l_{\mathrm{II}} - l_{\mathrm{I}} = (N + \tfrac{1}{2})\lambda \tag{25}$$

and for maximum intensity

$$p = l_{\mathrm{II}} - l_{\mathrm{I}} = N\lambda \tag{26}$$

where p is the optical path different in centimeters, N is an integer called the "band order," and l_{I} and l_{II} refer to the optical path lengths transversed by the two rays. The optical path length is defined as $l = \sum n_i d_i$, where n_i and d_i refer to the refractive index and thickness, respectively, of the ith medium traversed by the light ray; l is thus equal for two paths for which the light requires the same time to traverse. The wavelength λ is in centimeters.

There is one point in the focal plane for which $N = 0$, i.e., $l_{\mathrm{II}} = l_{\mathrm{I}}$. This point is labeled 0 in Fig. 16. Accordingly, 0 is a position of maximum light intensity called the zero order band. It can be seen that the position of this band in the focal plane will be a function of the media in the two optical paths. Measurement of refractive index is accomplished by measuring the displacement of the zero-order band from the position where the two optical paths are identical. This is often done for gas measuring devices by viewing the focal plane with a telescope and counting the number of fringes that pass the cross hairs as the sample is slowly introduced to the cells. For liquids, the number of bands that go past the reference cannot be counted because the introduction of the sample is necessarily discontinuous. A number of methods have been developed for determining the band position. One of the simplest methods is to change the optical path in one of the beams until the zero-order band comes back to the reference position. This is conveniently accomplished by tilting a glass plate in the path of one beam until the effective thickness of the plate is great enough to compensate for the presence of the sample in the other beam. The amount by which the compensator plate is turned is a measure of the difference in index, $n_2 - n_1$.

3. The Rayleigh–Haber–Löwe Interferometer

The design of the Rayleigh–Haber–Löwe interferometer is very similar to that shown in Fig. 16, except that the fringe pattern is viewed with a telescope. Rayleigh and his co-workers used the instrument for measurement of gases, and did so by allowing the gas pressures to vary slowly as the fringes were counted. Haber and Löwe (1910) fitted the instrument with an optical compensator of the type discussed above, so that by merely turning a micrometer drum any sudden displacement of the fringes, within range, could be restored gradually and thus counted. White light was also adapted as the source except for very precise determinations to a fraction of a wavelength, when a monochromatic source is used. When a white light source is used only the zero-order fringe is white and well defined. This is because all wavelengths are in phase at this point. As soon as one moves away from the zero-order fringe, some wavelengths will constructively interfere, but others will destructively interfere due to the difference in wavelengths. This phenomenon tends to blur all bands other than the zero-order band, making it easy to find.

The instruments that were marketed commercially consisted of two models, a small model called a portable gas and water interferometer, and a larger laboratory apparatus. The laboratory interferometer takes cells of various lengths from 10 to 100 cm for use with gases, and cells from 1 to 80 mm long for liquids. The tradeoff between accuracy and range is given in Table VII. This type apparatus is widely used for comparisons of dilute solutions, for tests of purity, and for rates of progress of certain reactions.

Many details regarding the calibration and use of an interferometer for the analysis of a mixture of gases, and for other problems involving gases,

TABLE VII
Accuracy and Range for the Rayleigh–Haber–Löwe Interferometer

	Cell length (cm)	Error (n in units of 1×10^{-7})	Range (n in units of 1×10^{-5})
Gas cells	100	0.2	5
	50	0.4	10
	25	0.8	20
	10	2.0	50
Liquid cells	8	2.5	63
	4	5.0	125
	2	10.0	250
	1	20.0	500
	0.5	40.0	1000
	0.1	200.0	2000

are given by Berl and Andress (1921). Much of the work done with inter-ferometers has been done with aqueous solutions. Cohen and Bruins (1921) have discussed the additional precautions that are necessary for obtaining comparable accuracy in work with organic solutions.

4. The Jamin Interferometer

Jamin (1856) made an interferometer with two thick glass plates which he cut from one large plane-parallel plate to ensure approximate equality of index and thickness. The first plate serves as an amplitude or beam splitter; the second plate is a beam uniter. Separation of the two paths of the coherent rays is accomplished by the oblique (near 40°) incidence and the thickness of the plates. For perfectly parallel adjustment of perfect plates, the path difference would be zero and no localized fringes would be observed, the field being uniformly illuminated as part of one very broad fringe. Slight rotation of one of the plates produces localized fringes which then may be used for refractive index measurement when a path difference is inserted. If L is the effective length of a cell containing a medium of index n and a like cell containing a medium of index n_0, then one can write

$$N\lambda = (n - n_0)L \tag{27}$$

where N is the number of fringes that traverse the field.

In the usual method, a parallel beam of monochromatic light is directed by a collimator through the mirrors and is received by a microscope or a telescope in which the localized interference fringes are observed with respect to cross hairs. Several observers have noted more or less "drift" of the fringes, and Cuthbertson and Cuthbertson (1932) and Huxley and Lowery (1943) have attempted to eliminate this problem. A proper remedy, it would seem, is division of the objectives so that automatically fiducial fringes may be available, just as is done for the Rayleigh instrument.

The cost of the fine quality glass that is required in a Jamin-type interferometer probably discourages its commercial use or manufacture for sale. The advantage of this instrument over the Rayleigh or other conventional types is that, because of the principle of superposition, wider sources or slits are usable and thus monochromatic sources of lower intensity are employable.

5. The Zehnder–Mach Interferometer

If an interferometer is to be used to study differential effects on a given medium, especially heat effects, the proximity of the cells may be disadvantageous, or very inconvenient, or both. Michelson (1890) indicated by one of his many diagrams and Zehnder (1891) showed in detail how Jamin's interferometer could be modified so that the separate paths of the coherent beams would be much farther apart. Mach (1892) built such an instrument. Here, by means of the additional freedom in independent adjustment of the plates and the mirrors, it is possible to gain additional control of fringe widths and locations.

H. Image Displacement Methods

The most direct method for determining refractive index, and the simplest in principle, is to measure the angles of incidence and refraction or, equivalently, to observe at a fixed distance from the sample the displacement of rays which pass through the sample at a fixed angle of incidence. There have been many different arrangements designed for this purpose, some of which are relatively easily constructed in the laboratory. Perhaps the most basic of the image displacement methods is the spectrometer.

1. The Spectrometer

In this method a source capable of producing a discrete line spectrum is used to illuminate a slit situated at the focal point of a collimator lens. The essentially parallel light from the source is passed through the sample in the form of a prism (axis vertical) placed on a rotatable table. The light is received in a telescope mounted on a vertical axis. The telescope is provided with a suitable horizontal scale to measure the angle of deviation of the light caused by the prism. The general method by which measurements are made is to measure the angle of minimum deviation. This is accomplished by rotating the prism until the minimum angle of deviation is observed. A different orientation of the prism and a correspondingly different angle of deviation will be observed for each spectral line used.

The refractive index is calculated from the equation

$$n = \sin[\tfrac{1}{2}(A + D)]/\sin\tfrac{1}{2}A \qquad (28)$$

where A is the prism angle and D is the angle of minimum deviation. Grange et al. (1976) found they could measure the angle of minimum deviation more accurately by illuminating two faces of the prism and measuring the angle of minimum deviation for each. They then used an average as their result.

The spectrometer as a refractometer is a very flexible instrument, but is not, perhaps, well adapted to routine work. However, there are some cases in chemical research for which the spectrometer is indispensable. Since any magnitude of angular deviation can be measured, there is no upper or lower limit to the refractive indices which can be covered. Thus it is possible to investigate certain fluorine compounds (e.g., n-C_7F_{16}, $n_D^{30} = 1.2572$) which are below the range of standard refractometers. Also, observations on less intense spectral lines can be made, and this is important for complete studies of dispersion. It is also possible to carry out investigations in the ultraviolet or infrared regions of the spectrum if the spectrometer is equipped with appropriate optics.

The spectrometer has been mainly used for the measurement of solids for which the refractive index need be known to a high degree of accuracy (1×10^{-6}). Solid samples must be transparent, reasonably free from occlusions, bubbles, or opacities, and formed into prisms with optically plane and polished faces. Liquid samples are contained in hollow prisms with polished plane-parallel windows. These prisms are available from a number of optics manufacturers. Tilton and Taylor (1938) have secured probably the best available refractometric data on water in this way; they have described the apparatus, including the means of temperature control.

A modification of the basic spectrometer described above incorporates the collimator and observing telescope into a single unit. The opposite face of the prism is silvered, and at minimum deviation the light beam returns to the collimator-telescope from whence it came. Such instruments are called autocollimating, and Eq. (28) is applied for the calculation of the results. An autocollimating spectrometer of special design has been described by Forrest et al. (1956).

The hollow prism spectrometer design for the measurement of refractive indices of liquids is a convenient refractometer for measurement of indices at temperatures substantially above or below ambient. The hollow prism can be conveniently heated, e.g., by means of a surrounding jacket through which the vapors of a boiling liquid are circulated. Lauer and King (1956) and Black *et al.* (1954) describe the construction and use of a hollow prism spectrometer at high temperatures. Grange *et al.* (1976) discuss the use of their spectrometer at temperatures near the freezing point of aqueous NaCl solutions.

2. Differential Instruments

The image displacement principle has been used as the basis of a number of useful differential instruments. They have the common characteristic that their refracting surfaces are made up of the equivalent of two adjacent prisms, one containing the liquid sample (n), the other consisting of or containing a reference substance (n_0). Most of the instruments designed differ mostly in the construction of the sample cell (prism). Some of the different designs that have been published are those by Hill and Jones (1955), Hadow *et al.* (1949), Körösy (1954), McCormick (1953), Sieflow (1953), and Svensson (1953).

The key feature of differential instruments is the ability to make measurements accurate to $\pm 3 \times 10^{-6}$. Most of the conventional techniques, Pulfrich or Abbe, require temperature control to $\pm 0.02°C$ to make measurements with a precision of one unit in the fifth place. It is difficult to control the temperature this precisely and for measurements into the sixth decimal place even more difficult. Differential instruments are superior to conventional refractometers not only in accuracy but also in simplicity of temperature control. Ambient temperature need not be closely controlled since the temperature coefficient of the difference in refractive index between a solution and its solvent is much smaller than that for the refractive index of the solution or solvent alone. It is essential, however, that solution and solvent in the differential cell have the same temperature to $0.01°C$ or better.

(*a*) *The Debye Refractometer.* The differential refractometer designed by Debye (1946) has been used for a number of studies involving the estimation of large molecular weights from data on light scattering and refractive index of dilute solutions. A diagram of his apparatus is shown in Fig. 17.

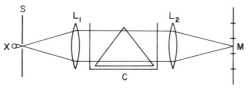

Fig. 17 Debye differential refractometer. X, light source; S, slit; L_1, collimating lens; L_2, focusing lens; C, cell containing the sample and prism; M, filar micrometer.

The apparatus is essentially a spectrometer with a prism containing the solution immersed in a cell C containing the solvent. The slit S is illuminated with monochromatic light and rendered parallel by lens L_1. Lens L_2 forms an image of the slit on a micrometer scale M. The position of the image is noted when both the prism and the surrounding cell contain the solvent. The image is displaced an amount proportional to $n - n_0$ when the solution is introduced to the prism, if the difference is not too large. For small image displacements, the displacement x is related to $n - n_0$ by the following equation:

$$x = 2f \tan(A/2)(n - n_0) \qquad (29)$$

where f is the focal length of L_2, and A is the refracting angle of the prism (generally about 125°). For temperature control the cells are placed in a jacket through which thermostated water can be circulated.

(b)   *The Brice–Phoenix Refractometer.*   Brice and Halwer (1951) published the design for a differential refractometer that has become very popular and is now available commercially. The Phoenix Precision Instrument Company markets the instrument and the G. N. Wood Manufacturing Company markets an attachment for their light scattering photometer that essentially makes it a refractometer of this design. A diagramatic sketch of the optical system of this differential refractometer is shown in Fig. 18. The light source is a mercury vapor lamp mounted in a housing that is equipped with filter turrent and filters to isolate the mercury lines. A semitransparent mirror is permanently mounted on the optical bench at an angle of 45° to the optical axis. This mirror permits the use of the standard mercury lamp or some accessory light source, such as a sodium lamp, maintained at 90° to the optical axis. The light beam then passes through an adjustable slit and enters the jacketed cell housing. The beam is deviated through an angle proportional to the refractive index difference between the two liquids and leaves the cell housing through an exit port. A projection lens images the slit onto the focal plane of the microscope objective fitted with a Filar micrometer eyepiece. This permits the displacement of the light beam to be determined within one micron. The iris diaphragm of the projection lens is useful to control the brightness of the slit image.

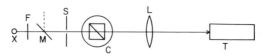

Fig. 18   The Brice–Phoenix Differential Refractometer. X, light source; F, filter, M, 45° semitransparent mirror; S, slit; C, jacketed cell; L, focusing lens; T, telescope.

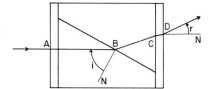

Fig. 19 The Brice differential cell indicating the optical path for a ray incident normal to the first surface of the cell.

A diagram of the differential cell used in the Brice–Phoenix refractometer, as well as the optical path of the incident beam, is shown in Fig. 19. The light beam is incident normal to the cell face at A, passes through the cell window and solvent, is incident at B with angle i on the thin partition, and undergoes deviations as shown on passing into the solution, the second glass window at C, and into the air at D, with a final angle of refraction r. The standard cell has a compartment capacity of 2 ml but only 1 ml is required in each compartment for measurement. The compartments are open at the top to facilitate filling and flushing. The angle of incidence to the cell partition is approximately 69°. The cell is suitable for solutions of refractive index up to 1.62 and for the determination of refractive index differences up to 0.01 to a sensitivity of $\pm 3 \times 10^{-6}$. Other cells are available equipped with permanent covers for the handling of volatile liquids or with a lower angle of incidence to extend the range of the instrument to a difference in refractive index of 0.07 with a corresponding decrease in sensitivity to about $\pm 2 \times 10^{-5}$.

Recently Scholte (1970) has described a modification to the Brice–Phoenix refractometer which makes possible a repeated removal and installation of the cell without the need for an exactly reproducible positioning, thus alleviating the need for repeated calibration. Cancellieri *et al.* (1974) describe an arrangement for the measurement of refractive index at any wavelength, including that corresponding to the HeNe laser.

The Brice differential refractometer has been used in many areas in the last two decades, but perhaps the area of greatest use is in the study of high molecular weight compounds. A few recent applications include; a characterization of functionally terminated polybutadienes by Ahad (1973), a dimensional study of polysulfone A in dimethylformamide by Allen and McAinsh (1970), a study of dilute solution viscosity and conformation of poly(*n*-hexylisocynanate) by Berger and Tidswell (1973), and the determination of the degree of aggregation of polyoxyethylene in dilute solutions by Carpenter *et al.* (1974).

(*c*) *Photometric Adaption.* In several situations, including the following of fractionation by distillation or monitoring a chromatographic process, the need for continuous measurement of refractive index has arisen. Even when making multiple determinations the Brice refractometer becomes very

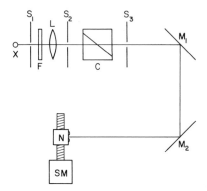

Fig. 20 Photoelectric modification of the Brice–Phoenix refractometer. X, light source; S_1, S_2, and S_3, slits; F, filter; L, collimating lens; C, cell; M_1 and M_2, mirrors; N, nut holding two photocells; SM, servomotor.

tedious to operate. The light beam configuration used in this instrument, a narrow vertical line, lends itself readily to automatic following. The optical arrangement used by Penther and Noller (1958) to convert a Brice refractometer to a photoelectric readout refractometer is shown in Fig. 20. The slits on either side of the cell are used to restrict stray light. A pair of photocells are mounted on a nut with a separation of about 0.005 in. between them. The nut is mounted on a screw connected to a servodrive that works to keep the light beam (line) between the two detectors. The position of the nut is read on a counter and converted to refractive index units from a calibration. The authors found, by carefully calibrating and by allowing sufficient time for the cell to reach thermal equilibrium, that the incremental differences of refractive index were reliable to $\pm 2.5 \times 10^{-7}$. Pittz and Bablouzian (1973) describe an instrument that is similar in optical arrangement but must be manually nulled.

3. Liquid Chromatography Detectors

High pressure liquid chromatography has developed into a very powerful tool for separating and quantifying certain classes of compounds. These compounds are generally too high in molecular weight or too labile to be handled by gas chromatography, but represent a formidable percentage of the chemical families currently of interest. The whole process of liquid chromatography depends on the presence of a detector capable of quantifying dilute solutions of these compounds as they come off the column. An ideal liquid chromatography detector has several properties: sensitivity to all compounds or at least those of interest; insensitivity to changes in solvent composition, temperature, and flow rate; reliability and simplicity in use; high sensitivity with a wide dynamic range; fast response; and low pressure drop across the detector. The refractometer comes close to being the ideal detector and is close to being a universal detector. For a refractometer to be universal, the only requirement is that the solvent refractive index be different from that of the sample. The optical principles involved allow for the construction of a very small differential refractive index cell which provides internal compensation for flow, pressure, and temperature changes.

Refractive index detectors for liquid chromatography are based primarily on two different measurement principles. Both types are made to be differential instruments and both types are available commercially. Most refractometers suitable for chromatography are of the beam-deflection type in which the bending of a light beam passing through a wedge-shaped sample is measured. Refractometers of this type have been discussed by Maley (1962), Vandenheuvel and Sipos (1961), and Trenner *et al.* (1953). The second approach makes use of the reflection principle first elucidated by Fresnel, in which the intensity of the reflected component of a beam of light impinging on the surface of the effluent stream changes inversely with the refractive index. Refractometers of this type have been discussed by Johnson *et al.* (1967), Jones *et al.* (1949), and Karrer and Orr (1946).

The angle of deviation technique utilizes the actual bending of a ray of monochromatic light as it passes through the effluent stream contained in a suitable optical cell. The bending angle changes as a function of sample concentration and is monitored by an electromechanical arrangement which moves either the source or photodetector to maintain optical alignment. A schematic diagram of a typical refractometer of this type is shown in Fig. 21. A light beam from an incandescent lamp X passes through the optical mask M which confines the beam within the region of the cell. The lens L acts as both the collimator and the focusing element. The beam passes through the cell containing both the sample S and reference R and is reflected by the mirror Y back to the detector D. As the beam changes location on the detector, an output signal is generated, which is then amplified and provided at an output for a recorder. Detectors based on this principle have been designed to detect changes of about $\pm 1 \times 10^{-7}$ refractive index units with a dead volume of less than 100 μl.

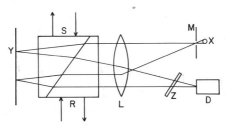

Fig. 21 Schematic drawing of a liquid chromatography detector based on the angle of deviation technique. X, light source; M, optical mask; L, lens; Y, mirror; S, sample side of differential flow-through cell; R, reference side; Z, glass plate for setting the zero; D, detector.

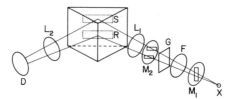

Fig. 22 Schematic drawing of a liquid chromotography detector based on Fresnel's laws of reflection. X, light source; M_1 and M_2, optical masks; F, filter; G, fine adjusting glass plate; L_1 and L_2, lenses; S, sample compartment of differential flow-through cell; R, reference compartment; D, dual detector.

The reflectance technique has certain advantages such as a smaller permissible flow and cell volume. The principle of this type system can be seen in Fig. 22, which is an optical diagram of a Fresnel-type detector. Rather than the original system of reflectance, the transmittance is measured which affords about an order of magnitude increase in sensitivity. Light from the source X passes through a mask M_1, an infrared absorbing filter F, a fine adjusting glass G, an aperture mask M_2, and a collimating lens L_1. The two beams defined by mask M_2 enter the prism and impinge on the two glass–liquid interfaces at the sample S and reference R compartments. The transmitted light is focused by lens L_2 onto the dual detector D. All components before the prism are mounted on a movable bar so that the angle of incidence can be set at slightly less than the critical angle. The detector is a dual photodetector and when the sample and reference compartments both contain the same material the amount of light impinging on each detector is equal and the output can be nullified. When the sample is changed the output signal changes indicating a difference in refractive index between the two liquids. Models of this type of detector have been constructed with a dead volume less than 10 μl and a sensitivity of about $\pm 3 \times 10^{-7}$.

4. *Schlieren Methods*

Longsworth (1946) and Antweiler (1951) have discussed the application and limitations of the schlieren method to the determination of refractive index. Any problem involving the formation of sharp concentration gradients in liquids, i.e., the moving boundary in an electrophoresis experiment, can benefit from this technique. As light passes through a cell any concentration gradients will cause the rays of the light beam to be deviated which results in the formation of shadows in the image of the cell. The deviated rays are detected by moving a sharp edged "schlieren diaphragm" into the optical path. The position of a shadow on the image can be correlated with the position in the cell of the concentration gradient. The corresponding refractive index gradient is proportional to the position of the schlieren diaphragm.

I. Handling of Special Sample Types

1. Liquids

(a) *Small Samples.* Often it is necessary to measure the refractive index of a very small sample. The standard Abbe refractometer requires only a very small sample, about 0.05 ml, and is thus directly suited for the measurement of small amounts. This amount can be further reduced by distributing the liquid on a piece of lens paper which is then clamped between the prisms. This method has been discussed by Alber and Bryant (1940). A Pulfrich refractometer can be converted for small samples by confining a thin layer of liquid on the prism surface with a glass plate. The refractometer then becomes equivalent to the Abbe type.

Two refractometers designed for use with very small samples, about 0.001 ml, are available commercially. In both of these instruments the liquid sample is made into a minute 45° prism by placing a drop of the sample between an appropriate set of glass surfaces. If the image of a slit is viewed through such a liquid prism, its apparent position is shifted from the true position by an amount proportional to the refractive index of the liquid. The position can be read on a scale and converted to the corresponding value of n via a set of calibrations with known liquids.

Jelley (1934) designed a microrefractometer that is marketed by the Fisher Scientific Company. It has a scale which gives values of n directly to ± 0.002 over the range of 1.30 to 1.90. A controllable electric heating device is attached to the cell, which shapes the liquid prism. In this way a variety of substances which are ordinarily solid may be measured in a molten state. The melting point of the substance can be measured at the same time.

The A. H. Thomas Company sells a microrefractometer based on the design of Nichols (1937). Two prismatic cells are mounted on a microscope slide and the cells are illuminated from below. A fine line just below the prisms appears doubled and the separation can be measured in the image plane of the microscope. Different prism sets are used to cover a range of 1.30 to 2.0 to an accuracy of ± 0.001 in n.

For a further discussion of micromethods see Wilson (1946).

(b) *Flowing Streams.* Many instruments have been produced for use as process monitors. The earliest recording refractometers were used during World War II to monitor the purity of butadiene and styrene streams. Applications in the food industry involve materials such as soybean and cottonseed oils. These oils have indices of approximately 1.47, whereas the finished products, shortenings and margerine, have indices of approximately 1.43. The process can be followed and the end point easily detected without the delays required in making spot checks with laboratory instruments.

Many other applications of process refractometers involve the concentration of solutions such as sugar solutions, salt brines, and acids. The degree of concentration can easily be followed by the change in refractive index. These instruments can be classified into three categories according to the method of operation: (i) based on the measurement of the total intensity of light; (ii) detection of the difference of position of a reference and a sample beam; (iii) tracking of a critical boundary with a servo mechanism. The details of operation of this type of instrument can be understood by studying the operation of liquid chromatography detectors (Section II,H,3) and photoelectric refractometers (Section II,H,2c).

2. Solids

(a) *Crystals.* Crystal refractometers, especially adapted for crystallographical and mineralogical investigations, have been marketed commercially. A similar instrument has been developed by Haake and Hartmann (1954). The optical block of the instrument is usually a hemisphere of high index glass, well polished on both plane and spherical surfaces. Either grazing incidence or critically directed (monochromatic) light can be radially thrown on or into the hemisphere by means of a mirror, but the reflected-light method is used almost exclusively on crystals because only one polished crystal surface is necessary in that case.

As with the Pulfrich refractometer, the angle through which the observing telescope revolves is measured on a vertical circle, and tables are provided for the corresponding index of refraction. The telescope objective can be removed and a microscope objective substituted in order to observe a crystal directly, or match it with an immersion liquid, as it lies on the plane surface of the hemispherical block. The hemisphere is measurably rotatable around a vertical axis.

By using a reducing telescope, and with suitable diaphragms, indices can be measured by reflection from surfaces less than 1 mm in area. This refractometer can be used on larger polished solids and for liquids in a special temperature-stabilizing cell with plane-parallel base and a thermometer.

A critical-angle instrument, similar to the crystal refractometer in that it utilizes a hemispherical block, has been described by Pfund (1949). This instrument makes use of a rutile hemisphere and permits measurements up to $n = 2.61$.

(b) *Thin Films.* Many optical systems in use today depend on components coated with thin dielectric films such as ZnS, ZnO, TiO_2, Al_2O_3, and CeO_2. The main parameters characterizing such films are the refractive index and the film thickness. Thus, several methods have been developed for the measurement of the refractive index of these thin films.

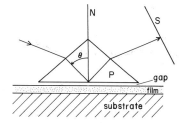

Fig. 23 Prism coupling technique for the measurement of the refractive index of a thin film.

One method of refractive index measurement is the prism-coupling technique which has been discussed in the literature by Ulrich and Torge (1973), Harris *et al.* (1970), and Tien and Ulrich (1970). In this method a laser beam is coupled to the thin film using a prism as shown in Fig. 23. The degree of coupling is determined by the angle θ. Strong coupling into the film only occurs at certain values of θ. The values of θ that give strong coupling lead to a determination of n. The accuracy obtainable by this method can be as good as $\pm 1 \times 10^{-4}$.

Daneu and Sanchez (1974) present a method for the measurement of the refractive indices of thin transparent films. This method consists of illuminating the film with a laser and observing the resulting interference pattern in the light scattered by the film. A measurement of the fringe spacing yields the refractive index of the film to ± 0.001. This method is particularly applicable to transparent films deposited on a transparent substrate such as glass. Harrick (1971) also discusses a method for the determination of refractive index based on the measurement of interference fringes. This method can be applied to free-standing films or to films deposited on a substrate of higher refractive index. He points out that there are a number of advantages to the measurement of the fringes in the reflected rather than in the transmitted light.

(*c*) *Powders.* The measurement of the refractive index of a powdered material can present difficult problems. Tilton and Taylor (1950) discuss a method applicable to very small grains of isotropic solids or powders of average cross-sectional extent greater than 0.01 mm. In this method the powder is immersed in a liquid of nearly identical or slightly lower refractive index and allowed to settle on the prism of a monochromatic critical angle refractometer. If the match between the refractive indices is good the critical boundary will be sharp and a precise reading can be made. More recently, Spitzer (1975) and Spitzer and Borsboom (1976) have proposed a new method for the study of powdered materials. In this method the sample is placed in a capillary and irradiated by a parallel beam of monochromatic light. The reflected radiation is then detected on a film strip. Fresnel's equations apply and thus the refractive index of the material can be found.

References

Ahad, E. (1973). *J. Appl. Polym. Sci.* **17**, 365.

Alber, H. K., and Bryant, J. T. (1940). *Ind. Eng. Chem. Anal. Ed.* **12**, 305.

Allen, G., and McAinsh, J. (1970). *Eur. Polym. J.* **6**, 1635.

Altmann, C. G. J. (1948). *Chem. Weekbl.* **44**, 708.

Antweiler, H. J. (1951). *Mikrochemie* **36/37**, 561.

Arshid, F. M., Giles, C. H., McLure, E. C., Ogilvie, A., and Rose, T. J. (1955). *J. Chem. Soc.* 67.

Arshid, F. M., Giles, C. H., Jain, S. K., and Hassan, A. S. A. (1956a). *J. Chem. Soc.* 72.

Arshid, F. M., Giles, C. H., and Jain, S. K. (1956b). *J. Chem. Soc.* 559.

Bauer, N., and Fajans, K. (1942). *J. Am. Chem. Soc.* **64**, 3023.

Bauer N., Fajans, K., and Lewin, S. Z. (1959). *In* "Physical Methods of Organic Chemistry" (A. Weissberger, ed.), Vol. 1, pp. 1169–1211. Wiley (Interscience), New York.

Berger, M. N., and Tidswell, B. M. (1973). *J. Polym. Sci. Part C* **42**, 1063.

Berl, E., and Andress, I. (1921). *Z. Angew Chem.* **34**, 370.

Berthelot, M. (1856). *Ann. Chim. Phys.* **48**, 342.

Black, E. P., Harvey, W. T., and Ferris, S. W. (1954). *Anal. Chem.* **26**, 1089.

Block, A. M. (1971). *Appl. Opt.* **10**, 207.

Brice, B. A., and Halwer, M. (1951). *J. Opt. Soc. Am.* **41**, 1033.

Browne, C. A., and Zerban, F. W. (1941). "Physical and Chemical Methods of Sugar Analysis," 3rd ed. Wiley, New York.

Campanile, V. A., and Lantz, V. (1954). *Anal. Chem.* **26**, 1394.

Cancellieri, A., Frontali, C., and Gratton, E. (1974). *Biopolymers* **13**, 735

Carpenter, D. K., Santiago, G., and Hunt, A. H. (1974). *J. Polym. Sci. Part C* **44**, 75.

Charles, D. F., and Meads, P. F. (1955). *Anal. Chem.* **27**, 373.

Cohen, E., and Bruins, H. R. (1921). *Proc. K. Akad. Wet.* **24**, 114.

Conrady, A. E. (1903). *Mon. Not. R. Astron. Soc.* **64**, 458.

Cuthbertson, C., and Cuthbertson, M. (1932). *Proc. R. Soc. London Ser. A* **135**, 40.

Daneu, V., and Sanchez, A. (1974). *Appl. Opt.* **13**, 122.

Debye, P. P. (1946). *J. Appl. Phys.* **17**, 392.

Doty, P., and Geiduschek, E. P. (1953). *In* "The Proteins" (H. Neurath and K. Bailey, eds.), Vol. 1A, pp. 399–402. Academic Press, New York.

Fajans, K. (1924). *Z. Phys.* **23**, 1.

Fajans, K. (1941). *J. Chem. Phys.* **9**, 281.

Fajans, K., and Johnson, O. (1942). *Trans. Electrochem. Soc.* **82**, 273.

Fajans, K., Hoelemann, P., and Shibata, Z. (1931). *Z. Phys. Chem. Abt. B* **13**, 354.

Forrest, J. W. (1956). *J. Opt. Soc. Am.* **46**, 657.

Forrest, J. W., Straat, H. W., and Dakin, R. K. (1956). *J. Opt. Soc. Am.* **46**, 143.

Geffchen, W., and Kruis, A. (1928). *Z. Phys. Chem. Abt. B* **1**, 456.

Gibson, R. E. (1938). *J. Am. Chem. Soc.* **60**, 517.

Grange, B. W., Stevenson, W. H., and Viskanta, R. (1976). *Appl. Opt.* **15**, 858.

Grosse, A. V. (1937). *J. Am. Chem. Soc.* **59**, 2739.

Grosse, A. V. (1946). Manhattan Project Declassified Document, P.B. 60787, Dec. 9 (M.P.D.C. 523).

Guild, J. (1917). *Proc. Phys. Soc. London* **30**, 157.

Haake, H., and Hartmann, J. (1954). *J. Opt.* **11**, 380.

Haber, F., and Löwe, F. (1910). *Z. Angew. Chem.* **23**, 1393.

Hadow, H. J., Sheffer, H., and Hyde, J. C. (1949). *Can. J. Res. Sect. B* **27**, 791.

Harrick, N. J. (1971). *Appl. Opt.* **10**, 2344.

Harris, J. H., Shubert, R., and Palby, J. N. (1970). *J. Opt. Soc. Am.* **60**, 1007.

Hill, R., and Jones, A. G. (1955). *Analyst (London)* **80**, 339.

Huxley, H., and Lowery, H. (1943). *Proc. R. Soc. London Ser. A* **182**, 207.

Icenogle, H. W., Platt, B. C., and Wolfe, W. L. (1976). *Appl. Opt.* **15**, 2348.

Jamin, J. (1856). *Pogg. Ann.* **98**, 345.

Jelley, E. E. (1934). *J. R. Microsc. Soc.* **54**, 234.

Jenkins, F. A., and White, H. E. (1957). "Fundamentals of Optics," 3rd ed. McGraw-Hill, New York.

Johnson, H. W., Jr., Campanile, V. A., and LeFebre, H. A. (1967). *Anal. Chem.* **39**, 32.

Jones, H. E., Ashman, L. E., and Stahley, E. E. (1949). *Anal. Chem.* **21**, 1470.

Karrer, E., and Orr, R. S. (1946). *J. Opt. Soc. Am.* **36**, 42.

Koenig-Gressmann, M. L. (1938). Thesis, Univ. of Munich, Munich.

Körösy, F. (1954). *Nature (London)* **174**, 269.

Kurtz, S. S., Jr., and Ward, A. L. (1937). *J. Franklin Inst.* **224**, 583.

Lauer, J. L., and King, R. W. (1956). *Anal. Chem.* **28**, 1697.

Lipson, H. G., Tsay, Y. F., Bendow, B., and Ligor, P. A. (1976). *Appl. Opt.* **15**, 2352.

Longsworth, L. G. (1946). *Ind. Eng. Chem. Anal. Ed.* **18**, 219.

McCormick, H. (1953). *Analyst (London)* **78**, 562.

Mach, L. (1892). *Z. Instrumentenkd.* **12**, 89.

Mair, B. J. (1932). *J. Res. Natl. Bur. Stand.* **9**, 461.

Maley, L. E. (1962). *ISA Trans.* **1**, 245.

Michelson, A. A. (1890). *Am. J. Sci.* **39**, 118.

Moilliet, A. (1955). *Rev. Sci. Instrum.* **26**, 519.

Nichols, L. (1937). *Natl. Paint Bull.* **1**, 12.

Penther, C. J., and Noller, G. W. (1958). *Rev. Sci. Instrum.* **28**, 43.

Perlmann, G. E., and Longsworth, L. G. (1948). *J. Am. Chem. Soc.* **70**, 2719.

Pfund, A. H. (1949). *J. Opt. Soc. Am.* **39**, 966.

Pittz, E. P., and Bablouzian, B. (1973). *Anal. Biochem.* **55**, 399.

Pockels, F. (1902). *Ann. Phys. (Leipzig)* **7**, 745.

Pulfrich, C. (1892). *Ann. Phys. (Leipzig)* **45**, 609.

Reed, J. O. (1898). *Ann. Phys. (Leipzig)* **65**, 707.

Richter, R. (1930). *Z. Instrumentenkd.* **50**, 254.

Röntgen, W. C., and Zehnder, L. (1891). *Ann. Phys. (Leipzig)* **44**, 49.

Roth, W. A., and Eisenlor, F. (1911). "Refraktometrisches Hilfsbuch," p. 37. Leipzig.

Scholte, T. G. (1970). *J. Polym. Sci. Part A-2* **8**, 841.

Sieflow, G. H. F. (1953). *J. Sci. Instrum.* **30**, 407.

Spitzer, D. (1975). *Appl. Opt.* **14**, 1489.

Spitzer, D., and Borsboom, P. C. F. (1976). *Appl. Opt.* **15**, 2301.

Straat, H. W., and Forrest, J. W. (1939). *J. Opt. Soc. Am.* **29**, 240.

Svensson, H. (1953). *Anal. Chem.* **25**, 913.

Swietoslawski, W. (1920). *J. Am. Chem. Soc.* **42**, 1945.

Tien, P. K., and Ulrich, R. (1970). *J. Opt. Soc. Am.* **60**, 1325.

Tilton, L. W. (1935). *J. Res. Natl. Bur. Stand.* **14**, 393.

Tilton, L. W. (1936) *J. Res. Natl. Bur. Stand.* **17**, 389.

Tilton, L. W. (1942a). *J. Opt. Soc. Am.* **32**, 373.

Tilton, L. W. (1942b). *J. Opt. Soc. Am.* **32**, 376.

Tilton, L. W. (1943a). *J. Res. Natl. Bur. Stand.* **30**, 311.

Tilton, L. W. (1943b). *J. Res. Natl. Bur. Stand.* **30**, 314.

Tilton, L. W. (1943c). *J. Res. Natl. Bur. Stand.* **30**, 319.

Tilton, L. W. (1943d). *J. Res. Natl. Bur. Stand.* **30**, 323.

Tilton, L. W., and Taylor, J. K. (1938). *J. Res. Natl. Bur. Stand.* **20**, 419.

Tilton, L. W., and Taylor, J. K. (1950). Refractive index measurement, *in* "Physical Methods in Chemical Analysis" (W. G. Berl, ed.), Vol. 1, pp. 411–462. Academic Press, New York.

Tilton, L. W., and Tool, A. Q. (1929). *J. Res. Natl. Bur. Stand.* **3**, 622.

Timmermans, J., and Delcourt, Y. (1934). *J. Chim. Phys.* **31**, 85.

Timmermans, J., and Hennaut-Rowland, M. (1932), *J. Chim. Phys.* **29**, 529.

Tool, A. Q., and Tilton, L. W. (1926). *J. Opt. Soc. Am.* **12**, 490.

Trenner, N. R., Warren, C. W., and Jones, S. L. (1953). *Anal. Chem.* **25**, 1685.

Ulrich, R., and Torge, R. (1973) *Appl. Opt.* **12**, 2901.

Vandenheuvel, F. A., and Sipos, J. C. (1961). *Anal. Chem.* **33**, 286.

Vogel, A. I., Cresswell, W. T., and Leicester, J. (1954). *J. Phys. Chem.* **58**, 174.

Waldmann, H. (1938). *Helv. Chim. Acta* **21**, 1053.

Ward, A. L., and Kurtz, S. S., Jr. (1938). *Ind. Eng. Chem. Anal. Ed.* **10**, 559.

Wiley, R. H., Brauer, G. M., and Bennett, A. R. (1950). *J. Polym. Sci.* **5**, 609.

Wilson, C. L. (1946). *Analyst (London)* **71**, 117.

Wright, F. E. (1921). *J. Opt. Soc. Am.* **5**, 389.

Yound, W. G., and Rule, J. M. (1954). *Anal. Chem.* **26**, 1393.

Young, J. C., and Finn, A. N. (1940). *J. Res. Natl. Bur. Stand.* **25**, 759.

Zehnder, L. (1888). *Ann. Phys. (Leipzig)* **34**, 91.

Zehnder, L. (1891). *Z. Instrumentenkd.* **11**, 275.

Index